Transport and Energy Processes

Presentations at the 2010 AIChE Annual Meeting

Salt Lake City, Utah, USA
7-12 November 2010

ISBN: 978-1-61782-153-0

Printed from e-media with permission by:

Curran Associates, Inc.
57 Morehouse Lane
Red Hook, NY 12571

Some format issues inherent in the e-media version may also appear in this print version.

Printed by Curran Associates, Inc. (2011)

For permission requests, please contact AIChE
at the address below.

AIChE
3 Park Avenue
New York, NY 10016-5991

Phone: (203) 702-7660
Fax: (203) 775-5177

www.aiche.org

Additional copies of this publication are available from:

Curran Associates, Inc.
57 Morehouse Lane
Red Hook, NY 12571 USA
Phone: 845-758-0400
Fax: 845-758-2634
Email: curran@proceedings.com
Web: www.proceedings.com

TABLE OF CONTENTS

Mitigation of heat transfer fouling in thin stillage evaporators

Kent D. Rausch[1,*], Amit Arora[2], Nancy Lin[1], Michelle Kwak[1], Vijay Singh[1], M. E. Tumbleson[1]
[1]University of Illinois, Urbana, IL 61801
[2]University of Wisconsin, Stevens Point, WI 54481
*corresponding author: krausch@illinois.edu, 217.265.0697, 1304 W. Pennsylvania Ave.

Introduction

From 2000 to 2009, US fuel ethanol production increased from 1.6 to 10.6 billion gallons per year [1]. The large majority of fuel ethanol is from the processing of corn starch using two processes: wet milling and dry grinding. Approximately 82% of ethanol production is from the dry grind process (Figure 1). During fuel ethanol production, whole corn grain is ground, slurried with water, cooked with enzymes to reduce viscosity and break down starch granules, and placed in a fermenter. Following simultaneous saccharification and fermentation, the resulting beer is stripped of ethanol and a portion of water, leaving behind a material called whole stillage. This process stream is fractionated into two constituents, thin stillage and wet grains. When thin stillage is subsequently concentrated using evaporators, deposits routinely form on evaporator surfaces through a process termed fouling. Heat transfer fouling is a major factor adversely affecting evaporator efficiency and increases water, chemical and energy use at bioethanol production facilities. Due to fouling deposits, evaporators must be cleaned at intervals of 1 to 2 weeks, thereby negatively impacting profitability and capacity of the ethanol facility at several levels. First, evaporator operations must be over-designed in order to compensate for decreased process throughput during cleaning, which leads to increased capital costs. As deposits form, evaporator efficiency is reduced, increasing energy requirements for evaporation. Further, cleaning of fouling deposits increase water used during ethanol production; following cleaning, water and cleaning chemicals must be either reused or treated.

Evaporator fouling is a common problem in bioprocess industries that reduces energy efficiency and thus the productivity of the plant [2-7]. Growth of deposits reduces evaporator capacity and performance of equipment to the point where it must be shut down and cleaned. Cost of cleaning evaporators has lead to study of the fouling problem [8-11]. In the petroleum, dairy and food processing industries, investigators have conducted research aimed at understanding and mitigating fouling. Evaporator fouling largely depends on composition of the input stream [12]. Particulate fouling is also one of the primary mechanisms of deposit buildup [13]. In the dry grind process, changes in thin stillage compositions and solids concentrations may alter fouling tendencies. Recovery of solids contained in thin stillage is a challenge due to relatively low concentrations (7% total solids) and larger volumes to be processed. Membrane filtration is a potential alternative technology to separate solids and alter thin stillage composition. In addition, permeate may be recycled within the process at higher rates than currently possible for unfiltered thin stillage, leading to potential reductions in fresh water requirements. A limited number of researchers [4, 6-7] have described heat transfer fouling tendencies of thin stillage.

More than 200 U.S. fuel ethanol plants use evaporators and are adversely affected by heat transfer fouling. Fouling and its associated maintenance increases the overall environmental footprint and costs of biofuel production, creating a critical need to more fully understand the causes of fouling and to develop strategies to mitigate fouling in thin stillage evaporators. The objectives were to (1) measure effects of solids concentration on heat transfer surface fouling tendencies of thin stillage and (2) determine effect of microfiltration on heat transfer surface fouling tendencies of thin stillage.

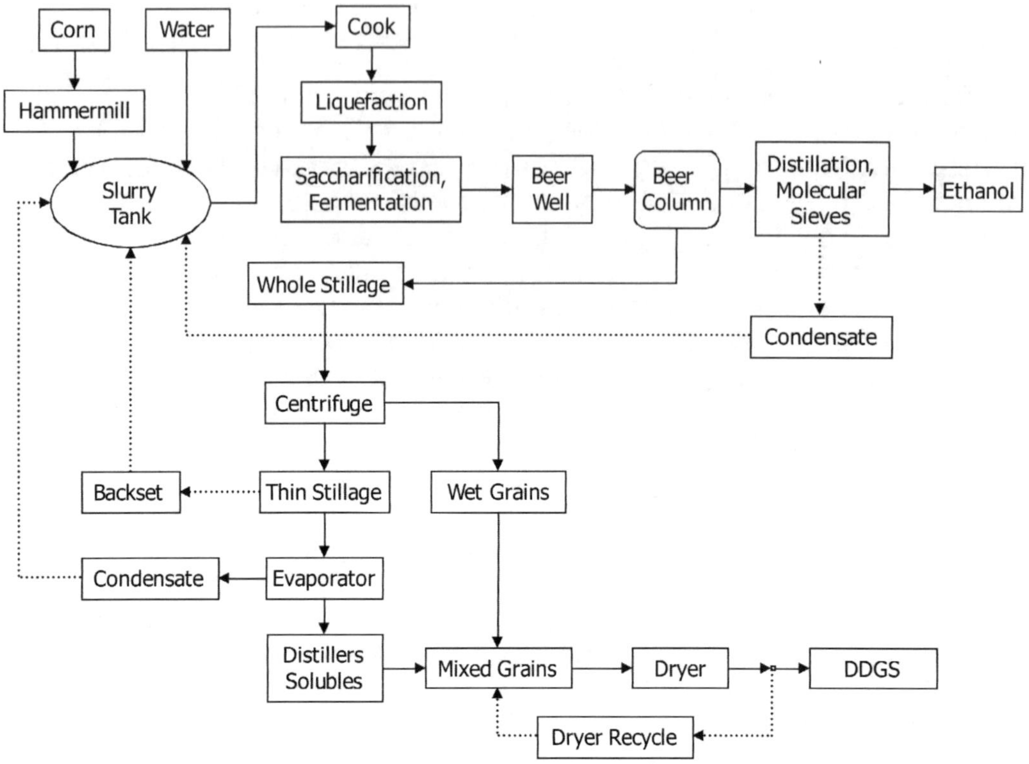

Figure 1. A simplified schematic of the corn dry grind process used to produce fuel ethanol.

Material and Methods

Thin stillage was collected from a commercial dry grind ethanol facility located in the Midwest US and stored at 4°C. The facility followed a typical dry grind process scheme (Figure 1) that ground whole corn kernels prior to fermentation (kernel components were not separated). Two batches (160 L) of thin stillage were collected from the facility during a period of several weeks. One 500 mL sample from each batch was analyzed for total nitrogen, ash and fat contents using standard methods [14]. Total solids contents of thin stillage were determined using a two stage oven method [15].

Three streams derived from corn based ethanol fermentation were evaluated: original thin stillage, diluted thin stillage (DTS) and permeate from microfiltration of thin stillage (MFP). Microfiltration was carried out using a bench top membrane unit. A tubular stainless steel module with six tubes having 0.61 m length, nominal diameter of 0.64 cm, 0.1 lm pore size, 0.31 cm wall thickness and 0.07 m^2 membrane area (Scepter model 2.5-250A-2P6, Graver

Technologies, Glasgow, DE) was used. The unit was equipped with a batch tank (15 L capacity). Permeate that passed through the membrane was termed MFP and the material that was retained and returned to tank was termed retentate. MFP was collected in a graduated cylinder during batch concentration. For each batch (15 L) experiment, permeate flux rate measurements (L/m^2/h) were taken during concentration until batch material was exhausted. Permeate flux rates were determined manually every 2 min, with a graduated cylinder and stopwatch. The selected transmembrane pressure (TMP) was 690 ± 13 kPa. A crossflow velocity of 4.75 m/s was selected according to the manufacturer's recommendations. In an ethanol plant, thin stillage temperatures range from 60 to 75°C. Therefore, the operating temperature chosen for the filtration study was 75 ± 2°C.

An annular probe was used to quantify fouling tendencies of fluids passing over heated surfaces. Two batches (160 L each) of thin stillage were collected from the same dry grind facility as in the membrane filtration experiments; batches were collected during a period of several weeks. Each batch was further divided into two equal parts (two replicates within each batch) to evaluate effects of thin stillage, solids concentration and composition on heat transfer fouling. Within each 80 L batch, 30 L was used for a fouling experiment and 35 L was used for producing 30 L permeate to be later used evaluating compositional effects on heat transfer fouling. The remaining thin stillage (15 L) was diluted with distilled water to reduce the solids level to equal that of permeate and used to evaluate effect of solids concentration on heat transfer fouling.

The same probe and associated equipment were used as in previous work [2, 4, 6-7]. To simulate thin stillage bulk temperatures of a commercial dry grind process, we chose T_b = 60°C which was near the upper operating limits of the fouling apparatus. Bulk fluid temperature of 60 ± 2°C and power input of 970 ± 15W were used throughout the experiment. Selected samples of fouling deposits were analyzed for total solids, protein and ash concentrations using standard methods [14]. The deposition of foulants on heat transfer equipment depends in part on the shape and structural material of the equipment. Surface temperature of the equipment, fluid velocity, residence time associated with the fluid and heated surface, and fluid material composition also contribute to fouling rate [16]. The principle of measuring heat transfer fouling resistance was to measure change in overall heat transfer coefficients due to fouling. The overall fouling effect is usually represented by fouling resistance (R_f) [3, 8]. R_f was obtained experimentally by determining the values of overall heat transfer coefficients for unfouled and fouled probe conditions [4, 6-7, 9]:

$$R_f = \frac{1}{U_{fouled}} - \frac{1}{U_{unfouled}}$$

where $U_{unfouled}$ is the initial overall heat transfer coefficient for a clean probe at time t = 0 and U_{fouled} is the overall heat transfer coefficient of a fouled probe at time t. U accounts for heat transfer from conductive and convective resistances. The heat transfer coefficient can be thought of as an inverse of thermal resistance and can be defined as the amount of heat which passes through a unit area per unit time when the temperature difference is one degree. As fluid flows through the annular space between two surfaces, a portion of fouling constituents of thin stillage accumulate on the heated surface of the probe. Foulant thicknesses increased with time which represented an additional resistance to heat flow from heater to fluid, causing increased probe surface temperature (T_s) [8]. Mean thermocouple temperature inside the probe (T_{tc}) was used to determine the outer surface temperature (T_s) of the probe using Fourier's law of conduction:

$$Q = \frac{kA(T_{tc} - T_s)}{x} \quad \text{or} \quad T_s = T_{tc} - \frac{Qx}{Ak}$$

where Q was the rate of heat loss (W/m^2), Q = P/A, calculated from P (electrical power input, watts), A was the total heated surface area (0.0034 m^2) of the probe. The parameter x/k was the distance of thermocouples from the surface divided by thermal conductivity of the probe metal. The values of x/k for each thermocouple were determined using a calibration procedure described by Fischer et al. [17]. T_s was used to calculate the heat transfer coefficient of the probe (U_{fouled}) as the testing period progressed. Four thermocouples located inside the probe measured temperature (T_{tc}); three were used to measure probe temperature change and the fourth was used to control an electric relay that protected the probe by turning off power at T_{tc} = 200°C. U_{fouled} and $U_{unfouled}$ were determined using:

$$U = \frac{Q}{A(T_s - T_b)}$$

where T_b was the bulk fluid temperature. R_f and t were plotted and a linear regression line was fitted to the data for each test. The fouling rate was defined as the slope of the R_f vs t regression line [2, 6-7]. Induction period was defined as the period during which the 1 min moving average of R_f was less than 0.01 m^2°C/kW, beginning with the start of the test at t = 0 [6-7]. After thin stillage temperature (T_b) was adjusted to 60°C, power was supplied to the fouling probe and was adjusted to give T_{tc} = 100°C, after which constant power was maintained (±15 W). As fouling progressed, T_{tc} increased; each test was stopped when T_{tc} = 200°C. The fouling resistance obtained at 200°C was considered the maximum fouling resistance. Fouling data were collected using each fluid (thin stillage, MFP and diluted thin stillage or DTS) using two replications. Fouling deposits were removed from the probe using plastic spatula and dried overnight at 49°C. Deposits were collected from each test using thin stillage and MFP. Deposits were analyzed for total solids, protein (total nitrogen × 6.25), ash and fat contents using standard methods [14].

Results and Discussion

Reductions in solids concentrations and changes in composition resulted in reduced fouling rates and fouling resistances (Table 1). At 10 h of fouling, 55 and 95% reductions in fouling resistance were observed in DTS and MFP streams, respectively, although both streams had decreased similarly from 7.2 to 3.5% solids. During DTS and MFP fouling, induction periods were prolonged by factors of 4.3 and 9.5, respectively, compared to the induction period for thin stillage fouling. Mean fouling rates were decreased by factors of 2.3 and 23.4 for DTS and MFP, respectively. Fouling of MFP took twice the time to reach a probe temperature of 200°C than did thin stillage (22 vs 10 h, respectively).

Table 1. Process stream composition, induction period, maximum fouling resistance (R_f) and fouling rates of original thin stillage (TS), microfiltration permeate (MFP) and diluted thin stillage (DTS). Adapted from [18].

	Composition				Induction		Fouling rate $\times 10^{-4}$ ($m^2 °C/kW/min$)	
	Dry Matter (%)	Protein** (% db)	Fat (% db)	Ash (% db)	period (h)	R_f* ($m^2 °C/kW$)	At t = 10 h	At T_{tc} = 200°C
TS	7.3 a	23.9 a	15.9 a	10.0 b	43.5 c	0.356 a	7.1 a	7.1 a
DTS	3.5 b	23.9 a	15.9 a	10.0 b	188 b	0.158 c	3.1 b	4.2 b
MFP	3.4 b	19.5 b	2.9 b	16.5 a	412.5 a	0.031 b	0.25 c	2.6 c

*Values shown for t = 10 h. For thin stillage, DTS and MFP, maximum R_f occurred at t = 10, 14 and 22 h, respectively (ie, at T_{tc} = 200°C).

**Protein = total nitrogen × 6.25.

[abc]Means followed by different letters in same column are different ($P < 0.05$). N = 4 observations.

Reduced solids content alone did not explain decreased fouling tendencies in microfiltered stillage (Figure 1). These results are indicative that selective separation of components from the thin stillage may improve evaporation efficiency and mitigate evaporator fouling. From these data, one could theorize that a membrane filtration scheme could be devised to selectively remove components from thin stillage prior to evaporation, rather than using membranes as a preconcentration step. Typical membrane applications in grain and starch processing have used the resulting retentate streams from membrane filtration as input streams to an evaporation step [19]. However, these data suggest that membranes may be useful in creating permeates that increase evaporator efficiency by recovering fouling components in the retentate. These data also illustrate that additional work is needed to determine which thin stillage components are responsible for accelerated rates of fouling.

There were negative fouling resistances initially in MFP (Figure 1). After an induction period of about 400 min (3.5 h), R_f increased with time in a linear fashion and attained maximum wall temperature (200°C) in 22 h. Other researchers [2, 4, 6, 20] also reported negative fouling resistance values. These negative resistances were attributed to the disturbance of the thermal boundary layer by particles and early formation of rough deposits, which increased the film transfer coefficient [20].

Fouling deposits obtained from thin stillage had reduced protein concentrations and lower fat concentrations; fouling deposits from MFP had increased ash contents (Table 2). This may have been due to removal of protein and fat content during MF, which increased the proportion of ash in permeate.

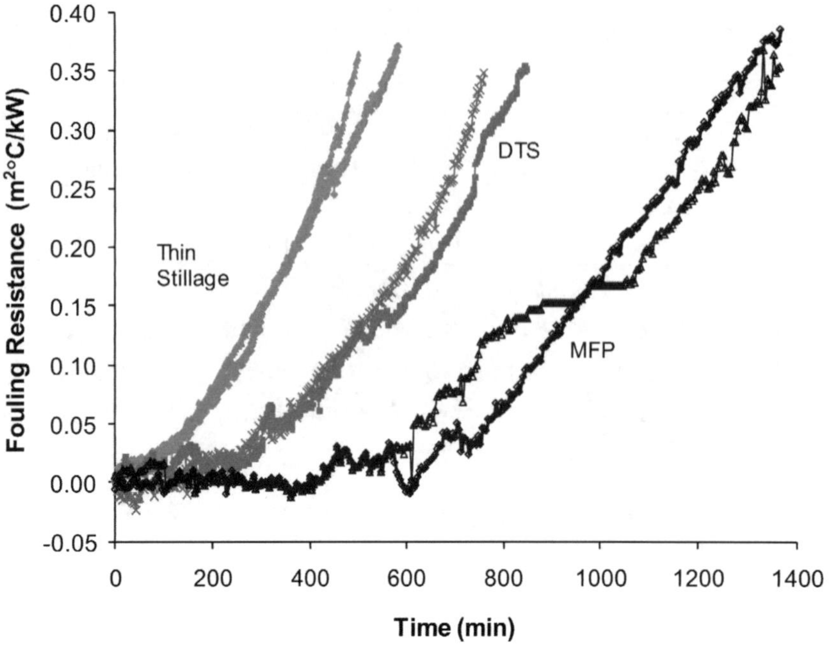

Figure 1. Fouling resistance (R_f, m^{2}°C/kW) of original, diluted (DTS) and microfiltered (MFP) thin stillage. Total solids content of original thin stillage = 7.2 to 7.4%; total solids contents of DTS and MFP = 3.5%. Adapted from [18].

Table 2. Compositions of thin stillage (TS), microfiltration permeate (MFP) and the resulting fouling deposits on an annular probe* (adapted from [18]).

Sample	Dry Matter (% db)	Protein (% db)	Fat (% db)	Ash (% db)
TS**	7.3 b	26.6 a	15.9 d	10.0 d
Deposit, TS	89.6 a	11.65 b	6.3 b	55.5 b
MFP	3.4 c	19.5 b	2.9 c	16.5 c
Deposit, MFP	ND+	ND	ND	92.2 a

*Mean ± one standard deviation. Observations for each mean = 2.
**TS: thin stillage.
+Not determined.
abcdMeans followed by different letters in the same column are different at $P < 0.05$. Protein = total nitrogen × 6.25.

Conclusions

Microfiltration of thin stillage resulted in a 95% reduction of the fouling resistance value for the same reduction in total solids content and at the same time interval. Reductions in solids concentrations and changes in composition resulted in reduced fouling rates and fouling resistances. At 10 h of fouling, a 55% reduction in R_f was observed when solids level decreased from 7.2% to 3.5%, with no change in composition. In DTS and MFP, the induction period was prolonged by factors of 4.3 and 9.5, respectively, compared to the induction period for thin stillage fouling. Mean fouling rates were decreased by factors of 2.3 and 23.4 for DTS and MFP, respectively. Fouling of MFP took twice the time to reach a probe temperature of 200°C than did thin stillage (22 h vs 10 h, respectively). A reduction in heat transfer fouling could be achieved by altering process stream composition using microfiltration. Additional research could find components that, when removed, reduce fouling rates and increase the efficiency of fuel ethanol production.

Acknowledgement

The authors gratefully acknowledge support from Graver Technologies, Glasgow, DE. Partially funded by the Illinois Council for Food and Agricultural Research (IDA CF07I-012-1-SEN), the National Center for Agricultural Utilization Research (Specific Cooperative Agreement SCA 58-3620-2-149), Agriculture Research Service, USDA, Peoria, IL and by the USDA National Institute of Food and Agriculture (Hatch project no. ILLU-741-324).

References

1. RFA. *Industry statistics.* 2010; Available from: http://www.ethanolrfa.org/pages/statistics.
2. Agbisit, R.M., et al., *Technique to measure surface-fouling tendencies of steepwater from corn wet milling.* Cereal Chemistry, 2003. **80**(1): p. 84-86.
3. Changani, S.D., M.T. BelmarBeiny, and P.J. Fryer, *Engineering and chemical factors associated with fouling and cleaning in milk processing.* Experimental Thermal and Fluid Science, 1997. **14**(4): p. 392-406.
4. Singh, V., C.B. Panchal, and S.R. Eckhoff, *Effect of corn oil on thin stillage evaporators.* Cereal Chemistry, 1999. **76**(6): p. 846-849.
5. Visser, J. and T.J.M. Jeurnink, *Fouling of heat exchangers in the dairy industry.* Experimental Thermal and Fluid Science, 1997. **14**(4): p. 407-424.
6. Wilkins, M.R., et al., *Analysis of heat transfer fouling by dry-grind maize thin stillage using an annular fouling apparatus.* Cereal Chemistry, 2006. **83**(2): p. 121-126.
7. Wilkins, M.R., et al., *Effect of pH on fouling characteristics and deposit compositions in dry-grind thin stillage.* Cereal Chemistry, 2006. **83**(3): p. 311-314.
8. Chenoweth, J.M., *Liquid fouling monitoring equipment*, in *Fouling Science and Technology*, L.F. Melo, T.R. Bott, and C.A. Bernardo, Editors. 1988, Kluwer Academic Publishers: Dordrecht, The Netherlands. p. 49-65.
9. Epstein, N., *Thinking about heat transfer fouling: a 5 x 5 matrix.* Heat Transfer Engineering, 1981. **4**(1): p. 43-56.

10. Kessler, H.G., *Fouling, cleaning, sanitizing, rinsing and associate processes at the interface*, in *Dairy Technology*, A. Kessler, Editor. 2002: Munich, Germany. p. 563-613.
11. Steinhagen, R., H. Mullersteinhagen, and K. Maani, *Problems and costs due to heat-exchanger fouling in New Zealand industries.* Heat Transfer Engineering, 1993. **14**(1): p. 19-30.
12. Bansal, B. and X.D. Chen, *A critical review of milk fouling in heat exchangers.* Comprehensive Reviews in Food Science and Food Safety, 2006. **5**(2): p. 27-33.
13. Bott, T.R., *General fouling problems*, in *Fouling Science and Technology*, L.F. Melo, T.R. Bott, and C.A. Bernardo, Editors. 1988, Kluwer Academic Publishers: Dordrecht, The Netherlands. p. 3-14.
14. AOAC, *Official Methods of the AOAC, 17th edition.* 2003, The Association of Official Analytical Chemists: Gaithersburg, MD.
15. AACC, *Approved Methods of the American Association of Cereal Chemists. 10th edition*. 2000, AACC International: St. Paul, MN.
16. Knudsen, J.G., *Apparatus and techniques for measuring of fouling of heat transfer surfaces*, in *Fouling of Heat Transfer Equipment. Hemisphere.* 1981, Hemisphere: Washington, DC. p. 57-81.
17. Fischer, P., J.W. Suitor, and R.B. Ritter, *Fouling measurement techniques.* Chemical Engineering Progress, 1975. **71**(7): p. 66-72.
18. Arora, A., et al., *Heat transfer fouling characteristics of microfiltered thin stillage from the dry grind process.* Bioresource Technology, 2010. **101**(16): p. 6521-6527.
19. Rausch, K.D., *Front end to backpipe: Membrane technology in the starch processing industry.* Starch/Starke, 2002. **54**(7): p. 273-284.
20. Panchal, C.B. and A.P. Watkinson, eds. *Chemical reaction fouling model for singlephase heat transfer.* AICHE Symposium Series. Vol. 89. 1993. 323-333.

(46b) Retrofit of Refinery Heat Exchangers to Mitigate Fouling Based On An Advanced Thermo-Hydraulic Model

Francesco Coletti, Chemical Engineering, Imperial College London, London, United Kingdom and Sandro Macchietto, Chemical Engineering, Imperial College, London, United Kingdom

Stringent environmental policies and tightening refining margins in a highly competitive market are forcing oil companies worldwide to increase their focus on the efficient utilisation of energy. Fouling, the deposition of unwanted material on heat transfer surfaces is a ubiquitous problem in oil refineries and a major cause of costly inefficiencies.

Traditional design and retrofit methodologies rely on highly empirical, fixed, TEMA fouling factors. The use of these 'safety factors' has been largely criticized [1, 2] as it often results in over-designed heat transfer units that, in theory, should compensate for reduced thermal efficiency given by fouling deposition. Ultimately, however, this approach turns out to exacerbate fouling rather than mitigating it as a consequence of producing higher temperatures at start-up and lower fluid velocities than those for which the unit was designed [3]. The ability to capture at the design stage the dependence of fouling on process conditions and fouling evolution over time becomes therefore pivotal to achieve fouling–resilient designs and retrofits of heat exchangers.

A novel, sophisticated thermo-hydraulic mathematical model for shell-and-tube heat exchangers that takes into account local fouling rates and deposit ageing as a function of exchanger geometry, oil characteristics, process conditions and time was recently proposed [4]. The model has been validated against real plant data from four units in two refineries operated by major oil companies showing for all units an accuracy in predicting outlet temperatures (in °C) over extended periods (i.e. 4–16 months) with an excellent accuracy of ±1% for the tube–side and ±2% for the shell–side [5]. The model has also been used to assess economic and environmental impact of fouling [6] and to evaluate retrofit configurations for the network structure [7] of refinery pre-heat trains.

In this paper, the model is used to assess alternative retrofit options for existing units. The thermo-hydraulic simulations allow capturing in a quantitative way the complex interactions between flow dynamics and fouling deposit growth over time, and establishing the trade-off between costs and benefits. Instant and integral energy losses caused by fouling can thus be accurately assessed for all the retrofit options proposed and configurations that mitigate fouling by design identified.

References

[1] Chenoweth, J. M., 1997, "The Tema Standards Fouling Section: Another Look," Proc. Fouling

Mitigation of Industrial Heat-Exchange Equipment, C. B. Panchal, ed. San Luis Obispo, California (USA), pp. 61-71.

[2] Bennett, C. A., Kistler, R. S., Lestina, T. G., and King, D. C., 2007, "Improving Heat Exchanger Designs," Chemical Engineering Progress, 103(4), pp. 40-45.

[3] Butterworth, D., 2004, "Process Heat Transfer 2010," Applied Thermal Engineering, 24(8-9), pp. 1395-1407.

[4] Coletti, F., and Macchietto, S., 2009, "A Dynamic, Distributed Model of Shell-and-Tube Heat Exchangers Undergoing Crude Oil Fouling," Submitted for publication.

[5] Coletti, F., and Macchietto, S., 2010, "Validation of a Novel Model for Shell and Tube Heat Exchangers Undergoing Crude Oil Fouling," Proc. 14th International Heat Transfer Conference (IHTC14), Washington, DC (USA).

[6] Coletti, F., and Macchietto, S., 2009, "Refinery Pre-Heat Train Network Simulation: Assessment of Energy Efficiency and Carbon Emissions," Proc. International Conference on Heat Exchanger Fouling and Cleaning VIII, H. Müller-Steinhagen, et al., eds., Schladming, Austria, pp. 61-68.

[7] Coletti, F., Macchietto, S., and Polley, G. T., 2010, "Effects of Fouling on Performance of Retrofitted Heat Exchanger Networks; a Thermo-Hydraulic Based Analysis," Ischia, Naples, Italy, accepted.

See more of this Session: Process Heat Transfer Technology
See more of this Group/Topical: Transport and Energy Processes

(46c) Integrating Crude Oil Fouling Modelling and Experiments On a New High Pressure Pilot-Scale Test Rig

Francesco Coletti, Geoffrey F. Hewitt, Denis R. Dugwell and Sandro Macchietto, Department of Chemical Engineering, Imperial College London, London, United Kingdom

Crude oil fouling is a long-standing problem in oil refineries which causes major energy efficiency and economic losses, extra environmental emissions and disruption of operations, with related health and safety hazards. Because of the complex and interacting phenomena involved, the underlying mechanisms leading to fouling depositions are not well understood. Acquiring experimental data in a controlled laboratory environment representative of realistic refinery conditions, through very precise measurements, is a required step for gaining a better understanding of the fouling mechanisms and for testing mitigation strategies. This is important for improving refineries operational efficiency. Moreover, data on crude fouling behavior provide sensitive and important commercial information for oil companies which often base buying, pricing and allocation strategies on the quality of the crudes, including the ease with which they can be processed in specific refineries.

A state-of-the-art high pressure oil rig (HIPOR) has been built at Imperial College London by a team under the lead of Prof. G. F. Hewitt [1]. This facility is designed to process crude oil in two test sections (a tube and an annulus) up to 30 bars and 300 °C and to measure with accuracy key variables such as inlet and outlet temperatures, flowrates, heat fluxes and pressure drops. Moreover, temperature profiles across the length of the annular test section can be measured through a radiation equilibrium thermometer whilst the actual thickness of the foulant layer can be measured simultaneously using a novel dynamic gauging technique [2].

The traditional experimental approach to crude oil fouling research focuses on the collection of fouling data which are typically interpreted through overall, lumped heat balances to yield overall fouling resistances. In this paper, an integrated approach is discussed which combines the detailed data from the above experimental apparatus with sophisticated thermo-hydraulic modelling to tackle the fundamental issues involved in crude oil fouling.

A distributed, dynamic model that takes into account local fouling resistance and the thickness of the deposit layer across the length of a tube was recently proposed [3]. The model also includes the structural changes of the deposit over time (ageing) as well as the impact of surface roughness dynamics produced by the initial deposition and the complex interactions between hydraulic and thermal effects of fouling. That model is modified here to match the specific operating conditions and geometries of the test rig.

The synergies achieved by combining experimental and modelling activities are discussed. They include a way to better interpret the data obtained, to identify key parameters (e.g. thermal conductivity of fouling layer, sheer stress, etc) and to validate the mathematical model. This integrated approach also allows the rapid testing of alternative models of fouling, ageing and roughness dynamics and help explain the complex phenomena leading to deposition. Moreover, the use of model-based experiment design techniques to plan the experiments is discussed. The goal here is to devise effective experiments that maximize the information content and minimise experimental costs and time. References

[1] Macchietto, S., Hewitt, G. F., Coletti, F., Crittenden, B. D., Dugwell, D. R., Galindo, A., Jackson, G., Kandiyoti, R., Kazarian, S. G., Luckham, P. F., Matar, O. K., Millan-Agorio, M., Müller, E. A., Paterson, W., Pugh, S. J., S.M., R., and Wilson, D. I., 2009, "Fouling in Crude Oil Preheat Trains: A Systematic Solution to an Old Problem," Proc. International Conference on Heat Exchanger Fouling and Cleaning VIII, H. Müller-Steinhagen, et al., eds., Schladming, Austria.

[2] Tuladhar, T. R., Paterson, W. R., and Wilson, D. I., 2002, "Investigation of Alkaline Cleaning-in-Place of Whey Protein Deposits Using Dynamic Gauging," Food and Bioproducts Processing, 80(3), pp. 199-214.

[3] Coletti, F., Ishiyama, E. M., Paterson, W. R., Wilson, D. I., and Macchietto, S., 2010, "Impact of Deposit Ageing and Surface Roughness on Thermal Fouling: Distributed Model," AIChE J. Accepted for publication.

See more of this Session: Process Heat Transfer Technology
See more of this Group/Topical: Transport and Energy Processes

Eddy Impaction as an Ash Deposition Mechanism: An Theoretical and Experimental Investigation

Minmin Li, Larry Baxter, Sayantan Ghosh.
Chemical Engineering Department, Brigham Young University,
Provo, UT 84601

Abstract

The eddy impaction ash deposition model predicts eddy impaction rates as a function of turbulence intensity, boundary layer thickness, and gas velocity. The experimental apparatus utilizes the blender method to introduce toner particles (5µm diameter) into a gas stream flowing through a horizontal pipe (Re 2,300-8,000). The toner particles impact onto the pipe wall and the total mass of impacted particles is measured. Experimental relationship between the deposition velocity and Reynolds number shows increasing trend, which follows the deposition model predict that eddy impaction increases with the increasing the turbulence.

Introduction

Ash deposition represents a long-standing operational and design research issue. Ash deposits contribute to corrosion and decrease heat transfer by decreasing thermal conductivity and emissivity and increasing effective path length. They also increase weight and stress on components, radically increasing maintenance associated with ash removal. Therefore, ash formation, transformation and deposition models can provide valuable information to help reduce corrosion, fouling and emission problems [1]. Computer simulations and fuel performance indicts help select proper fuels and operating conditions, reducing ash deposition problems. Relatively little quantitative experimental data exist for validating predictions of ash deposition simulations. Detailed and quantitative ash deposition data and models could significantly improve thermal processor (combustor, gasifier, etc.) designs and performance. While many of the mechanisms of ash deposition are well established, such as inertial impaction and condensation, eddy impaction rate expressions and understanding remains largely empirical.

Eddy impaction describes impaction rates associated with particles with too little momentum to transit average fluid boundary layers around surfaces but within the range that they can transit this boundary layer given the extra energy associated with, and the change in boundary layer thickness associated with, turbulent eddy velocity fluctuations. This investigation will seek to determine the fundamental mechanisms and theoretical descriptions of eddy impaction, including original model development and experimental investigations. The goal is to provide a sound fundamental basis to replace the existing largely empirical description of this phenomenon.

The dual focus of this study is to present an analytical eddy impaction ash deposition model and validate obtained results via comparison with experimentally gathered data. This work will largely deal with monodisperse particles in ideal, isothermal flows for which eddy impaction is the only deposition mechanism.

Model description

Start from the differential velocity equation and force balance on a particle, use the initial condition to derivate, finally find the relationship between the eddy impaction efficiency and the dimensionless parameter.
Beginning with a force balance on a particle, where the primary forces are drag and weight,

$$\sum \vec{F} = \frac{d\, m_p \vec{v}_p}{dt} = \vec{F}_D + \vec{F}_g$$

For small particles (small Reynolds numbers),

$$C_D = \frac{24}{Re} = \frac{24\mu_g}{\rho_g d_p |\vec{u}_g - \vec{u}_p|}$$

Leading to

$$\frac{d\,\vec{u}_p}{dt} = \beta(\vec{u}_g - \vec{u}_p) + \vec{g}$$

where

$$\beta = \frac{18\mu_g}{\rho_p d_p^2 C_C}$$

The Cunningham correction factor for small particles has been added,

$$C_c = 1 + \frac{2\lambda}{d}\left[A_1 + A_2 \exp\left(-\frac{A_3 d}{\lambda}\right)\right]$$

and A_1, A_2, and A_3 = 1.257, 0.4, 0.55 in air, respectively.
Assuming the initial position is at the edge of the boundary layer and is taken as zero (which is an arbitrary value, so this does not decrease the general application of this approach), fluctuations sufficient to deposit a particle on the surface are

$$u'_p > \beta\delta$$

The magnitudes of particle and gas velocity fluctuations are related by

$$\frac{u'^2_{p,rms}}{u'^2_{g,rms}} = \frac{\beta^2}{(\beta + r)^2 + s^2}$$

where the ratio is between root-mean-square fluctuations. The large particle fluctuation may actually exceed that of the gas:

$$u'_{g,rms} = \frac{\sqrt{(\beta + r)^2 + s^2}}{\beta} u'_{p,rms}$$

where s is the gas fluctuation characteristic time constant and r is the gas turbulence decay rate.
Assuming a Gaussian distribution of gas velocity fluctuations, that is, that the probability density function for the gas fluctuations has is given by

$$g(u_g) = N(0, u'_g) = \frac{1}{\sqrt{2\pi} u'_g} \exp\left[-\frac{u_g^2}{2u'^2_{g,rms}}\right]$$

The fraction of particles depositing on the surface is the same as the probability of x_p^∞ exceeding the boundary layer thickness δ, and is given by

$$\eta = p(x_p^\infty > \delta) = \frac{1}{2}\mathrm{erfc}\left[\frac{\beta^2\delta}{u'_{p,rms}\sqrt{2[(\beta + r)^2 + s^2]}}\right] = \frac{1}{2}\mathrm{erfc}\left(\frac{\beta\delta}{u'_{g,rms}\sqrt{2}}\right)$$

defining an eddy impaction Stokes number as the inverse of the argument of the complementary error function above.

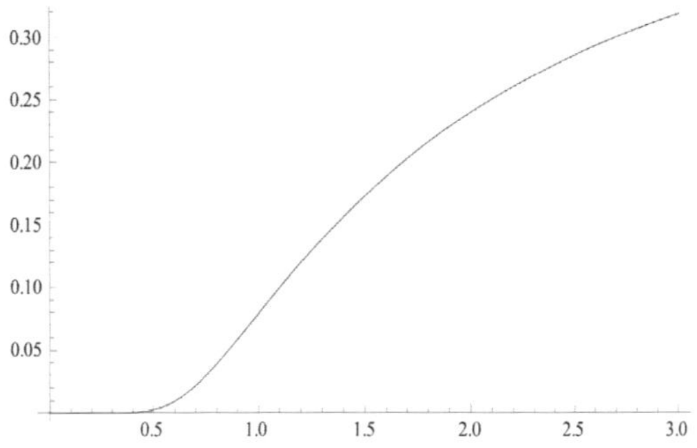

Figure 1 Plot of eddy impaction efficiency vs. the dimensionless parameter ——— or, equivalently, ——— up to a value of 2.0.

As graph indicates, the eddy impaction efficiency increases with increasing particle/gas turbulent intensity and decrease with increasing boundary layer thickness.

Experimental

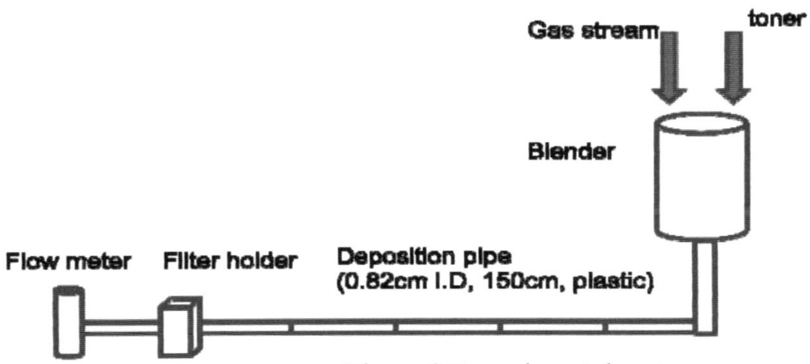

Figure 2 Experimental system

Figure 2 is a schematic diagram of the system used. In this investigation, HP color toner was used as a surrogate for ash. The toner particles are uniform in size with a diameter of 5μm. The deposition experiment was run at room temperature and ambient atmosphere pressure conditions. A long and thin deposition pipe generates a fully developed flow. Blender (Rival 6-spend blender) is help to mix the particles. Gas stream transports particles through the tube over a range of Reynolds numbers. Sulfur hexafluoride (SF_6) is used because it's the densest gas which can make the higher Reynolds number. The deposition tube is separated into five 30 cm sections and they are connected by larger flexible tubing which creates and air tight seal. Separate sections make it possible to determine deposition in each individual section of pipe. A filter holder containing a 42.5mm diameter filter paper (Whatman GF/A 42.5mm, Cat No1820

042) collects the particles that don't deposit on the tube surface. The flow transducer was previously calibrated with a standard precision rotameter. Generally, all the particles deposit on the tube prior to exiting the flow system. The amount of toner in each section is monitored over a range of flow rates.

Results

Table 1. Sample particle deposition data for deposition pipe

	Deposited in section(g)	Penetrated in section(g)	Fractional penetration in section	Deposition velocity for section(cm/s)	Average deposition velocity calculated from sections1-4
Connection	0.1241				
Section5(30cm) (Entrance)	0.2087	1.2040	0.852	0.246	
Section4(60cm)	0.3507	0.8533	0.709	0.540	
Section3(90cm)	0.2566	0.5967	0.699	0.552	0.611
Section2(120cm)	0.2023	0.3944	0.661	0.671	
Section1(150cm) (Exit)	0.1383	0.2561	0.649	0.684	
Filter paper	0.1320				
sum	1.4127	1.4127			
	Flow rate=	122.67cm^3/s	Re=	7300	
	D=	0.82cm	L=	30cm	

Table 1 shows the result for a typical experimental run. The quality of HP laser printer toner on the individual section is shown in column 1 and column 2. From these values, the fraction of particles penetrating through each individual section is calculated. The deposition velocity is calculated from the equation given by Sehmel [2]:

$$V = \frac{Q}{\pi DL}\ln\left(\frac{1}{P}\right)$$

Where *Q* is the volumetric flow rate, *D* is the i.d. of the deposition pipe, *L* is the length of the section, and *P* is the fractional penetration through the individual section.

Since the fractional penetration between section 1 and 4 are essentially constant, that the corresponding deposition velocities based on these penetration values vary by not more than 10 percent from the average deposition velocity of 0.611cm/sec. The lower deposition velocity for section 5 near the entrance to the deposition pipe is expected because of the finite transition length needed for the flow to become fully turbulent.

In order to consider the entrance length effect, Reynolds number will be focus on because it is the only parameter affecting that [3]. If,

$$Le = f(d, V, \rho, \mu) \quad V = \frac{Q}{A}$$

then $\frac{Le}{d} = g\left(\frac{\rho v d}{\mu}\right) = g(Re)$

For laminar flow, the accepted correlation is

$$\frac{Le}{d} \approx 0.06Re$$

The maximum laminar entrance length, at $Re_{d,crit}$=2300, is Le=138d, which is the longest development length possible, where d is diameter of the tube. In this system, d=0.82cm, thus Le=113cm. In turbulent flow the approximation for smooth walls:

$$\frac{Le}{d} \approx 4.4Re_d^{1/6}$$

Some computed turbulent entrance lengths are calculated in table 2.

Table 2 some computed turbulent entrance length based on Reynolds number

Re	4170	5210	6250	7300	8350
Le	17.65	18.32	18.88	19.38	19.82

Table 2 indicates that the entrance effect may be only effect the section near the entrance. However, even in laminar flow, section 1 and 2 which are the farthest sections away from entrance are fully developed. So the section 1 and 2 will be investigated. Figure 3 is the deposition velocity changed with Reynolds number in section1 and 2.

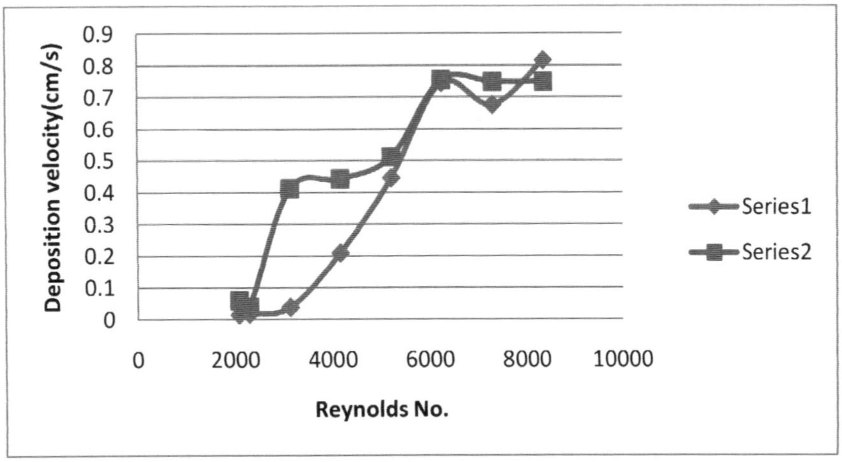

Figure 3 Experimental relationships between deposition velocity and different Reynolds number

In figure3, there very a rapid increase in the deposition through the pipe with increasing the flow rate which increases the Reynolds number before it reaches 6000. Then the deposition velocity moderate changes with Reynolds number. They almost look same when Re lower than 2300. In transit region, they are apart. In turbulence region starts from 5000, they begin to close again. That indicates the system is stable due to the section 1 and 2 are at different places. However, the deposition velocities in transit area are random. Generally, this increasing trend follows the figure 1, eddy impaction increases with the increasing the turbulence.

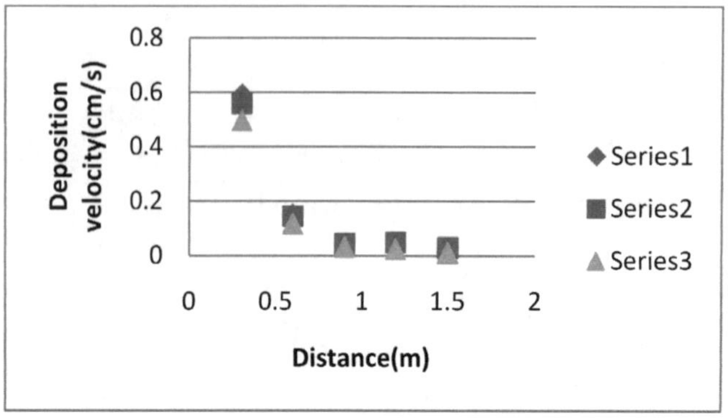

Figure 4 Plot of deposition velocity at different sections in different runs when Re=2300

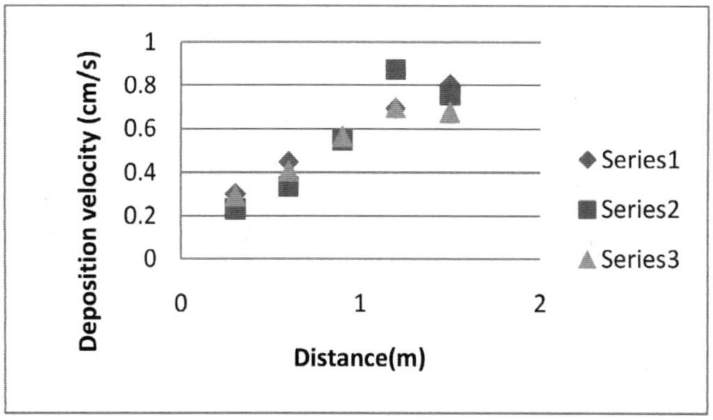

Figure 5 Plot of deposition velocity at different sections in different runs when Re=6250

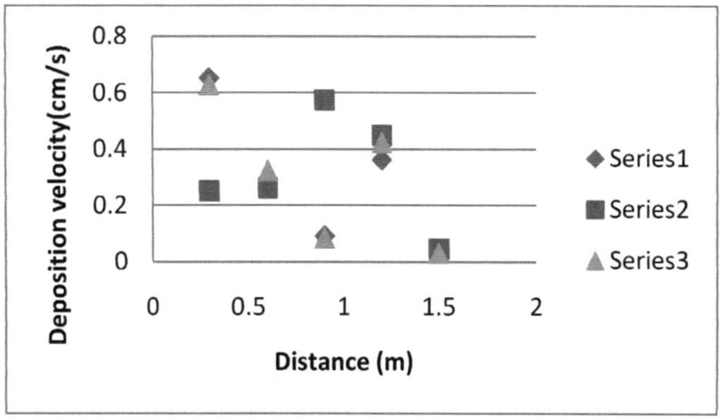

Figure 6 Plot of deposition velocity at different sections in different runs when Re=3130

Several duplicate runs were made throughout the course of these experiments to determine the reproducibility of the experiments and to discover whether all interfering variables had been brought under experimental control.

The results (Fig.4 and 5) tell us that the maximum difference at duplicate runs is about 10 percent which suggested that the system is stable and only small variations in laminar and turbulent region. However, Figure 6 shows more random because when Re=3130 is in transit region that the particles' movements are more complex.

Table 3 Particle deposition in 0.82cm i.d plastic pipe

Reynolds No.	Average velocity of gas	Fraction velocity(v*)	Dimensionless particle relaxation time, τ+	Average deposition velocity,V(cm/s)	Dimensionless deposition velocity, V+
2085	0.664	0.051	0.002	0.037	0.736
2300	0.730	0.055	0.002	0.028	0.513
3130	0.996	0.072	0.003	0.225	3.109
4170	1.328	0.093	0.006	0.326	3.501
5210	1.660	0.113	0.008	0.478	4.223
6250	1.992	0.133	0.012	0.750	4.594
7300	2.324	0.152	0.015	0.712	4.690
8350	2.656	0.171	0.019	0.782	4.582

Table 3 is the summary of the experimental results for average deposition velocity at different Reynolds numbers. A dimensionless relaxation time of the particles, τ_+, is defined by Liu and Agarwal [4] as

$$\tau_+ = \frac{\tau v_*^2}{v}$$

and a dimensionless deposition velocity defined as:

$$V_+ = \frac{V}{v_*}$$

Here v is the kinematic viscosity of air, τ is the particle relaxation time, and v_* is the friction velocity of the turbulent flow. For spherical particles of a radius a, the relaxation time is given by Liu and Agarwal [4] as:

$$\tau = \frac{2\rho_P \, a^2 C}{9\mu}$$

where ρ_P is the particle density, μ is the viscosity of the gas, and C is the slip correction. The friction velocity defined by Liu and Agarwal [4]:

$$v_* = (f/2)^{\frac{1}{2}} \bar{u}$$

where $\bar{u}$ is the average velocity of the fluid in the pipe. The friction factor, f, for the smooth pipe, has been calculated from the Blasius formula, valid for Re<100,000, given by Schlichting[5]:

$$f = \frac{0.316}{4(Re)^{1/4}}$$

where Re is the pipe Reynolds number.

Figure 7 is a plot of the experimental results given in Table 3. The dimensionless deposition velocity, V+, is seen to increase rapidly with increasing particle relaxation time, $\tau+$, reaching a peak at a $\tau+$ value of approximately 0.003. After that, V+ shows decreasing slightly at higher $\tau+$.

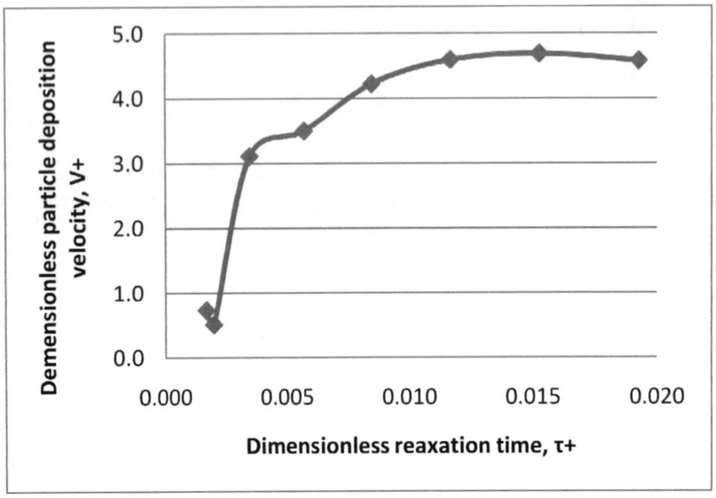

Figure 7 Experimental relationship between the dimensionless particle deposition, V+ , and the dimensionless particle relaxation time, $\tau+$.

Comparison and conclusion

Several of the existing theories on turbulent deposition are based on the so-called "diffusion free-flight" model. In this model, particles are assumed to be transported by turbulent diffusion from the turbulent core through the boundary layer to within one "stopping distance" from the wall, at which point the particles make a "free flight" to the wall.

In the theories of Friedlander and Johnstone [6], Davies [7] and Beal [16], the particle diffusivity in the boundary layer is assumed to be the eddy momentum diffusivity of the fluid. In the theory of Davies [7], the initial free-flight velocity of the particle is assumed to be the root-mean-square fluctuating velocity of the fluid as the free flight begins. Liu and Agarwal [4] (see Fig 8) have reported some new experiments for deposition on the walls of a vertical pipe. They also compared their experimental data with other theories (see Figure 8). The theory of Liu and Ilori[8] shows equally good agreement with experimental data between τ_+= 2 and 20.The deposition velocities estimated by Davies' model are considerably lower in the τ_+<30 region. Sehmel [2] measured the variation in particle deposition velocity with size for particle diameters over the range estimated between 10^{-4} and 10^{2} μm. His experimental data for large particles gave deposition rates for the floor and ceiling of a wind tunnel, providing an evaluation of the influence of gravity on the effective deposition velocity.

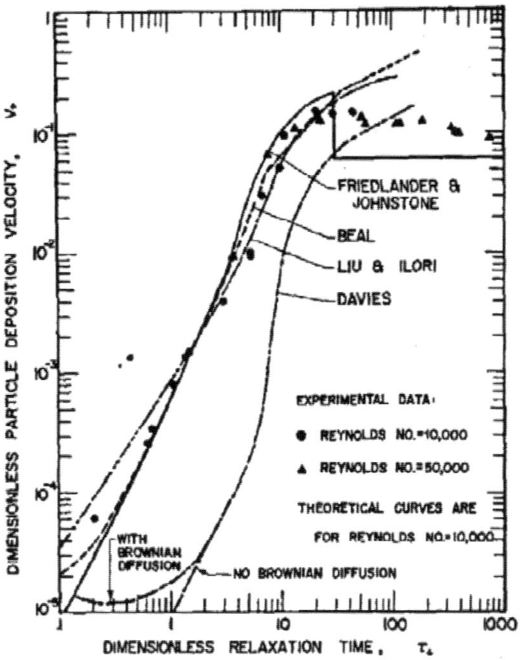

Figure 8 Comparison of theories and experiments by Benjamin Y.H. Liu and Jugal K.Agarwal

Figure 7 shows the tiny particles deposition behaviors in a short time. When τ_+<<2, the dimensionless deposition velocity increased as the dimensionless relaxation time increasing. Compared with Figure 8, this trend follows the other theories' results. It indicates my experimental results have good agreement with existing theories and also can extend their model.

Conclusion:

- With the Reynolds number increased, more ash particles will deposit further sections.
- Blender can largely improve the stable of the system
- In the transition region, the deposition velocity distributes a little random, however, with the Reynolds number increased from 6000, ash prefer to deposits further place.
- More ash deposits with increased the Reynolds number.

References

[1] Joller M, Brunner T, Obernberger I, *Modeling of aerosol formation during biomass combustion for various furnace and boiler types*. Fuel Processing Technology,2007. **88**: p.1136-1147.

[2] Sehmel, G.A, *Particle eddy diffusivities and deposition velocities for isothermal flow and smooth surfaces*. J.Aerosol Sci., 1973. **4**: p.145.

[3] Frank M.White *Fluid mechanics fifth edition*, p.350.

[4] B.Y.H. Liu and J.K.Agarwal, *Experimental observation of aerosol deposition in turbulent flow*. J. Aerosol Sci., 1974. **5**: p.145.

[5] Schlichting, H, *Boundary layer theory*. McGraw-Hill, New York, 1968. Ch.20.

[6]Friedlander, S.K and Johnstone, H.F, *Deposition of suspended particles from turbulent gas streams*. Ind. Engeg Chem, 1957. **49**: p.1151.

[7]Davies, C.N, *Deposition from moving aerosols*. In Aerosol Science Academic Press, New York,1966.
[8]Liu, B.Y.H and Ilori,T.A, *Inertial deposition of aerosol particles in turbulent pipe flow*. Presented at the ASME Symposium on flow studies in air and water pollution, Altanta, Georgia, 1973: p.20-22.

(46e) Marangoni Condensation of Steam-Ethanol Mixtures On a Horizontal Tube at Sub-Atmospheric Pressure

Hassan Ali, School of Engineering and Material Science, Queen Mary University Of London, London, United Kingdom

Recent experimental studies (Utaka and Wang (2004), Murase et al. (2007), Li et al. (2008) have found that very small concentrations of ethanol in the liquid phase can increase significantly heat-transfer coefficients compared to those with pure steam.

Condensation of steam-ethanol mixture on a horizontal tube at atmospheric pressure was reported in Murase et al. (2007). In view of the possible application of steam-ethanol mixture condensation in power plant condensers where condensation takes place below atmospheric pressure, accurate experimental data at sub-atmospheric pressures become necessary. The present paper reports experimental results for steam-ethanol mixture condensation on a horizontal tube over a wide range of vapour-to-surface temperature difference and composition at sub-atmospheric pressures.

See more of this Session: Process Heat Transfer Technology
See more of this Group/Topical: Transport and Energy Processes

Concentration and Temperature Gradient Measurements at Binary Mixture Boiling Bubbles

Oliver S. Knauer, Andreas Braeuer, and Alfred Leipertz
Lehrstuhl für Technische Thermodynamik and Erlangen Graduate School in Advanced Optical Technologies (SAOT),
Universität Erlangen-Nürnberg, Erlangen, Germany

Introduction

Nucleate Boiling is commonly used in many engineering applications, as due to the phase change huge amounts of heat can be transferred. This is quantified by the heat transfer coefficient, which is the transferred heat per heater surface area divided by the temperature difference between the heated wall and the bulk liquid. In many configurations, the working fluid is often a mixture of liquids with different saturation temperatures and a gap between the bubble point line and the dew point line in the vapor-liquid-equilibrium (VLE)[1]. In these cases, the heat transfer coefficient is significantly reduced compared to the value of the pure fluids or the volume fraction averaged value of the mixture[2,3]. The main reason for this effect is the preferential evaporation of the more volatile component, causing composition gradients in the liquid. Though many correlations and models are reported in the literature[4,5], the occurring heat and mass transfer processes and especially their close and complex linking are not fully understood. To contribute to knowledge about the physical phenomena and to validate models for boiling of mixtures, non-invasive measurements with high spatial and temporal resolution were carried out. Therefore, a laser-based measurement technique as one-dimensional Raman spectroscopy was applied to measure composition and temperature gradients in the liquid surrounding boiling bubbles of binary mixtures simultaneously.

Raman spectroscopy for simultaneous Composition and Temperature Measurements

The Raman effect occurs when light is scattered inelastically from a molecule. During the scattering process, the energy level of the molecule changes and causes a shift in the wavelength of the scattered photon. Due to energy conservation, this so called Raman shift (denoted in wavenumbers [cm^{-1}]) is related to the difference of the involved energy levels of the molecule and is therefore species specific. When a monochromatic light source as a laser is used for excitation, the Raman shift can be calculated, when the scattered light is spectrally resolved. As the intensity of the scattered light is linear to the number of scattering molecules, quantitative composition measurements can be carried out[6,7]. Figure 1 shows calibration spectra of acetone and isopropanol mixtures for different preset volume fractions x_{IPP} at a constant temperature of 308 K acquired with the setup described later. It can be seen that the intensity of the OH-peak between 3100 and 3700 cm^{-1} increases as the mixture gets richer in isopropanol.

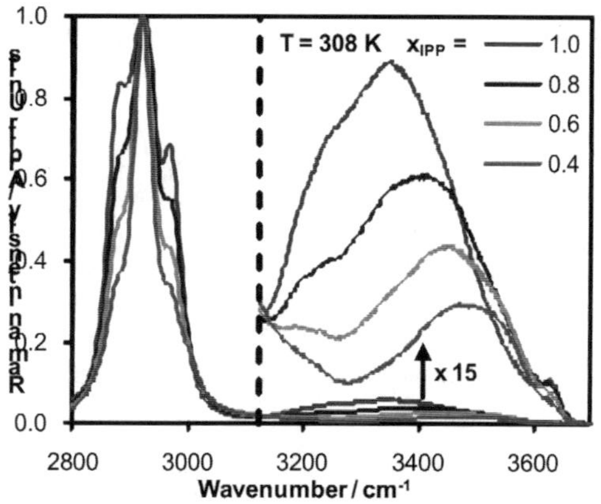

Figure 1 Raman spectra of acetone-isopropanol mixtures at different volume fractions

In liquids containing water or alcohols, hydrogen bonds between the molecules occur, which influence the shape of the OH-peak in the Raman spectrum. Figure 2 shows Raman spectra of a mixture with a constant isopropanol volume fraction $x_{IPP} = 0.8$ at different temperatures. With increasing temperature of the liquid, more hydrogen bonds break up due to collisions and the OH-peak changes. Temperature measurements based on these changes in the Raman spectra were carried out successfully for different kinds of liquids[8-10]. Details on a novel strategy to evaluate composition and temperature of the studied mixture of acetone and isopropanol from the Raman spectra simultaneously are reported elsewhere[11].

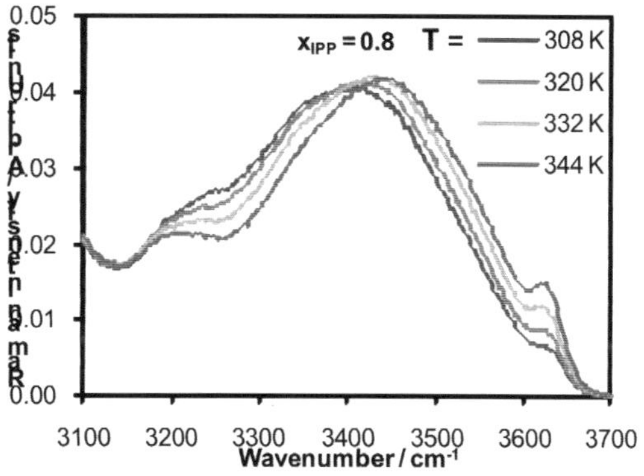

Figure 2 Raman spectra of an acetone-isopropanol mixture at different temperatures

Experimental Setup

Figure 3 shows the measurement chamber to investigate individual boiling bubbles. The inner reservoir is filled with approximately 100 ml of a preset mixture. The outer reservoir surrounding the mixture reservoir can be purged with water from a temperature-conditioning circulator to temper the mixture. A flatted hemisphere of teflon with a hole of 0.5 mm in the center is fixed at the bottom of the inner chamber. An electrical heater is placed inside this hemisphere, so that the small tip of the heater is located just underneath the small hole in the teflon, providing an artificial bubble site. If the temperature of the mixture is set a few Kelvin below its saturation temperature and the heater temperature is set a few Kelvin above, single boiling bubbles begin to grow and detach with a certain frequency, depending on the heater temperature, the bulk mixture temperature and composition. Three windows allow optical access to the location of the bubble nucleation.

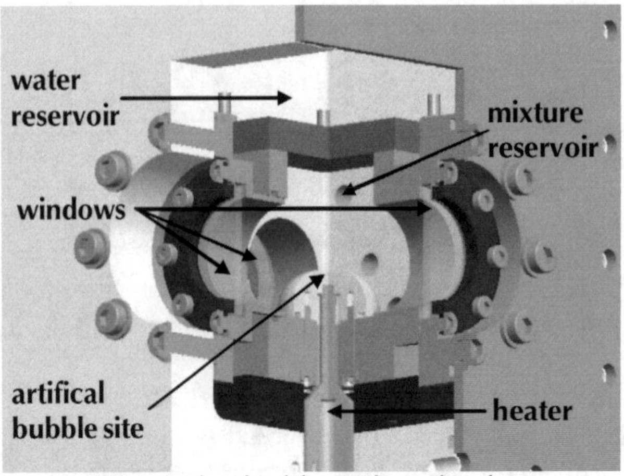

Figure 3 Sketch of the Boiling Chamber

Figure 4 shows the optical setup to acquire Raman spectra from the liquid surrounding the boiling bubbles. A pulsed and frequency-doubled Nd:YAG laser with a repetition rate of 10 Hz is used as excitation source. As Raman scattering cross sections are very low, high pulse energies are required to produce sufficient signal intensities. To avoid laser-induced window damage or breakdown in the liquid, two precautions are carried out to reduce the pulse power per area but keep the total pulse energy constant. Firstly, an optical pulse stretcher expands the temporal width of the laser pulse from approximately 10 ns to 100 ns. Secondly, a system of lenses (L) enlarges the diameter of the laser beam, so that it passes the windows in a larger area, when focused into the measurement chamber (MC). The laser is blocked behind the measurement cell with a beam dump (BD). An achromatic lens (AC) collects and collimates the light scattered from the measurement volume (MV) in a solid angle as large as possible. A dichroitic mirror reflects the elastically scattered light to a charge-coupled-device (CDD) camera, taking a photo of the measurement volume. The inelastically scattered light is transmitted and remaining laser light is blocked by filters (F), before the Raman signal is focused onto the entrance slit of an imaging spectrometer (IS) by another achromatic lens. An electron-multiplying charge-coupled-device (EMCCD) with a quantum efficiency of more than 90 % in the required wavelength range detects the Raman spectra, resolving even the minor differences in the shape of the OH-peak. The imaging spectrometer disperses the elastically scattered light from the measurement volume into its spectral components while conserving the spatial information. The width of the spatial axis of the EMCCD camera and the magnification factor of the optical setup define a 3.2 mm long measurement volume of 20 increments, resulting in a spatial resolution of 160 µm. Spectra from all measurement points can be acquired simultaneously and composition and temperature can be evaluated from each spectra.

Acetone and isopropanol were chosen as constituent parts of the mixtures for several reasons. They are suited for Raman measurements, as they are transparent, completely miscible, and the OH-bond of isopropanol enables temperature measurements. As they form a large gap in the binary vapor-liquid-equilibrium (VLE), the mixture impact on the boiling process is distinct and therefore large composition and temperature gradients are expected. For the calibration measurements, the mixture in the chamber was preset to a certain composition and temperature and Raman spectra were averaged over 1000 acquisitions. A large range of isopropanol volume fraction (x_{IPP} = 0.4, 0.5, 0.6, 0.7, 0.8, and 0.9) and temperature (from T = 308 K up to the saturation temperature of the corresponding mixture in steps of 4 K) was covered for calibration. For the measurements on boiling bubbles, the EMCCD and the CCD camera were temporally synchronised to the pulses of the laser. With the photos of the CCD camera, the measurements, where a boiling bubble was inside the measurement volume, could be selected for evaluation of composition and temperature.

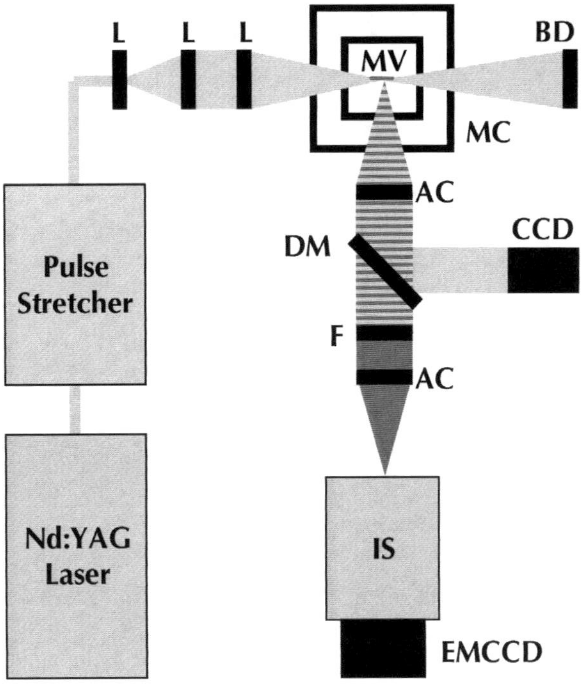

Figure 4 Sketch of the optical setup

Results

Measurements were carried out at boiling bubbles in a mixture of a bulk isopropanol volume fraction x_{IPP} = 0.4. The bulk temperature was set to T_{bulk} = 333 K and the heater was set to a temperature of T_{heater} = 336 K. Approximately 85 acquisitions, where a boiling bubble was located inside the measurement volume and hit by the laser beam in the same height as depicted in Figure 5, were evaluated. In the photo, the scale and the location of the measurement volume is indicated by the white axis.

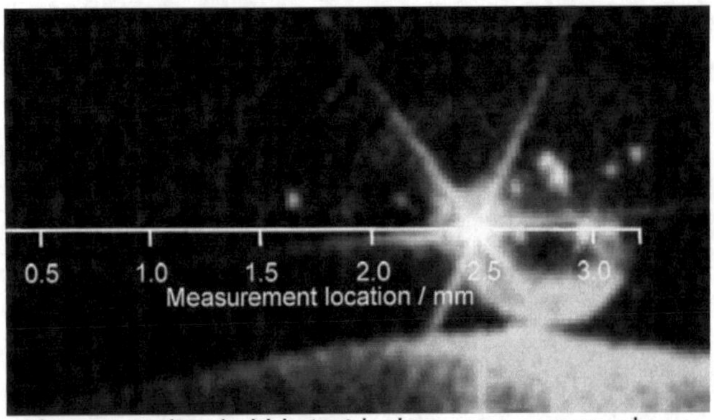

Figure 5 Boiling bubble inside the measurement volume

Figure 6 shows the spatially resolved averaged values of temperature and composition together with the corresponding standard deviation bars. As the Raman signal intensity in the gas phase is much weaker than in the liquid phase, spectra from beyond the phase boundary (dotted line) located at a measurement position of approximately 2.4 mm are not reliable and therefore not shown. The results show an accumulation of isopropanol near the phase boundary, due to preferential evaporation of acetone, which is the more volatile component. The temperature increase near the phase boundary indicates a heat transfer from the boiling bubble to the surrounding liquid due to evaporation of vapor from inside the boiling bubble at the phase boundary. Isopropanol volume fraction as well as temperature for measurement locations smaller than 1.4 mm hold the preset bulk values of $x_{IPP} = 0.4$ and respectively $T = 333$ K. This indicates that a single boiling bubble influences the surrounding liquid not further than approximately 1 mm away from the phase. A species conservation calculation[11] supports this assumption: It shows that the composition profile can be explained by a demixing process inside a sphere covering the boiling bubble and 1 mm of surrounding liquid. The minima in both composition and temperature at a measurement location of 1.7 mm, which is 0.7 mm away from the phase boundary, were not expected and could not be explained up to now.

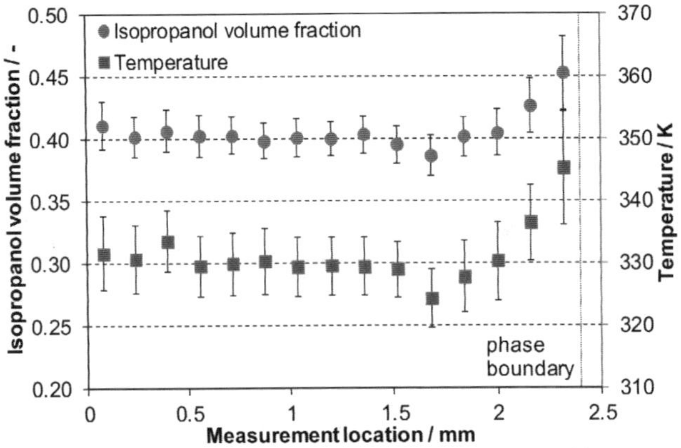

Figure 6 Composition and temperature profile

Conclusions

For the first time, one-dimensional Raman spectroscopy was applied at boiling bubbles of acetone-isopropanol mixtures. Extensive calibration measurements and a novel evaluation strategy enabled non-invasive measurements of composition and temperature profiles with high spatial and temporal resolution. Results from bubbles produced in a temperature adjustable boiling chamber were reported. Further measurements under systematic variation of the different experiment parameters, as mixture composition and temperature or heater temperature, will be carried out and help to get new insights in the heat and mass transfer phenomena of binary mixture boiling.

Aknowledgements

The authors gratefully acknowledge funding of parts of the project by the German Research Foundation (DFG), which also founds the Erlangen Graduate School in Advanced Optical Technologies (SAOT) in the framework of the German excellence initiative.

References

1. Gültekin N. (1989), "Vapor-liquid equilibria for binary and ternary systems composed of acetone, 2-propanol, and 1-propanol", *Journal of Chemical & Engineering Data,* 34(2), 168-171.
2. Shock R. A. W. (1982), "Boiling in Multicomponent Fluids", *Multiphase Science and Technology,* 1(1-4), 281-386.
3. Collier J. G., Thome J. R. (1996), "Multi-Component Boiling and Condensation", in *Convective boiling and condensation,* 3rd ed., Oxford; New York: Clarendon Press ; Oxford University Press.
4. Inoue T., Kawae N., Monde M. (1998), "Characteristics of heat transfer coefficient during nucleate pool boiling of binary mixtures", *Heat and Mass Transfer,* 33(4), 337-344.
5. Celata G. P., Cumo M., Setaro T. (1994), "A review of pool and forced convective boiling of binary mixtures", *Experimental Thermal and Fluid Science,* 9(4), 367-381.
6. Braeuer A., Leipertz A. (2009), "Two-dimensional Raman mole-fraction and temperature measurements for hydrogen-nitrogen mixture analysis", *Applied Optics,* 48(4).
7. Dowy S., Braeuer A., Schatz R., Schluecker E., Leipertz A. (2009), "CO2 partial density distribution during high-pressure mixing with ethanol in the supercritical antisolvent process", *Journal of Supercritical Fluids,* 48(3), 195-202.
8. Hopkins R. J., Reid J. P. (2006), "A Comparative Study of the Mass and Heat Transfer Dynamics of Evaporating Ethanol/Water, Methanol/Water, and 1-Propanol/Water Aerosol Droplets", *The Journal of Physical Chemistry B,* 110(7), 3239-3249.
9. Müller T., Grunefeld G., Beushausen V. (2000), "High-precision measurement of the temperature of methanol and ethanol droplets using spontaneous Raman scattering", *Applied Physics B-Lasers and Optics,* 70(1), 155-158.

10. Risovic D., Furic K. (2005), "Comparison of Raman spectroscopic methods for the determination of supercooled and liquid water temperature", *Journal of Raman Spectroscopy,* 36(8), 771-776.

11. Knauer O. S., Lang M. C., Braeuer A., Leipertz A. (2010), "Simultaneous determination of the composition and temperature gradients in the vicinity of boiling bubbles in liquid binary mixtures using one-dimensional Raman measurements", *Journal of Raman Spectroscopy,* http://dx.doi.org/10.1002/jrs.2689.

(99a) Barriers and Challenges to the Selective Removal of Vanadium From the Asphaltene Fraction

Greg P. Dechaine and Murray R. Gray, Chemical and Materials Engineering, University of Alberta, Edmonton, AB, Canada

Most crude oils contain traces of vanadyl porphyrins within their asphaltene fraction. Although these metals are only present in trace quantities, they have a significant detrimental impact on crude oil processing units; therefore, their selective removal is highly desirable. The fact that the majority of the vanadium is contained within the highly aromatic, highly polar asphaltene fraction also poses additional roadblocks to their selective removal. This fraction has been shown to associate/aggregate significantly in most (if not all) solvents. Recent work on the nature and possible mechanisms of the molecular association of asphaltenes in solution can be extended to help elucidate the molecular interactions occurring between asphaltenes and metalloporphyrins, and hence the nature of the inclusion of metalloporphyrins within the asphaltene fraction. Solubility measurements with simple vanadyl porphyrins showed that they fit the operational definition for asphaltenes, exhibiting negligible solubility in n-heptane and orders of magnitude higher solubility in toluene. Further studies were also carried out using a stirred diffusion cell. These experiments determined that the size of the asphaltene aggregates in toluene at 25°C and 1 g/L were between 5 and 9 nm. These aggregated structures were extremely stable, remaining intact for periods of >7 days. Neither an increase in temperature nor a decrease in concentration (to 0.1 g/L) reduces the size of these structures below 5 nm, and the origin of the asphaltenes (Athabasca, Safaniya, Venezuela) did not have a significant impact. A significant amount (>85%) of the native petroporphyrins present in the asphaltene fraction remain bound within the above aggregated structures and hence are not susceptible to selective removal. The native petroporphyrins that are free to diffuse appear to be larger than model vanadyl porphyrins as indicated by pore hindrance effects. Neither an increase in temperature nor a decrease in concentration results in an increase in the amount of petroporphyrins available for diffusion. The selective separation of metalloporphyrins is clearly hampered by gaps in the basic understanding of the aggregation behavior of the asphaltene fraction, as well as gaps in understanding metalloporphyrin properties and behavior in solution.

See more of this Session: Heavy Oil and Other Challenging Resources II
See more of this Group/Topical: Topical D: Chemical Engineering in Oil and Gas Production and Other Complex Subsurface Processes

SOLUBILITY AND MASS TRANSFER PARAMETERS OF HYDROGEN IN VACUUM RESIDUES UNDER HYDROCONVERSION CONDITIONS

Mariela Sanoja, Laurent Sehabiague, Yannick Heintz, Rachid Oukaci* and Badie I. Morsi
Chemical and Petroleum Engineering Department
University of Pittsburgh, Pittsburgh, PA 15261

**Energy Technology Partners, LLC*
Pittsburgh, PA 15238

The volumetric liquid-side mass transfer coefficient (k_La) and equilibrium solubility (C^*) for H_2 were measured in two different vacuum residues (Zuata and Merey-Mesa) using one-liter agitated autoclave operating in a Gas-Inducing mode under the typical conditions of hydroconversion of heavy residues. The effects of operating variables, including pressure (up to 55 bar), temperature (up to 300 °C), mixing speed (up to 2000 rpm) and solid concentration (up to 40 wt%), on k_La and C^* were investigated using the Central Composite Statistical Design and Analysis technique. The transient physical gas absorption method and a modified Peng-Robinson Equation-of-State were used to obtain k_La and C^*, respectively.

Under the operating conditions investigated, k_La values for H_2 were found to increase with increasing pressure, temperature and mixing speed, and to decrease with increasing solid loading. Also, the equilibrium solubilities for H_2 in the two vacuum residues appeared to increase with pressure at constant temperature and increase with temperature at constant pressure. These data are indispensable for proper design and scaleup of slurry reactors heavily employed in hydro-processing of vacuum residues.

(99c) On Specific Partial Molar Volume of Asphaltenes in Organic Liquids and Their Solubility Parameters

Kasra Nikooyeh and John M. Shaw, Chemical and Materials Engineering, University of Alberta, Edmonton, AB, Canada

Similarity of Hildebrand solubility parameter is frequently used in petroleum science, as a measure of compatibility of constituents and is commonly used for interpreting and correlating asphaltene-diluent interactions. Specific partial molar volume at near infinite dilution is a sensitive measure of solute-solvent interactions derived from high precision density measurements of dilute mixtures. It equals the partial molar volume of a solute divided by its molar mass, but is obtained directly from the density measurements without invoking a molar mass for the solute. Such a measurement is appropriate here because the mean molar mass of asphaltenes is not known a priori. In this contribution, the specific partial molar volumes of Athabasca and Maya heptane asphaltenes at near infinite dilution are reported for organic liquids over the temperature range 20 - 80 C. The organic liquids comprise pure compounds and binary mixtures including: heptane, toluene, 1-methylnaphthalene, quinoline, anisole, 2,6-lutidine, pyridine, methylene chloride, tetrahydrofuran, decane). At 20 C, these organic liquids possess solubility parameters ranging from 15 MPa0.5 (heptane) to 22 MPa0.5 (quinoline and methylene chloride). Surprisingly, the solubility parameter of the organic liquids, in this range, does not appear to correlate with the specific partial molar volume of these asphaltenes at near infinite dilution. Estimation of the solubility parameter of asphaltenes is discussed.

See more of this Session: Heavy Oil and Other Challenging Resources II
See more of this Group/Topical: Topical D: Chemical Engineering in Oil and Gas Production and Other Complex Subsurface Processes

(99d) Upgrading of Athabasca Oilsands Bitumen by Low Temperature Natural Zeolite Cracking

A.S.M. Junaid[1], W. Wang[2], C. Street[2], M. Rahman[2], W. McCaffrey[2] and S. Kuznicki[2], (1)Chemical and Materials Engineering, University of Alberta, Edmonton, AB, Canada, (2)University of Alberta, Edmonton, AB, Canada

Synthesized or acidified minerals, particularly clays, have been found to effectively break heavy petroleum resources such as gas oil. However, the application of natural zeolites (NZs) as cracking catalysts and/or adsorbents for oilsands bitumen is a novel approach. Canadian oilsands bitumen, the majority of which is in Alberta, represents approximately 15% of the proven global crude oil resources, and is an extremely valuable source of energy for North America. However, the density, viscosity, contaminant content, and residue fraction of the bitumen recovered from these sands are much higher than those of conventional crude oil due to the presence of large molecules such as asphaltenes. Our studies show that acidified NZs such as clinoptilolite from Saint Clouds, New Mexico and Ca-chabazite from Bowie, Arizona can be developed as very effective reagents for cracking of oilsands bitumen. Low-cost NZs, particularly clinoptilolites, are present globally in large natural deposits. NZ-cracked oilsands bitumen products are highly recoverable (up to ~84%) using light hydrocarbons such as pentane, which indicates substantial conversion of heavier fractions to lighter components. Boiling point distribution and viscosity measurements of the products demonstrate that NZ catalyzed cracking results in much less viscous low boiling point products compared to those produced from pure thermal treatment, or analogous catalytic treatment using the standard petroleum cracking agent, zeolite Y. An additional benefit of using NZs for oilsands bitumen processing is that they function as strong adsorbents to facilitate removal of contaminant heteroatoms such as vanadium (up to ~80%) and nitrogen (up to ~38%). Based on our findings we envision a NZ-based alternative to traditional primary upgrading processes for bitumen, that operates at lower temperatures, improves the transportability of bitumen, and simultaneously removes contaminants, reducing processing steps. This study investigates the efficacy of NZ catalyzed cracking for conversion of heavy fractions of Athabasca oilsands bitumen. Boiling point distribution, average molecular weight and viscosity of the catalytically cracked samples are compared with baseline values for raw and thermally cracked bitumen samples.

See more of this Session: Heavy Oil and Other Challenging Resources II
See more of this Group/Topical: Topical D: Chemical Engineering in Oil and Gas Production and Other Complex Subsurface Processes

SUPERCRITICAL WATER TO UPGRADE PETROLEUM FEEDSTOCK

Ki-Hyouk Choi, Ali M. Al-Somali, Mohammad F. Aljishi, Joo-Hyeong Lee, Mohammed R. Al-Dossary, Ashok K. Punetha, Bader M. Al-Otaibi, and Mohammad Z. Al-Bahhar
Research & Development Center, Saudi Aramco, Dhahran, Saudi Arabia

Introduction

Depleting good-quality crude oil necessitates the reliance on the abundant resources of non-conventional hydrocarbons and heavy & sour crude oils to meet the steadily increasing oil demand while supplying "clean" fuel to the market. However, these resources require major treatments since they are associated with low API gravity and high sulfur content, which present many challenges for refineries to manage these crude oils until final products. Alongside these obstacles, environmental regulations are getting more stringent to minimize air pollution. Growing cleaner alternative energy sources, biofuels for example, exert pressure on crude oil products, which are considered environmentally harmful. This presents an additional challenge in upgrading heavy and sour hydrocarbons. One of the most important processes for refining petroleum based fuels to meet the regulations on transportation fuels is desulfurization.

Available Technologies

At present, there are two major existing technologies to upgrade heavy and sour oil: carbon rejection and hydrogen addition [1]. Carbon rejection results in breaking down hydrocarbons into smaller chains. However, it results in the formation of coke, hence, a product loss. The product of carbon rejection is often unstable and requires further treatment. In general, carbon rejection is an inefficient process that does not lead to substantial results in upgrading heavy and sour oil. These factors make carbon rejection an unattractive option. The alternative, hydrogen addition, make use of hydrogen and a catalyst to upgrade heavy and sour oil. Coke formation is mostly avoided in this process. Although this process is preferred over carbon rejection, the use of hydrogen and a catalyst binds the process to a high cost which presents economical limitations. These factors call for innovative ideas to upgrade heavy and sour oil. A lot of effort is done to obtain new processes as alternative technologies to the existing ones for upgrading and desulfurization of heavy and sour oil. Different non conventional methods have been proposed and evaluated. For example, ultrasonic irradiation was reported to show certain performance in upgrading and desulfurizing heavy oil. A method is also discussed and introduced in this paper which is using supercritical water , which acts as an organic solvent at the critical point and pressure, to reduce the amount of sulfur in the heavy and sour oil.

Properties of SCW

Hydrothermal treatment of heavy and sour oil through supercritical water (SCW) is a viable option. SCW, which is water at temperatures over 374 °C and a pressure over 218 atm, exhibits distinctive chemical and physical behavior [2]. At supercritical conditions, hydrogen bonding in water is substantially weakened, which results in a significant increase in water ionic properties and a decrease in its dielectric constant. The ionic product of water could increase as much as 3 orders of magnitude, which results in an excess of $[H^+]$ or $[OH^-]$. This makes reactions involving SCW, whether acidic or basic, self-catalyzed. At supercritical conditions, water density is substantially reduced as well as viscosity, which minimizes the barrier for rotational and translational motion, enhancing water transport properties as a reaction stimulating solvent. Through cage effect, which stabilizes free radicals, coke formation is avoided. In addition, supercritical water stabilizes the transition state by modifying the geometry, easing the path toward upgraded heavy and sour oil. These unique chemical and physical behaviors could be utilized to upgrade heavy and sour oil. This is achieved through homolytic cleavage of covalent bonds which produces free radicals. Supercritical water-treatment of heavy oil achieves many goals at once; significant reduction in sulfur content, increase in API gravity, and reduction in asphaltene and metal content. The graph below shows the relationship between the temperature and dielectric constant at different pressures. Dielectric constant of supercritical water is very near to that of dichloromethane(DCM) and toluene[3].

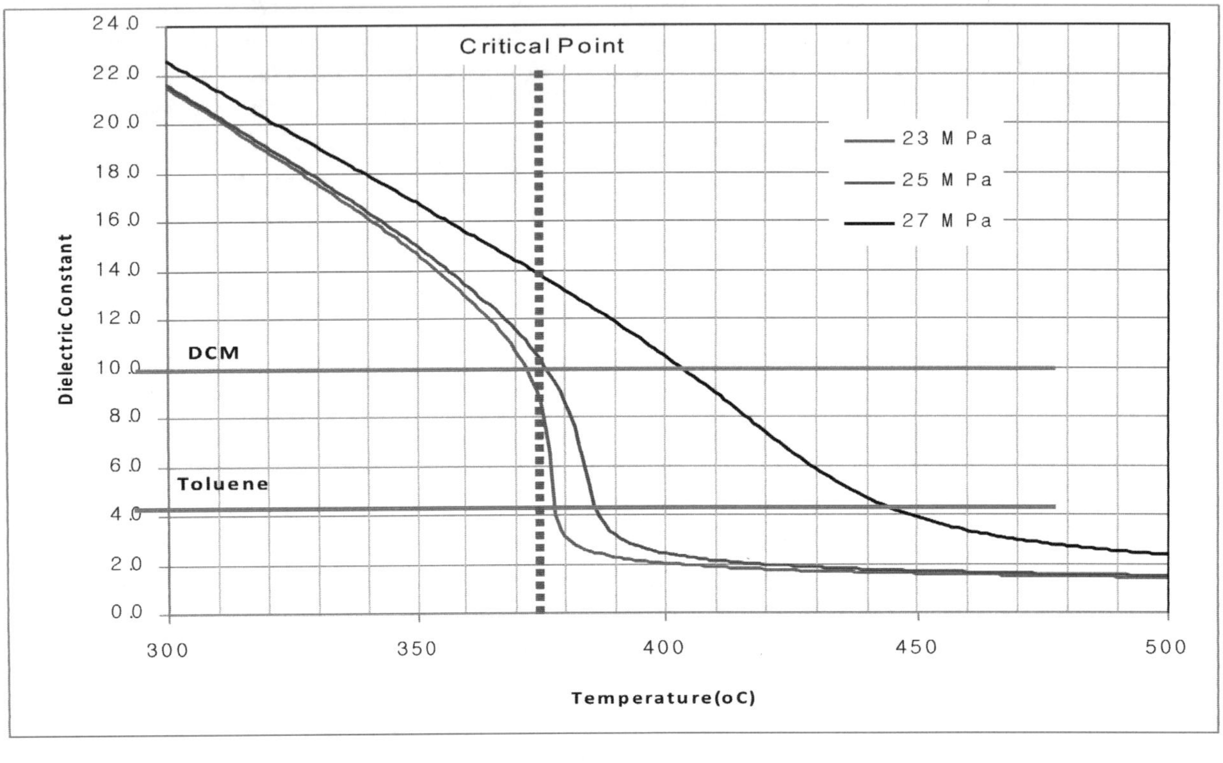

Fig. 1. Dieletric constant of supercritical water.

Experimental

A series of model compounds was tested to elucidate probable mechanism of reactions occurring in supercritical water environment. As a representative sulfur compounds included in heavy oil, thiophene, benzothiophene, and hexyl sulfide were selected for testing. Also, tetralin, decalin, dodecane, and toluene were used to simulate matrix of heavy oil derived from petroleum. Content of sulfur compounds was adjusted to between 0.5 – 2.2 wt% sulfur. All chemicals used for model compound study were obtained from Aldrich and had high purity. **Table 1** summarized some properties of heavy oil used for this study.

Table 1. Properties of Heavy Oil

Property	Measured Value	
Gravity	27.0 API Degree	
Total Sulfur Content	3.0 wt% Sulfur	
Distillation Property (GC-SIMDIS)	5 vol%	92 °C
	10 vol%	152 °C
	20 vol%	249 °C
	30 vol%	333 °C
	40 vol%	424 °C
	50 vol%	515 °C
	60 vol%	628 °C
Kinematic Viscosity(SUS)	38 SUS at 100 F	

Testing was conducted by using a reactor made of stainless steel tube reactor (0.25 inch or 0.5625 inch O.D., 0.083 inch or 0.188 inch I.D.) maintained at 380 - 485 °C by electric heater. Deionized water and oil were pumped into the reactor by high pressure metering pumps through separate lines which were heated to certain temperatures. The oil feed including model compounds, was not pre-heated to avoid conversion before combing with supercritical water. Water to oil ratio was kept at 1 or 2 by vol/vol. Residence time of feed mixture in the reactor was calculated by assuming that the density of model feed was the same to that of water in the reaction condition. Product stream from the reactor was cooled to room temperature before depressurization to atmospheric pressure. The liquid phase product is collected during certain periods to measure total liquid mass balance which was almost 100 wt %. The gaseous product was analyzed by on-line mass spectrometer to obtain qualitative composition. No solid deposit was observed in the reactor after the reaction. GC-MS(Agilent 7890 and MSD), GC-FID(Agilent 6890), and GC-SCD(Agilent 6890 coupled with Sievers 355) were used to identify and quantify product molecules. Total sulfur content and SIMDIS were measured by relevant ASTM methods. Also, 1H and 13C-NMR were taken to analyze change of hydrocarbons during supercritical water treatment.

Results and Discussion

Upgraded oil results indicate that a 15-30% reduction in sulfur content was achieved while increasing API gravity without generating coke at the reaction conditions of 380-450 °C and 3,200 – 3,600 psig.. **Table 2** shows the amount of sulfur removed during the tests.

Table 2. Properties of Feed and Product Oil.

	Feed	Product	Conversion
Total Sulfur	1.27%	1.16%	8.9%
Thiophene	0.67%	0.59%	11.5%
Benzothiophene	0.31%	0.30%	5.3%
Hexyl Sulfide	0.29%	0.28%	5.1%

Reaction temperature, pressure and residence time used for the testing of Table 2 were 413 °C, 1.53 seconds, and 3,600 psig, respectively.

Study with model compounds was conducted to investigate the effect of matrix hydrocarbon on desulfurization of thiophene. As shown in Fig. 1, dodecane showed the highest desulfurization activity than that of tetralin and decaline. Conversion of thiophene in decalin was much lower than that in tetralin. In general, decomposition of thiophene was understood to proceed through a radical mechanism.[3] The role of supercritical water in such a radical reaction has been proposed as the hydrogen donor, stabilizer, and dilution medium.[2] Impact of matrix hydrocarbon on the conversion of thiophene was understood to be related to participation of the hydrocarbon in the radical reaction network. As noted above, n-dodecane generated olefin and paraffin at the reaction conditions employed in this study while decalin and tetralin were relatively stable during reaction. From such observations, it was suggested that n-dodecane facilitated radical reactions which were responsible for thiophene conversion, by generating large amount of radicals in the reactor. In contrast, tetralin was known to stabilize radical by hydrogen donation and inhibit further decomposition or combination of the radical[4] although hydrogen donation might facilitate thiophene conversion. It must be noted that the difference of thiophene conversion was significant, what indicates the important role of matrix hydrocarbons in desulfurization.

Products from heavy oil showed the increase of hydrogen to carbon ratio as shown in Fig. 2. Increase hydrogen amount of product seems to be from hydrogen donation from water or steam reforming reaction. Products having lower sulfur content had higher hydrogen to carbon ratio. The amount of saturate and aromatics increased while the amount of asphaltene and resin has decreased by supercritical water treatment as shown in Fig. 3. Also, it was observed that the significant amount of asphaltenic sulfur was changed to maltenic sulfur which is shown in Fig. 4. The graph shows the amount sulfur transferred in the feed and in the product. Gas-phase product contained hydrogen, light hydrocarbons such as ethane, carbon dioxide and also hydrogen sulfide as in Fig. 5. Although mass spectra of Fig. 5 doesn't have quantitative information, it is obvious that some amount of hydrogen was generated in the reactor and sulfur was transformed into hydrogen sulfide. Source of hydrogen gas is not identified yet. However, steam reforming reaction can be probable source of hydrogen and also carbon dioxide

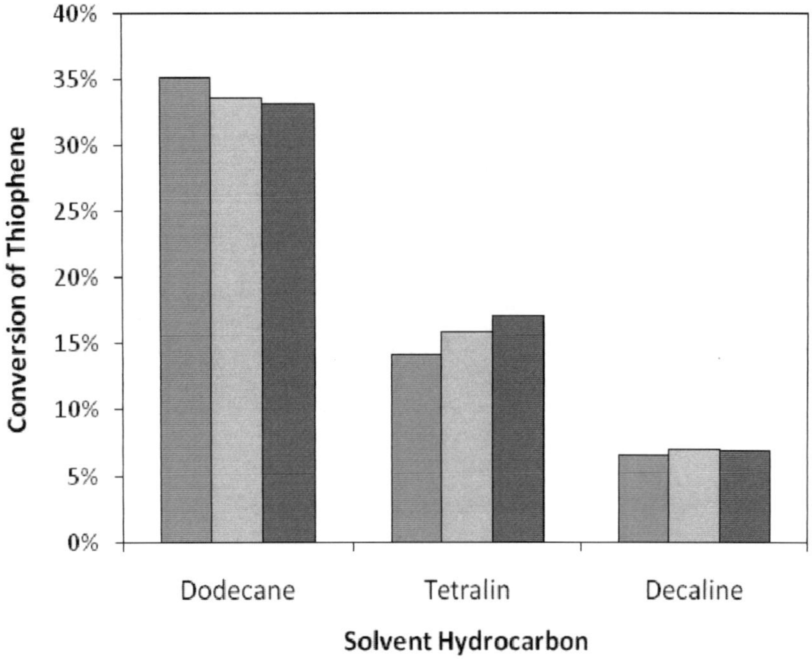

Fig. 1. Thiophene conversion in various matrix hydrocarbon. Reaction temperatures were 403 °C, 457 °C, 485 °C(from left to right). Residence time of feed mixture was in the range of 1.6 – 1.9 seconds at the operating condition.

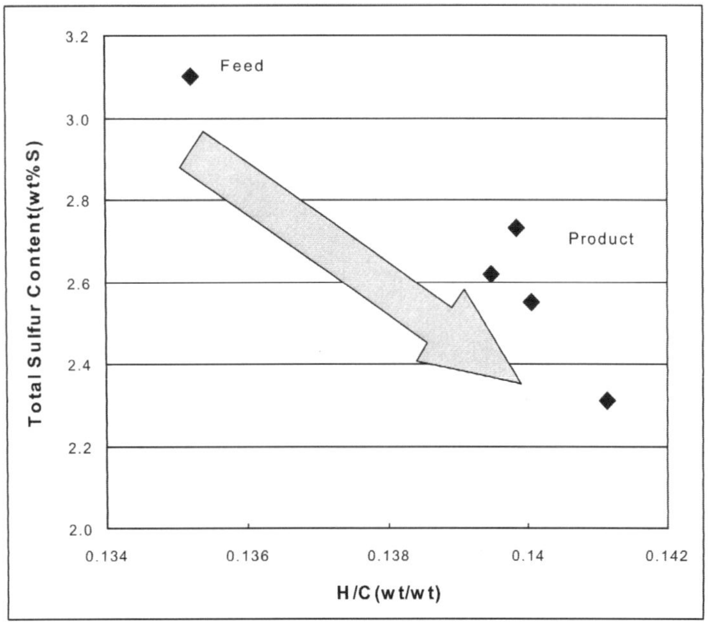

Fig. 2. Hydrogen to carbon ratio of product vs. total sulfur content of the product.

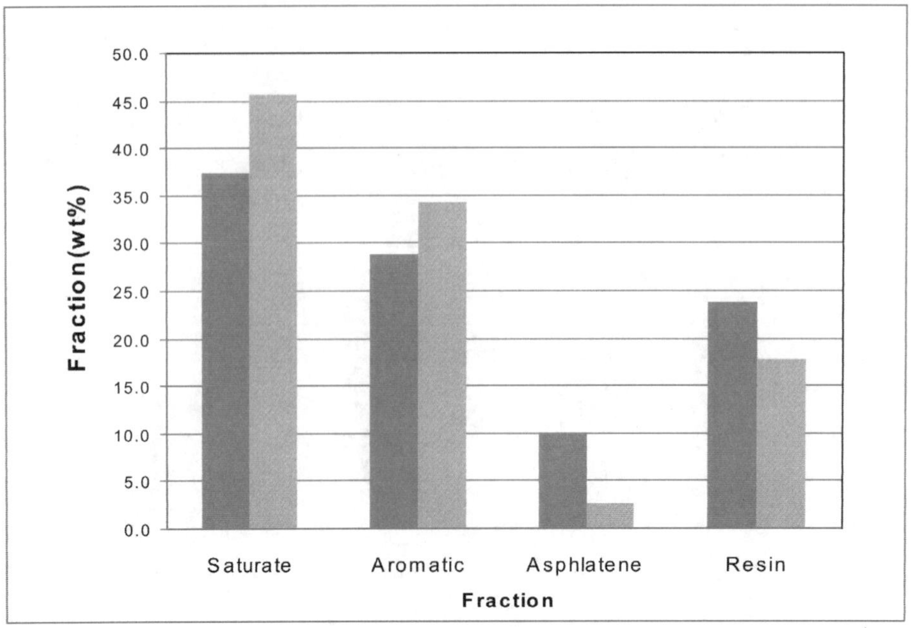

Fig. 3. Saturate, Aromatic, Asphaltene, and Resin amounts of feed(left) and product(right).

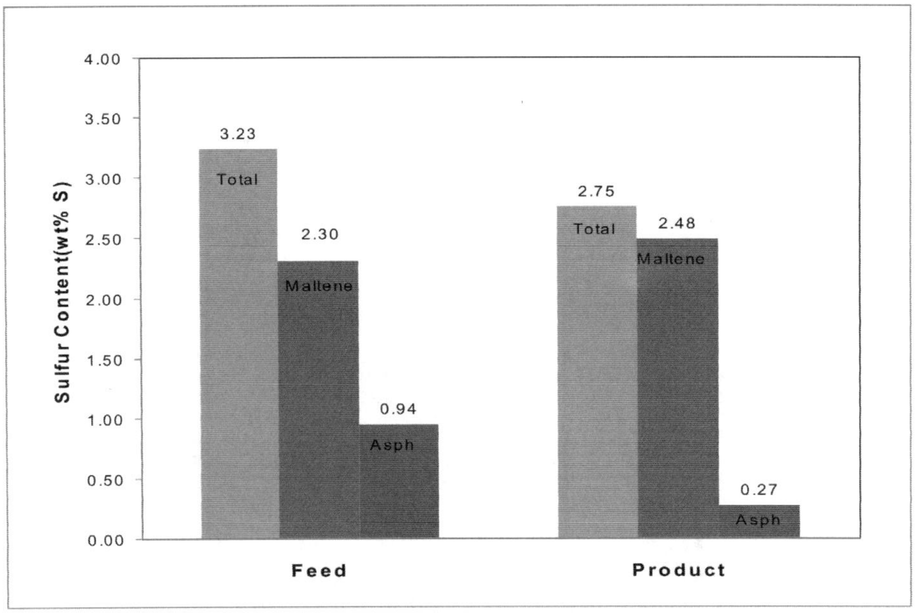

Fig. 4. Sulfur contents of Saturate, Aromatic, Asphaltene, and Resin fractions of feed and product.

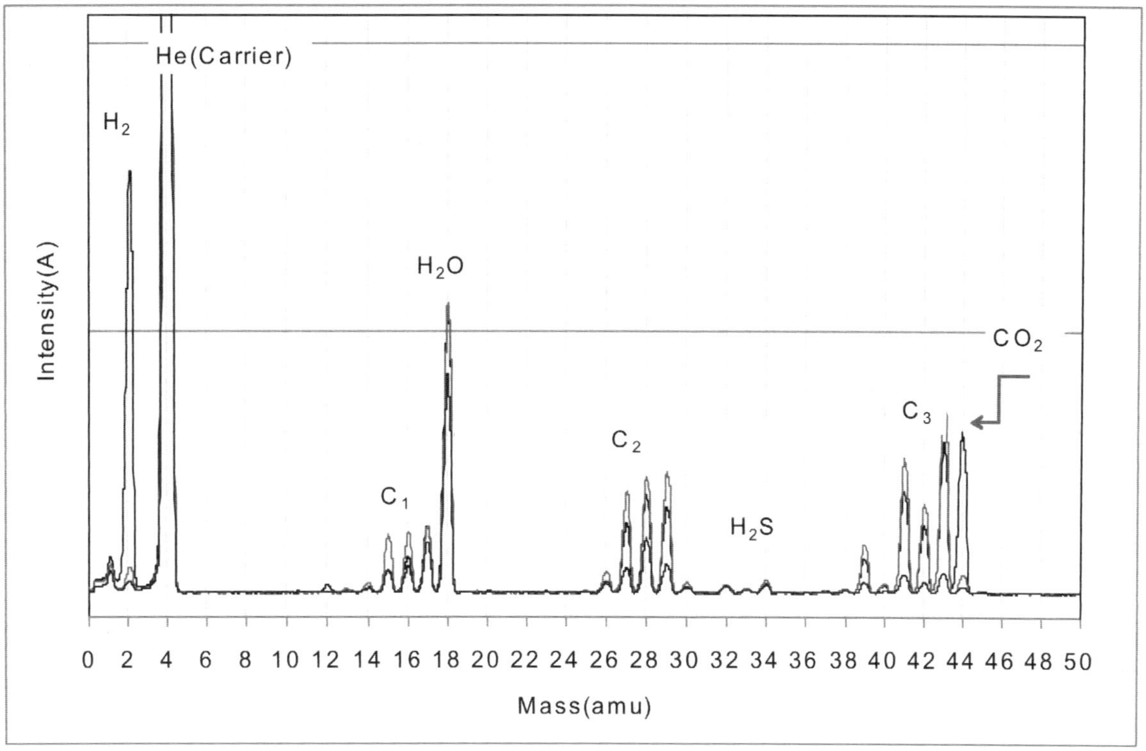

Fig. 5. Mass spectra of product gas. Black and red spectra were taken from different operating codntions. Helium was used as a carrier gas due to small amount of gas-phase product.

Conclusion

This study presents hydrothermal treatment as an innovative technology for upgrading heavy and sour oil. It discusses the unique properties of supercritical water and its role in upgrading heavy and sour oil as well as its applicability in the mid-stream operations sector. Upgrading reactions occurring in the supercritical water conditions were related to radical-mediated network and strongly influenced by co-present hydrocarbons. This process takes advantage of water, which is abundant in upstream operations, in an environmentally friendly manner and self-catalyzed conditions. It also eliminates the need of hydrogen, thus overcoming the economic limitation of hydrogen addition, and avoids product loss through coke formation. In spite of very promising results shown in this study, many technical hallenges still exist to improve the quality of upgraded heavy and sour oil. Challenges, such as corrosion, high energy requirement, and reactor design need to be resolved. Although corrosion is very notorious for supercritical water process, a Japanese group suggested that conventional stainless steel due to absence of oxygen and halogen in feedstock [4]. More comprehensive experiments should be carried to determine optimized conditions for oil-to-water ratio, reaction temperature, and residence time.

Acknowledgement.

Authors are grateful to Dr. Esam Z. Hamad for fruitful discussion.

References

(1) I. V. Babich and J. A. Moulijn, *Fuel*, **82**, 607(2003).
(2) N. Akiya and P. E. Savage, *Chem. Rev.*, **102**, 2725(2002).
(3) M. Uematsu and E. U. Franck, *J. Phys. Chem. Ref. Data*, **9**, 1291(1980).
(4) R. Fujisawa, M. Sakaihara, Y. Miyashita, W. Saiki, and R. Mori, *J. Japan Inst. Energy*, **88**, 180(2009).

.

(99f) Separation of Heavy Oil Emulsions

Jayant P. Rane, Energy Institute, Department of Chemical Engineering, City College of New York, New York, NY, David Harbottle, The Energy Institute, The City College of The City University of New York, New York, NY and Sanjoy Banerjee, Chemical Engineering, Energy Institute, City College of New York, New York, NY

As the conventional light oil is replenishing, the focus has been shifted to the recovery of heavy oil. Almost half of the world's reserves are classified as heavy or extra-heavy. For better understanding of the heavy oil recovery process, it is important to understand the gravity separation process using heavy oil system. Heavy oils have density typically in the range of 10-30 API causing much lower separation rate than conventional crude oil due to very low density difference between two immiscible liquids i.e. oil and water. To enhance the separation rate and achieve similar processing throughputs as observed in conventional crude oil processing, water droplet-droplet coalescence and droplet-homophase coalescence should be maximised. Thus it is important to understand droplet-droplet and droplet-homophase interactions. In a typical industrial process, much of the droplet-droplet coalescence occurs within the process pipeline leading into the gravity settler. However, there is still great uncertainty regarding droplet-droplet coalescence during separation. Aim of the current study is to understand these coalescence mechanisms in the continuous gravity settler and also, to study the droplet-droplet and droplet-homophase interactions independently in the microfluidic cell.

Laboratory scale vertical gravity settlers (height: 1m, diameter, 0.075m) have been constructed to study the coalescence mechanics during separation. To reduce the number of variables that are to be considered, model emulsions have been developed using mineral oil and de-ionised water, stabilised by Span 80. Also, some experiments have been carried out using asphaltenes instead of SPAN80. For the heavy-oil emulsion investigation, mineral oil is premixed with Halocarbon oil (ρ = 1.95g/cc) to increase the oil density to 0.94g/cc. To allow for measurements under steady state conditions a closed loop set-up has been utilised. The droplet size distribution and the emulsion density are measured along the length of the column at the steady state. From the drop size distribution it is possible to track the changes in size as a function of position. Such information is important when trying to understand coalescence mechanics. Through a combination of both measurements, and measurement of the water retrieval flow rate, a coalescence rate at the emulsion-water interface can be inferred. The experimental program considers both the separation of light-oil and heavy-oil systems as a function of surfactant concentration. At steady state, water outlet in the gravity settler corresponds to droplet-homophase coalescence rate. It is found that, this water outlet decreases with increase in surfactant concentration.

Microfluidic setup is used to generate water droplets in the range 20-100 microns in oil using flow-focusing geometry made up of PDMS using soft lithography. In order to understand how droplets interact with each other, droplets are brought into contact with the range of different velocities and surfactant concentrations (Capillary number). Nature of interaction and interaction time is recorded to get a broad spectrum of data. Such data is relevant in the sedimentation zone of the gravity settler where droplet-droplet interactions are predominant. For droplet-homophase interactions, generated droplets are brought in contact with the pool of water homophase at T-junction and droplets are stacked at particular pressure over the pool of water homophase. Such kind of interactions is observed near the packed bed-homophase interphase in the gravity settler.

See more of this Session: Heavy Oil and Other Challenging Resources II
See more of this Group/Topical: Topical D: Chemical Engineering in Oil and Gas Production and Other Complex Subsurface Processes

(135a) Biomass Pretreatment to Improve Bio-Oil Stability1

Najeeb M. Kuzhiyil[1], Dustin Dalluge[2] and Robert C Brown[2], (1)Mechanical Engineering, Iowa State University, Ames, IA, (2)Mechanical Engineeering, Iowa State University, Ames, IA

Storage stability is one of the most important properties of a fuel. Bio-oil produced from biomass by the fast pyrolysis process is known for its poor storage stability which makes it a less convenient fuel. The aldehydes and carboxylic acids contained in the bio-oil react with each other and with other species that causes the instability of the oil. Pretreatments are the first opportunity to improve the quality of bio-oil by selectively controlling the impurities in the feedstock that influence the pyrolysis reactions. Potential biomass pretreatment techniques were studied in order to understand their effect on the stability of the bio-oil produced from the treated feedstocks. Switchgrass was pretreated by washing with hot water and dilute acids separately, and also by infusing various acids at different concentrations. Pyrolysis-GC-MS set up was used to analyze the composition of the bio-oil produced from these treated feedstocks. For the purpose of predicting the stability of the oil produced, the pyrolysis products were categorized broadly into anhydrosugars, light oxygenates, furans and phenolics. Hot water and dilute acid wash improved the yield of levoglucosan and reduced the yield of carboxylic acids and aldehydes while the acid wash increased the yield of furans also. However, the overall effects were not good enough to justify the extra energy spent on drying the washed feedstock. In order to avoid the need for drying, acids were infused into switchgrass and tested. It was found that phosphoric acid at 2 wt% of the biomass produced the maximum yield of anhydrosugars and lowest yield of light oxygenates. Sulfuric acid was slightly better than phosphoric acid while nitric and hydrochloric acids were less effective. Acetic and formic acids were not effective at all. The oils produced from phosphoric and sulfuric acid treated switchgrass were found superior in terms of its storage stability when compared to the control oil.

1,This work was funded by the U.S. Department of Energy and the ConocoPhillips Company

See more of this Session: Biomass Pyrolysis
See more of this Group/Topical: Fuels and Petrochemicals Division

Process Design Data for Liquid Phase Pyrolysis of Biogenic Feedstock

Verena Mertlitz*, Nikolaus Schwaiger, Edgar Ahn, Matthäus Siebenhofer
BDI-BioDiesel International AG, Parkring 18, 8074 Grambach/Austria,
verena.mertlitz@bdi-biodiesel.com

This project focuses on pyrolytic conversion of CHO-feedstock such as fir chips under liquid phase operation conditions. The investigation of the reaction kinetics showed a change in the reaction mechanism at a reaction temperature of 350 [°C]. The activation energy for a reaction temperature below T = 350 [°C] was found to be 18 [kJ/mol]. At higher temperatures an activation energy of 105 [kJ/mol] was observed.
The temperature has a significant influence on the product distribution and the energy balance.

1. Introduction

Pyrolytic conversion of biogenic feedstock has become a high priority task due to declining crude oil reserves.
Subject of this project is the determination of design parameters for industrial scale plant design, including reaction mechanisms, thermodynamics and kinetics of liquid phase pyrolysis of CHO-feedstock. The results are needed for knowledge transfer from lab scale to technical scale.
From liquid phase pyrolysis improved convective heat transfer is expected, enabling fast pyrolysis conditions at ambient pressure. Operation temperature was limited to a range of less than 650 [K]. Investigations were carried out under isothermal conditions.

1.1 Experimental

The process was operated in a lab scale reactor in semi-batch mode. The rate of degradation for nearly monodisperse feed was recorded. The feed quantity was limited to an amount which causes a deviation of the reactor temperature of less than 2 [K] each feed cycle. The effect of temperature on product distribution was investigated. The influence of biomass composition on the mechanism and rate of pyrolytic degradation were analyzed. As feedstock wood, straw, pulp and bran were used.
The liquid and solid products were characterized by GC-MS, NMR-spectroscopy and elemental analysis. Reproducibility was checked with replicates.

1.2 Results and Discussion

The ratio of water formed during pyrolytic degradation increases with increasing temperature, passing a maximum, while formation of CHO-products linearly correlates

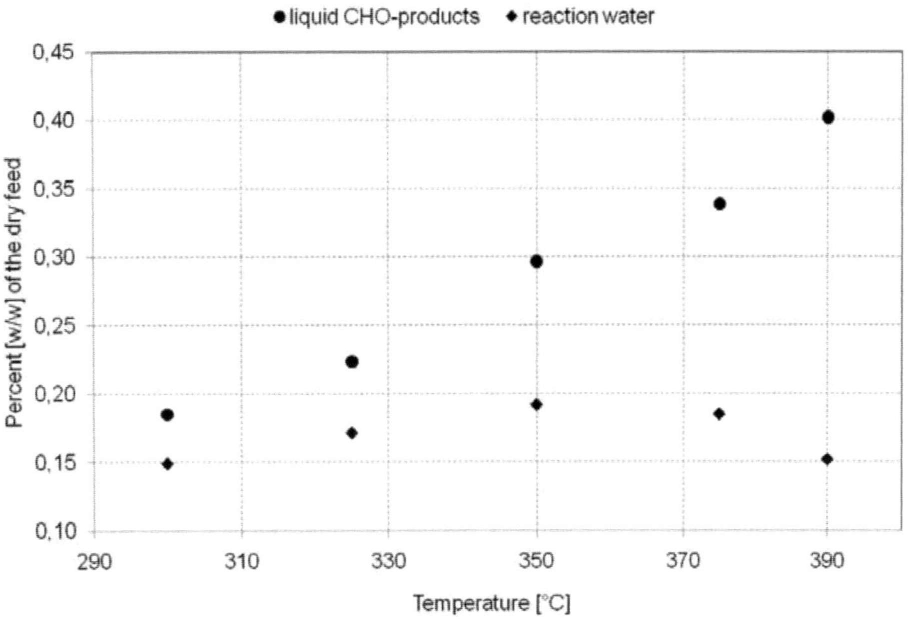

Figure 1: Temperature dependency of liquid products

with the operation temperature (Figure 1). Carbon transfer from feed to gaseous and liquid product phase also correlates linearly with the temperature.

Figure 2 shows the influence of the feedstock on the product distribution.

Figure 3 shows the formation of liquid products (liquid CHO-products plus water formed by chemical reaction) based on dry feed at different reaction temperatures. The feedstock was wood. The rate of formation of liquid products was best modelled by 2nd order kinetics. A shift in the reaction mechanism of the formation of liquid products was observed.

The activation energy for a reaction temperature below 350 [°C] was found to be 18 [kJ/mol]. Above 350 [°C] an activation energy of 105 [kJ/mol] was observed.

The reaction temperature also has an effect on the distribution of product fixed energy (Table 1). At lower temperatures (300 [°C], column 1) the energy of the biomass is mainly stored in the produced char.

Table 1: Energy content of products (based on the original biomass energy content) for different reaction temperatures

product	300 [°C]	350 [°C]	390 [°C]
Char	79%	53%	35%
Gas	1%	3%	12%
liquid CHO-products	19%	39%	47%
Loss	1%	5%	6%

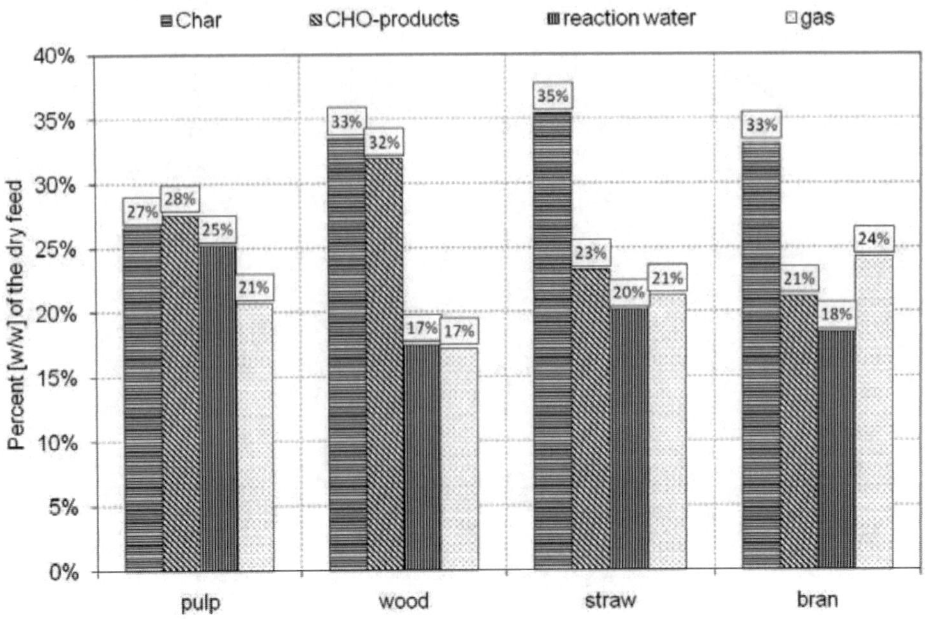

Figure 2: Influence of feedstock on the product distribution (Char, liquid CHO-products and reaction water).

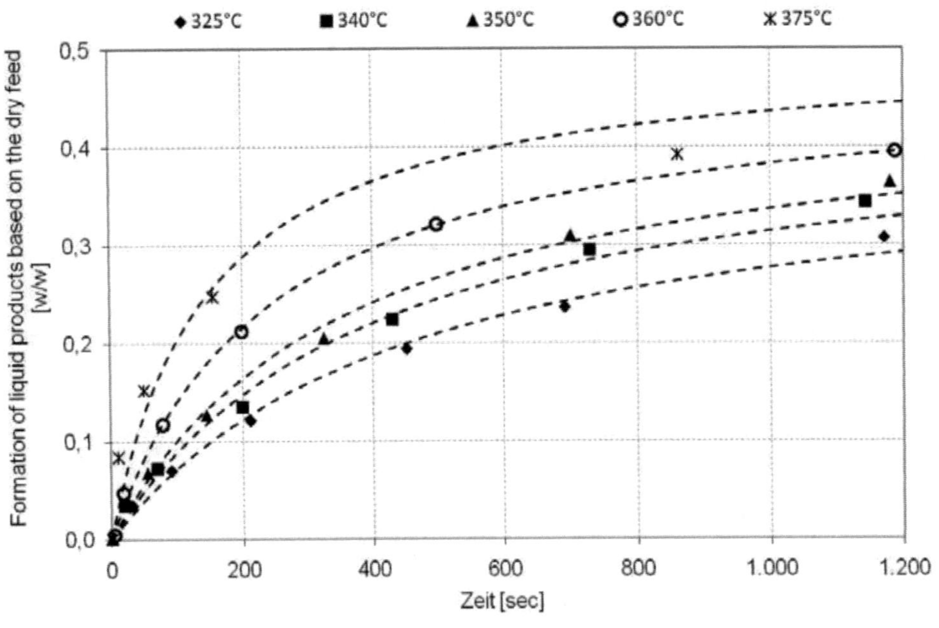

Figure 3: Formation of liquid CHO-products and reaction water based on dry feed.

At higher temperatures (390 [°C], column 3) the liquid-CHO-products are the fraction with the highest energy content. Different biogenic feedstock does not have a significant effect on energy distribution.
According to process thermodynamics liquid phase pyrolysis of wood based feedstock is a moderately exothermic process. The heat of reaction is less than the needed energy for vaporization. According to Figure 4 the overall process needs energy supply.

1.3 Conclusions

Based on the experimental results and thermodynamics it is possible to convert biogenic feedstock such as fir chips and straw into products with higher energy density by liquid phase pyrolysis.
The char left after separation of volatiles is highly reactive. Due to the elevated carbon content compared with the feed the char provides interesting application opportunities for further processing.
The macro-kinetics of liquid phase pyrolysis showed a shift in the reaction mechanism. Further investigation is needed to clarify the influence and interaction of the biomass composites cellulose, hemicelluloses and lignin on the reaction mechanism in detail.
Liquid phase pyrolysis of wood is moderately exothermic reaction energy cannot cover the demand for evaporation of condensable products.

1.4 Acknowledgments

This project is funded by the Climate and Energyfund Austria and is carried out under the program "Energy of the Future".
K. Zahel and A. Prettner are gratefully acknowledged for help and support. P. Pucher and G. Witek are acknowledged for helpful discussions.

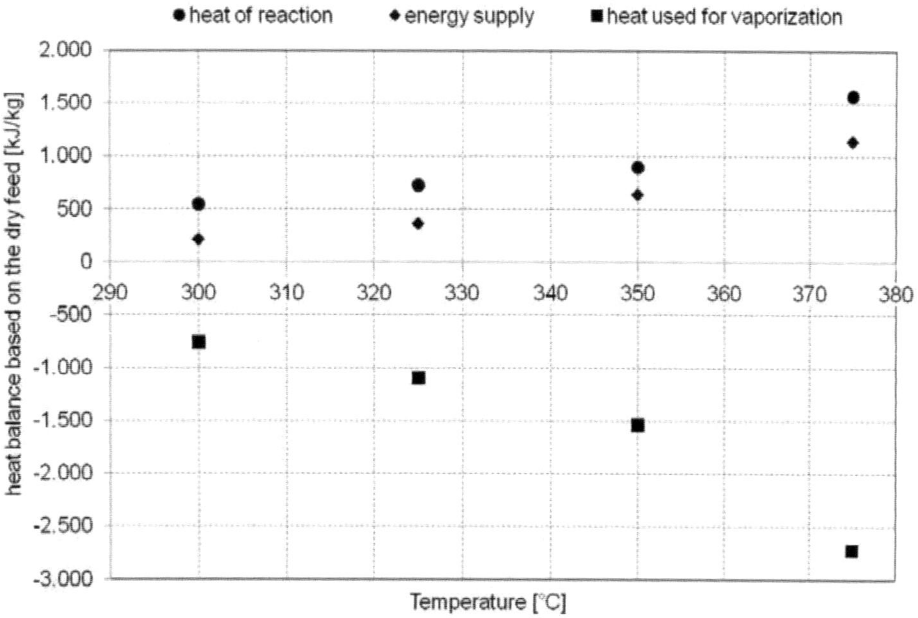

Figure 4: Results of the thermodynamic analysis

New method of total acid number (TAN) determination in biooils

FA Agblevor and J. Shao

Biological Systems Engineering, Virginia Polytechnic Institute and State University, Blacksburg, VA 24061

Abstract

Fast pyrolysis oil is one source of renewable liquid fuels, which is promising in satisfying environmental concerns over fossil fuels. The ASTM D664 method is used to determine the acidic constituents in petroleum products, lubricants, and biodiesel requiring mixtures of 2-propanol and toluene as solvent. The ASTM method also uses large quantities of solvents for the total acid number (TAN) determination. The ASTM D664 method has been adopted for the determination of TAN for bio-oils data published in literature. However, the properties of fast pyrolysis bio-oils are different from those of petroleum products and biodiesel, because bio-oils are soluble in polar solvents and partially soluble in water whereas petroleum products and biodiesel are insoluble in water and are more non-polar. An alternative method which is more suitable for TAN determination of pyrolysis bio-oil has been developed and validated. Aqueous potassium hydroxide (0.1 M KOH) was used as standard titrant instead of 0.1M KOH in alcoholic solution, and acetone was used as titration solvent instead of mixtures of toluene and 2-propanol as specified in the ASTM method. Four bio-oil samples were analyzed for determination of TAN with the new method and the ASTM method, respectively. The results showed that TAN of bio-oil samples obtained by the new method are in good agreement with the one obtained by the ASTM method. It was confirmed that the new method was simpler and relatively cheaper and more suitable for TAN determination of the pyrolysis oils than the ASTM D664 method.

Introduction

Pyrolysis of biomass is one of the most promising methods of producing liquid fuels from lignocellulosic biomass feedstocks. There are several variations of the pyrolitic conversion method including rapid pyrolysis, ultrafast pyrolysis, vacuum pyrolysis and fractional catalytic pyrolysis. All these methods produce liquid products that are acidic. The level of acidity varies with the pyrolysis method. The pH of liquids produced using these methods vary from 2 to 4. Because of the acidity of these oils, they could potentially cause corrosion in processing equipment or equipment using the oils either as fuel or other applications. The total acid number (TAN), which measures both the weak and strong acids in the oil is the traditional method of assessing the corrosivity of petroleum oils. The TAN is defined as the quantity of potassium hydroxide required to neutralize the acid in one gram of sample. The TAN is different from the pH in that whereas the pH measures the strength of the acidic or alkaline constituents, the TAN measures the concentration of the acidic constituents. TAN is also useful in detecting the presence of minor weak acids. For petroleum products, the ASTM standard method D664 is usually used. This method has been adapted to determine the TAN for biooils. However, both the physical and chemical properties of biooils are quite different from those of petroleum products. Whereas most petroleum products are hydrophobic, the biooils are hydrophilic and in some cases can dissolve substantial amounts of the biooil. Because of the hydrophobicity of the petroleum products, a mixture of solvents such as iso-propanol and toluene are used to ensure that the all the acidic components of the substrate are dissolved in the solvent prior to the titration. Large quantities of solvents are also normally required per determination of the TAN. A typical titration will require as much as 150 mL of solvent mixture; thus for a triplicate run, about 450 mL of solvent mixture will be required. For a large number of samples, the amount of solvent required becomes excessive, time consuming and not sustainable.

Biooils on the contrary are hydrophilic, easily dissolve in common solvents such methanol and acetone and are partially soluble in water. However in addition to weak organic acids such as acetic and propionic acids, biooils also contain phenols which are also weakly acidic. Because of the hydrophilicity of biooils, we proposed a new and simpler method for determining the acid number of the biooil. In this paper we report results of our new method of TAN determination for biooils.

Materials and methods

All chemicals used for these experiments (toluene, acetone, benzoic acid, potassium hydroxide, iso-propanol) were reagent grade obtained from Sigma- Aldrich (sigma-aldrich.com). the biooils used for the analysis were produced in our laboratory using both fractional catalytic pyrolysis method (Agblevor et al, 2010) and rapid pyrolysis using sand as the pyrolysis medium.

The biooils and benzoic acid standard were titrated using 798 MPT Titrino (Metrohm USA, Riverview, FL). One set of titration was carried out using the ASTM standard D 664 and another set was carried out using the new method. In the new method, 50 mL and 125 mL acetone were used to dissolve the biooil samples instead of the toluene iso-propanol mixture. The o.1M KOH solution was standardized with 0.1M hydrochloric acid solution. For each sample 0.1-0.5 g biooil was dissolved in 50 mL acetone in the first set of experiments and in the second set 125 mL of acetone was used to dissolve the biooil. The samples were titrated with 0.1 M KOH to the end point. Each sample was titrated in triplicate and the average and standard deviations calculated. Additionally each method was used to titrate benzoic acid standard to the end point. After the titration the total acid number was calculated using the following formula:

$$TAN = mgKOH/g = (A-B)*M*56.1/W$$

Where A = volume (mL) of KOH solution used to titrate the sample to end point; B = volume (mL) of KOH used for blank titration; M = molar concentration of KOH solution (mol/L); W = mass of biooil sample used.

Results and discussions

The TAN of benzoic acid standard using the two methods are similar to the theoretical calculations (Table 1). Clearly the accuracy of the new method is similar to that of the ASTM D 664 method. For the ASTM method, the volume of solvent mixture recommended was usually 125 mL. For the new method

50 mL and 125 mL of acetone were investigated. The results showed that there was no difference in the TAN using either 50 mL solvent or 125 mL. Thus, for subsequent analysis, only 50 mL solvent was used to dissolve the samples. In the ASTM, alcoholic KOH is used for the titration, but in the new method, 0.1 M KOH in water was adequate for the titration.

Table 1. Comparison of ASTM D 664 standard method with new acetone method

Method	ASTM D664	New acetone method
TAN (mg KOH/g benzoic acid)	456.81±2.61	461.39±3.52
Calculated TAN (mgKOH/g benzoic acid)	459.38	459.38
Repeatability	1.58%	2.11%
Accuracy	99.4%	100.4%

Repeatability= (2.77*SD)/(average)*100; Accuracy = Average value/calculated value

The comparison of the benzoic acid results showed similar TAN number, so the method was applied to both rapid pyrolysis biooil samples and catalytic pyrolysis biooil samples. The rapid pyrolysis biooil sample was supplied by NREL. The NREL biooil was prepared from hybrid poplar wood. The TAN number for bioooil samples determine by the two methods are shown in Table 2.

Table 2. Comparison of TAN for biooil samples using both ASTM D664 and new acetone method.

Biooil	ASTM D664, TAN (mgKOH/g)	New acetone method, TAN (mgKOH/g)
NREL (non-catalytic)	95.00±1.40	98.85±2.69
Poplar catalytic	86.92±0.55	89.93±1.81
Poplar catalytic	76.60±1.05	82.58±2.49
Pine catalytic	61.51±0.25	63.35±0.69

In all cases, the acetone method was slightly higher than the ASTM method. We attributed the differences to the improved solubility of the biooil in the acetone compared to the toluene/iso-propanol mixture. However, the TAN determined by the two methods are comparable. Since the new method is much simpler and does not require the use of large solvent volumes, and alcoholic potassium hydroxide, we recommend this approach as being more suitable for TAN determination in biooils.

(135d) Fractional Catalytic Pyrolysis of Biomass for Stable Pyrolysis Oils and Hydrocarbon Fuels Production

Foster A. Agblevor, Department of Biological Systems Engineering, Virginia Polytechnic Institute and State University, Blacksburg, VA, Ofei Mante, Biological Systsems Engineering, Virginia Polytechnic Institute and State University, Blacksburg, VA, S. Ted Oyama, Chemical Engineering, Virginia Polytechnic Institute and State University, Blacksburg, VA, Francine Battaglia, Mechanical Engineering, Virginia Polytechnic Institute and State University, Blacksburg, VA and Ron McClung, BASF Catalysts LLC, Iselin, NJ

Pyrolysis of biomass is developing rapidly and could play a very important role in the future renewable energy equation. However, rapid pyrolysis oils are usually acidic, unstable and have high viscosities that increase with storage time. We have developed fractional catalytic pyrolysis process that produces stable, low viscosity biomass pyrolysis oils that can be stored at ambient conditions without any significant increase in viscosity. The oils can be distilled at both atmospheric pressure and under vacuum without char or solid formation. Fractional catalytic pyrolysis oils produced from hybrid poplar wood was stored at ambient laboratory condition for more than 10 months and the dynamic viscosities changed from 11.2 ±0.45 cP to 12.7 cP ± 0.56 The oils were also characterized using the high temperature simulated distillation method and this showed 100% distillation at up to 640 oC (1200 F). About 15% of the stable pyrolysis oils were blended with Standard Gulf Coast gasoil and cracked in fluid catalytic cracking (FCC) unit into various fuel fractions. There was no difference in both conversion and yields of various products from the blend and the standard Gulf Coast gasoil. The pyrolysis oil/gasoil blend produced slightly less coke than the standard gasoil feedstock. 13C-NMR and FTIR analysis of the cracked fractions did not show any oxygenates in the products.

See more of this Session: Biomass Pyrolysis
See more of this Group/Topical: Fuels and Petrochemicals Division

IN SITU CATALYTIC UPGRADING OF BIOOIL WITH NOVEL CATALYSTS, FROM BENCH SCALE TO PILOT PLANT

*K.G. Kalogiannis[1], S. Stephanidis[1,2], S.S. Voutetakis[1], and A.A. Lappas[1]**

1. Chemical Process Engineering Research Institute (CPERI)/
Center for Research and Technology Hellas (CERTH)

2. Department of Mechanical Engineering, University of Western Macedonia

*Corresponding author: Tel: +0030-2310-498305, Fax: +0030-2310-498380,

email: angel@cperi.certh.gr, kkalogia@cperi.certh.gr

Introduction

Biomass flash pyrolysis is a very promising thermochemical process for the production of bio-liquids and/or chemicals [1, 2]. The catalytic pyrolysis of biomass has recently gained a lot of research interest for the production of bio-liquids from biomass flash pyrolysis in a single catalytic stage. However, early works have shown that the use of catalysts usually leads to additional water and coke production and to a decrease in the yield of organic products [3, 4]. Additionally, due to the fast catalyst deactivation during the biomass catalytic pyrolysis a catalyst regeneration step is necessary. Consequently, the fully circulating fluid bed (CFB) technology is the best technology for this process.

CPERI has extended experience in CFB technology for the FCC process. For biomass pyrolysis, CPERI has constructed a once through transported fluid bed unit. A detailed description of the unit and its operation has been given elsewhere [5]. The main disadvantage of the once through configuration is that it cannot run for a long time since spent catalyst should be reloaded manually into the regenerator. For the above reasons CPERI recently revamped this once through unit to a fully circulating system. CPERI has also experience in the catalytic pyrolysis of biomass for the production of an upgraded biooil. Catalyst evaluation takes place in a bench scale unit, while commercial catalysts have also been evaluated in the Pilot Plant unit to correlate the two units. This allows the fast screening of different catalytic materials.

The characterization of the produced biooil, the assessment of the achieved deoxygenation and the distribution of liquid products into different groups such as aromatic and aliphatic hydrocarbons, phenols, carbonylic compounds, furans, acids etc. is very important for both the understanding of the catalytic pyrolysis mechanisms and the evaluation of the effect of different catalytic materials on the end products. For these reasons several techniques have been developed that allow the determination of the elemental analysis of the produced biooil, its water yield and the qualitative and quantitative evaluation of the distribution of the different chemical groups. In this paper, the effect of several catalysts on the quality and yield of biooil will be investigated in the bench scale unit. Deoxygenation and GCMS analysis of the produced biooil are two important factors that will be thoroughly reviewed. In addition, data from the bench scale unit is compared to those received from the pilot plant unit so that a correlation between the two units is achieved.

Experimental

Prescreening experiments were performed on a bench scale fixed bed reactor made of stainless steel 316. The reactor was heated by a three zone furnace and was purged with nitrogen to ensure an oxygen-free atmosphere and to carry the pyrolysis vapours through the catalyst. The experimental setup is described in detail elsewhere [6]. Figure 1 presents a schematic diagram of the bench scale unit. The feed was Beechwood sawdust and the temperature of pyrolysis was 500 ^{0}C for all runs.

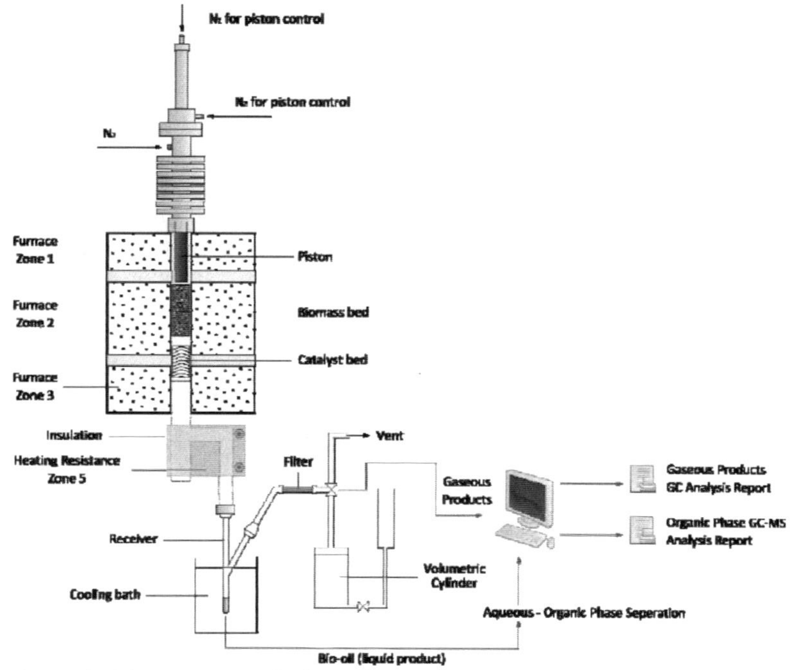

Figure 1. Bench scale biomass pyrolysis unit

In addition, some of the commercial catalysts were tested in the pilot plant unit. Mass balances were received in both pilot plant and bench scale units and the received biooils were analysed for oxygen and water content and with the help of GCMS a semiquantitative analysis for fuels and chemicals in the biooil took place.

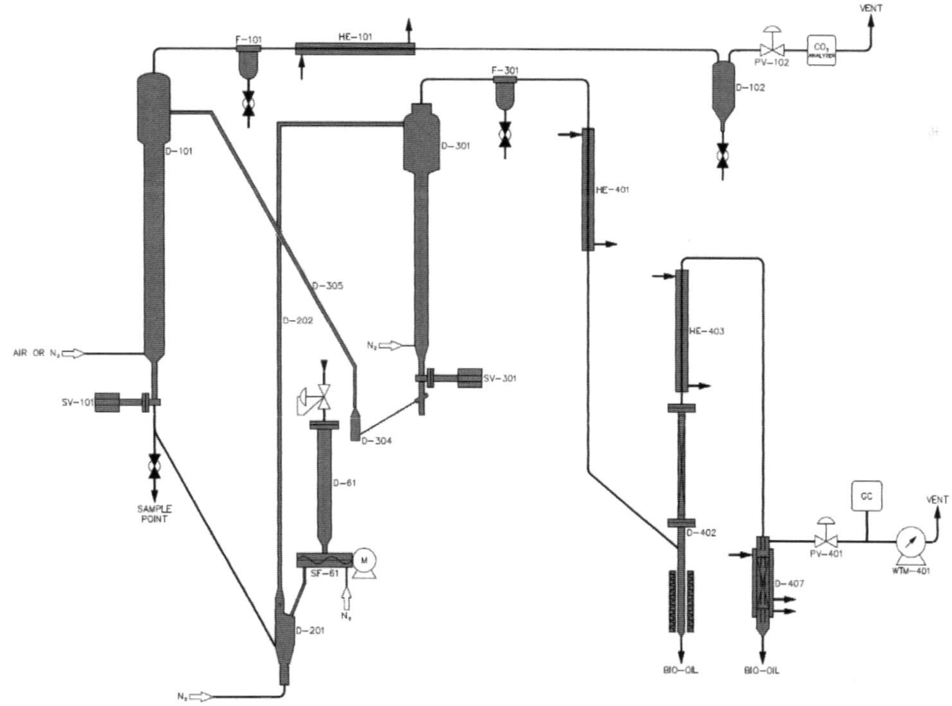

Figure 2. Schematic diagram of the biomass pyrolysis pilot plant in the fully circulating configuration

Figure 2 presents the biomass pyrolysis pilot plant unit in the fully circulating configuration. The unit can be continuously operated for long runs achieving mass balances always between 92-98 wt%. In the paper detailed results of tests will be presented.

Results and Discussion

All tests described in the current study for the bench scale unit were performed at similar experimental conditions Thus; the only independent variable of the work was the type of the catalyst. The mass balances of all tests which were satisfactory and within the range of 90-96 wt% have been normalized to 100%. Figure 3 presents the liquid, gas and coke yields of all the catalysts tested.

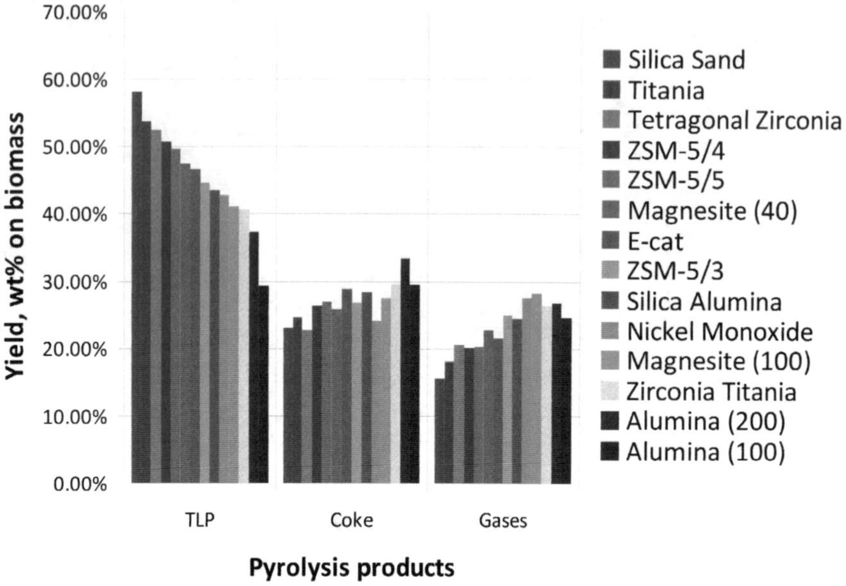

Figure 3. Liquid, Coke and Gases mass balance for catalytic pyrolysis runs of Beech sawdust at 500 ^{0}C in the bench scale unit.

Depending on the catalyst, different cracking takes place yielding different product yields and different biooil qualities. Figure 4 presents the organic biooil yield versus the oxygen content of the organic biooil. It is clear that different deoxygenation levels can be achieved with different catalysts. The lesser the oxygen content of the biooil the greater the decrease of the biooil yield. This is due to the fact that oxygen is removed from biooil in the form of CO, CO_2 and H_2O.

However, Figure 4 also reveals that different biooil yields may be achieved for a given deoxygenation level depending on the catalyst used. This way Silica-Alumina with a surface area of 400 m^2/g has an organic biooil yield of 16.5 wt% on biomass with an oxygen content of 30.5 wt% on biooil while the ZSM-5/3 has a yield of biooil 20.6 wt% on biomass with an oxygen content of 31.14 wt% on biooil respectively. Therefore, the ZSM-5/3 seems to be better suited for the catalytic upgrading of biooil compared to Silica-Alumina. The same is noted if a comparison between a USY Zeolite is made and ZSM-5/3 catalysts. Depending on the surface area of each catalyst different biooil yields and oxygen contents are achieved, however the ZSM-5 catalysts deoxygenate the biooil more effectively than the USY Zeolite. All catalysts produce less biooil of higher quality, less oxygen content, compared to the base cases of Silica sand that acts as inert material with no surface area and yields around 38 wt% organic biooil with an oxygen content of 38 wt%.

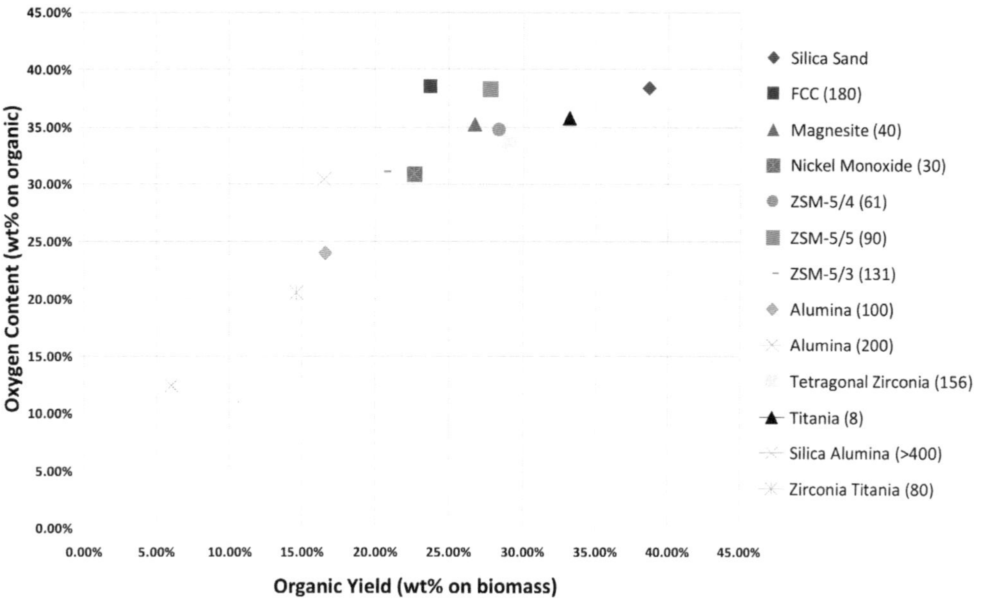

Figure 4. Oxygen content of organic biooil vs organic biooil yield.

GCMS analysis of the produced biooils is presented in table 1. The groups presented are Aromatic Hydrocarbons (AR), Aliphatic Hydrocarbons (ALI), Phenols (PH), Furans (FUR), Acids (AC), Esters (EST), Alcohols (AL), Ethers (ETH), Aldehydes (ALD), Ketones (KET), Polyaromatic Hydrocarbons (PAH), Sugars (SUG), Nitrogen compounds (NIT), Heavy compounds (HV) and Unidentified (UN).

Table 1. GCMS analysis of organic fraction (wt % on organic liquid) of biooils produced from different catalysts at 500 ^{0}C.

	Silica Sand	Tetr. Zirc.	FCC	ZSM-5/3	Magnesite (40)	Zirconia - Titania	Alumina (200)
AR	1.11	3.02	2.82	7.77	0.84	8.45	10.98
ALI	0.18	0.00	0.43	1.02	1.05	0.20	1.16
PH	29.62	32.03	36.15	37.02	39.12	32.76	25.76
FUR	1.04	0.37	1.47	1.67	0.83	1.43	1.91
AC	19.64	14.38	12.45	5.24	9.30	8.65	3.88
EST	4.18	3.29	0.64	1.12	6.12	3.91	0.89
AL	0.30	0.52	1.74	0.54	1.59	1.02	4.75
ETH	1.76	0.43	4.32	2.74	0.91	0.55	0.72
ALD	1.00	1.23	2.34	0.16	0.68	1.26	2.58
KET	13.62	11.42	8.41	6.03	14.36	10.65	4.18
PAH	0.00	0.00	1.79	2.74	0.23	4.87	10.48
SUG	1.77	0.50	0.00	0.00	0.00	0.45	0.00
NIT	4.58	3.50	3.60	6.10	1.84	2.39	4.05
HV	9.67	13.02	8.45	10.03	13.44	12.84	17.58
UN	11.53	16.29	15.39	17.82	9.69	10.57	11.08

From Table 1 it can be deduced that different catalysts enhance the production of different chemical groups. FCC, ZSM-5, Alumina and Zirconia-Titania produce more aromatics and PAHs while FCC, ZSM-5 and Magnesite also produce high Phenol yields. All catalysts reduce the Acids and Ketones yields that are largely responsible for the corrosivity and instability of biooils [7].

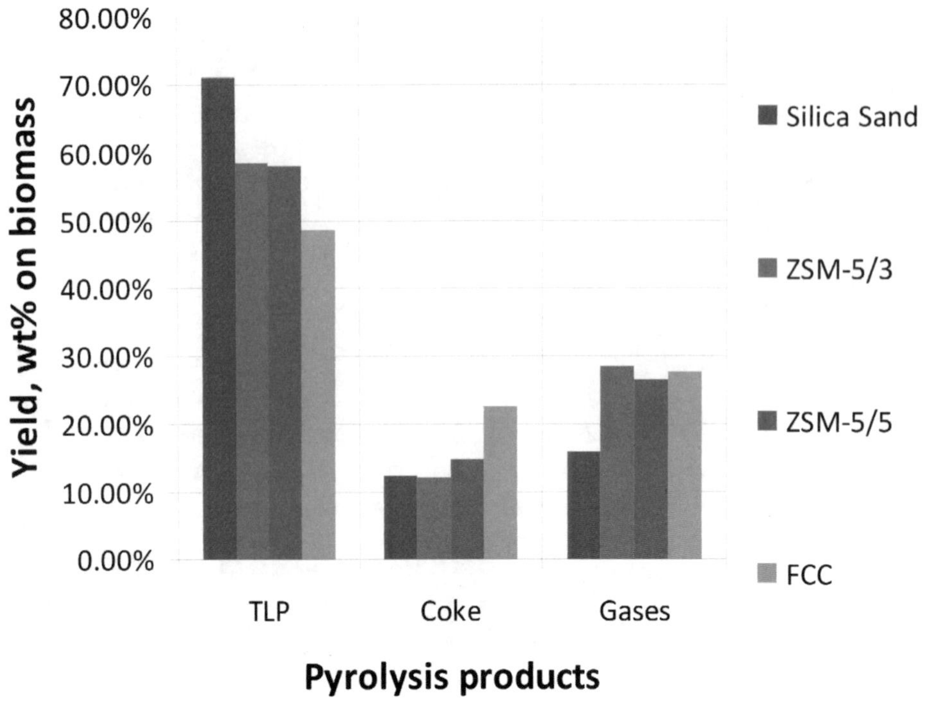

Figure 5. Pyrolysis product yields for in Pilot plant unit.

In the pilot plant two ZSM-5 additives with different surface areas (named ZSM-5/3 and ZSM-5/5), an FCC catalyst were also tested along with silica sand for the base case tests. Mass balances are presented in figure 5.

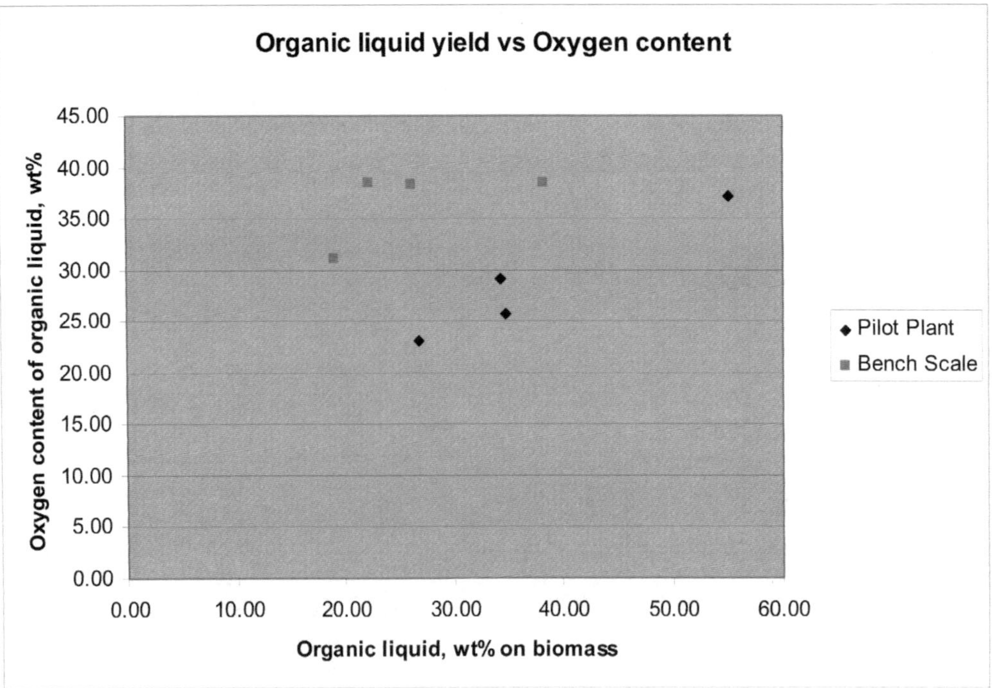

Figure 6. Comparison of oxygen content in biooils vs biooil yield produced in bench and pilot plant scale.

Figure 6 presents a correlation between the bench scale and the pilot plant data. In all cases, the pilot plant unit yields more organic biooil with less oxygen content. This is attributed to three main factors. Firstly, the heat transfer in the pilot plant unit where the biomass particles are directly mixed with hot catalyst results in the better mixing and cracking of the biomass. This way more liquid product and less coke and char are produced. In addition, the catalyst to biomass ratio in the pilot plant is 15:1, 30 times higher compared to the 1:2 ratio of the bench scale unit. The higher catalyst to biomass ratio along with the direct contact and better mixing of biomass particles with hot catalyst results in the further deoxygenation and cracking of the biomass. Finally, in the case of the pilot plant the catalyst is constantly regenerated by burning off coke in the regenerator vessel which allows it to retain its activity in contrast to the fixed bed catalyst that is deactivated due to coke formation on the catalyst. The general trend is that the pilot plant unit produces biooil in greater yields and with lower oxygen contents stressing the efficiency of the circulating fluid bed technology.

Conclusions

Catalytic biomass pyrolysis was investigated both at bench and pilot plant scale. Several new experimental and commercial catalysts were tested and the deoxygenation along with the selectivity for chemicals and fuels was evaluated. In addition a correlation between bench and pilot plant scale was achieved that allows for the fast screening of catalysts and the determination of the most suitable ones for the biomass catalytic cracking process with circulating fluid bed technology.

Acknowledgements

The work presented was funded by the ACENET HECABIO project contract no. ACE.07026

References

[1] A. Demirbas, Biomass resource facilities and biomass conversion processing for fuels and chemicals, Energy Convers. Managem. 42 (2001) 1357-1378..

[2] D. Meier, O. Faix, State of the art of applied fast pyrolysis of lignocellulosic materials- a review, Bioresource Technol. 68 (1999) 71-77.

[3] A. A. Lappas, M. C. Samolada, D. K. Iatridis, S. S. Voutetakis and I. A. Vasalos, Biomass pyrolysis in a circulating fluid bed reactor for the production of fuels and chemicals, Fuel, 81 (2002) 2087-2095.

[4] Adam, E. Antonakou, A. Lappas, M. Stöcker, M.H. Nilsen, A. Bouzga, J.E. Hustad, G. Øye, In situ catalytic upgrading of biomass derived fast pyrolysis vapours in a fixed bed reactor using mesoporous materials, Micropor. Mesopor. Mater. 96 (2006) 93-101.

[5] Lappas A., Dimitropoulos V., Antonakou E., Voutetakis S., Vasalos I., 7th World Congress of Chemical Engineering, (2005) 26.

[6] E. Antonakou, A. Lappas, M.H. Nilsen, A. Bouzga, M. Stöcker, Evaluation of various types of Al-MCM-41 materials as catalysts in biomass pyrolysis for the production of bio-fuels and chemicals, Fuel, 85 (2006) 2202.

[7] M.N. Islam, M.R.A. Beg, The fuel properties of pyrolysis liquid derived from urban solid wastes in Bangladesh, Bioresource Technol. 92 (2004) 181-186.

Title: Pyrolysis of algal biomass

Pyrolysis experiments with algal biomass were performed in a laboratory fixed-bed reactor. These experiments were carried out at 450, 500, 550 and 600°C to study the effects of temperature on biomass conversion to gas, oil and char products. These conditions were chosen based on thermo-gravimetric analyses of the feed stocks which showed that most of the thermal decomposition of algal biomass occurred within this temperature range. The reactor was purged with helium to maintain inert atmosphere for pyrolysis reactions. During our experiments, liquid products were collected by condensing with cold water; gas formation was monitored using mass flow meters and gas composition was measured using an online gas chromatograph equipped with a pulsed discharge helium ionization detector. Solid residues (or char) were recovered from the reactor at the end of the experiment. Proximate- (volatile matter, moisture content, ash and fixed carbon) and ultimate- analyses (C, H and N) of the feedstock and products were performed using various standard methods to close overall and element-specific mass balances. Additionally, bio-oil was analyzed using mass spectroscopy to identify specific compounds and the overall molecular weight distribution of this product. The heating values of biomass, bio-oil and bio-char were calculated. The results of these investigations will be presented and discussed.

Catalytic Upgrading of Switchgrass Pyrolysis Bio-oil Vapors

Jonathan E. Peters, Xiaohan Zhang, Trung T. Pham, Anirudhan Gangadharan, Xinli Zhu, Roberto Galiasso, Daniel E. Resasco and Richard G. Mallinson,
School of Chemical, Biological and Materials Engineering,
University of Oklahoma, Norman, OK, 73019, USA
Center for Biomass Refining,
University of Oklahoma, Norman, OK, 73019, USA

ABSTRACT

Catalytic upgrading of bio-oil from switchgrass pyrolysis was investigated over $Ce_{0.5}Zr_{0.5}O_2$ mixed oxides to produce higher carbon chain molecules for use as fuel additives from short aldehydes and carboxylic acids found in bio-oil mixtures. The experiments were carried out in a single two-stage process where the gaseous products from the pyrolysis reaction were fed directly to a fixed catalyst bed in a separate reactor in series with the pyrolysis reactor. The final products were condensed to the liquid phase and collected for analysis. GC by GC analysis of products shows evidence of catalytic activity. Acetic acid and propanoic acid found in the bio-oil were converted to acetone and 3-pentanone, respectively, among other products, when reacted over the catalyst. This is consistent with our work on model compounds that short aldehydes and carboxylic acids are converted to higher chain aldehydes and ketones by way of aldol condensation and ketonization over mixed oxides catalyst. Reacting over mixed oxides catalyst resulted in lower liquid yields from pyrolysis when compared to experiments without catalyst. The char yields remained the same while the gas yields, by difference, appear to have increased under these conditions. The coated monolith catalysts seem to have higher activity than the pelletized catalysts for the same amount of ceria and zirconia loading.

1. INTRODUCTION

There is strong motivation to produce renewable fungible fuels for transportation and power generation.[1,2] One way is by a high temperature thermo chemical breakdown of biomass called pyrolysis.[3] The liquid from pyrolysis is a complex mixture of oxygenated compounds called "bio-oil."[4] The light oxygenates in bio-oil can be upgraded to higher chain oxygenates for use as a fuel or fuel additive.[5] In this study, catalytic upgrading of bio-oil produced from switchgrass pyrolysis was investigated over $Ce_{0.5}Zr_{0.5}O_2$ mixed oxides to produce higher carbon chain molecules from short aldehydes and carboxylic acids found in bio-oil mixtures. This is consistent with our work on model compounds that short aldehydes and carboxylic acids are converted to higher chain aldehydes and ketones by way of aldol condensation and ketonization over mixed oxides catalyst.[5]

2. EXPERIMENTAL

2.1 Catalyst Preparation

A $Ce_{0.5}Zr_{0.5}O_2$ catalyst was prepared by a co-precipitation method with aqueous solutions of cerium (IV) ammonium nitrate and zirconium nitrate as described in the studies by Hori et al [6] and Noronha et al [7]. The aqueous solutions were mixed at a 1:1 ratio, co-precipitated with ammonia hydroxide, rinsed, dried, and calcined at 500°C for one hour.

2.2 Pyrolysis/Catalysis

The pyrolysis and the catalysis reactions were carried out in a single two-stage process in a nitrogen environment at atmospheric pressure. The pyrolysis stage consisted of a 0.870 in. I.D., stainless steel, counter-flow reactor with nitrogen fed in the bottom at 0.550 SLPM and pre-heated to 500 °C. The catalysis stage consisted of a 0.495 in. I.D., stainless steel, fixed-bed reactor maintained at 400 °C. The switchgrass, 180-250 micron diameter particles, was fed with a manual, auger-type feeder from the top of the first reactor by gravity. Later, some experiments were also carried out with 150-600 micron diameter switchgrass particles. The stages were operated in series where the nitrogen carried the pyrolysis vapor products to the catalytic reactor, and the catalysis products were condensed to the liquid phase for collection and analysis. Figure 1 shows the schematic of the two-stage process. Two sets of experiments were conducted; one set with catalyst and one without. In the first set, the catalytic reactor was charged with the $Ce_{0.5}Zr_{0.5}O_2$ mixed oxides fixed bed using pelletized catalyst with particle size 400-700 microns. The linear velocity of the gaseous products in the carrier gas is about 40 cm/s, and the catalyst bed is 1 cm long. The second set of experiments was conducted with ground glass of particle size 400-700 microns in place of the catalyst to show the changes from the catalyzed reactions while maintaining the fluid dynamics. The liquid products collected existed in two distinguishable phases, the aqueous fraction and the heavy organic fraction. In this study, the aqueous fraction was the focus and was analyzed separately by GCxGC, GC-MS, carbon NMR, free fatty acid number, and water content.

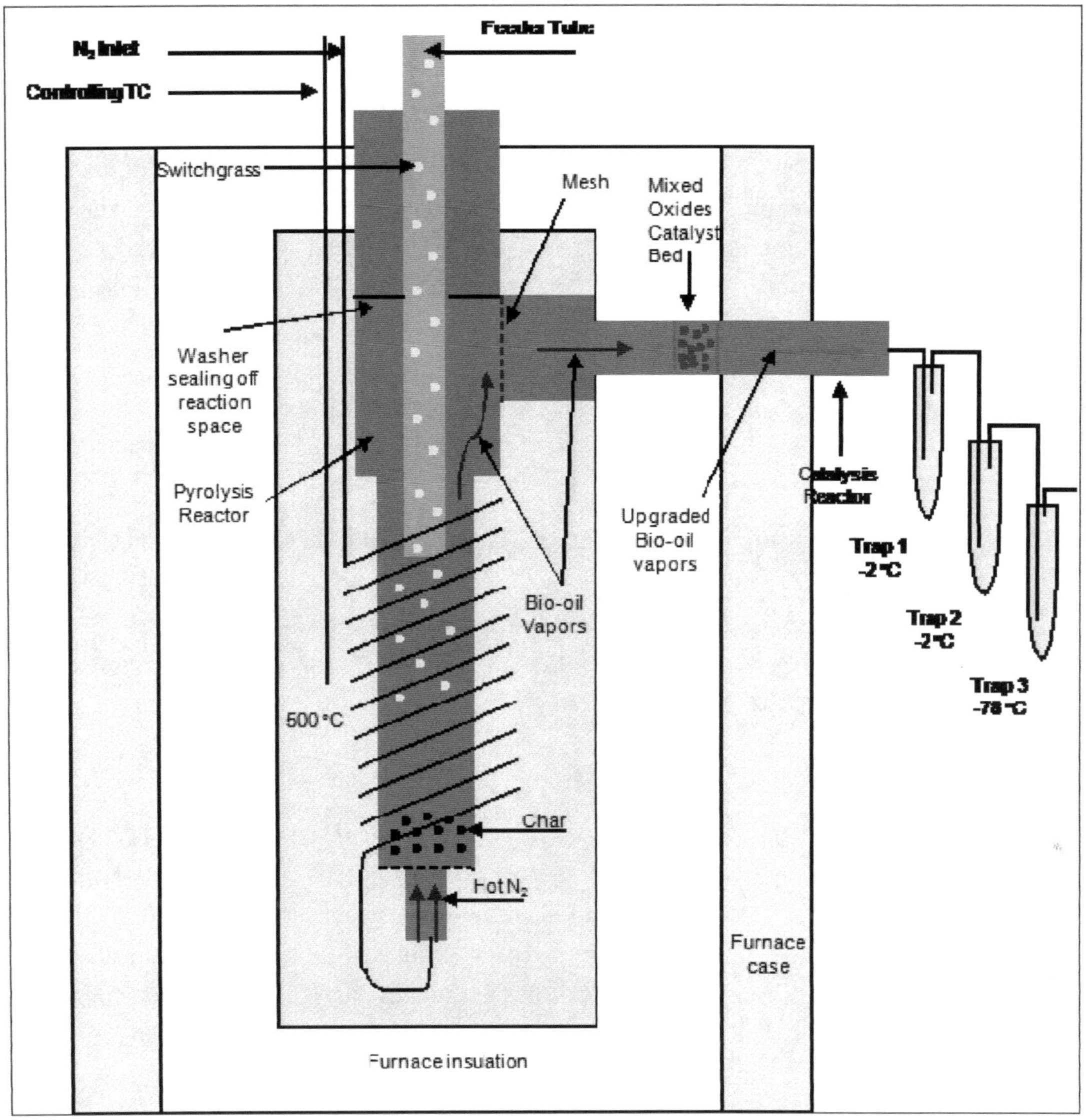

Figure 1: Pyrolysis-Catalysis Two-Stage Process Design Schematic

2.3 Two dimensional GCxGC

A GCxGC system was set up by modifying the GCMS-QP 5000 (Shimadzu), equipped with EI detector, with a two-stage thermal modulator (Zoex Corporation) and a secondary oven installed inside the main oven of the GCMS. A low temperature valve mounted outside the GCMS oven controls the cold jet from -80 °C to -120 °C, that produces narrow chemical pulses. A hot jet at 300°C releases the trapped solutes into the second column mounted inside the second oven. The modulator accumulates samples eluting from the first column for a period of 12 seconds, typically equal to one third of the duration of a chemical peak. However, some peaks may not completely elute within this modulation period, i.e. acetic acid, and showing smearing/tailing behavior in the 2D-chromatogram. An auto-sampler injects undiluted samples with a 10 micro-syringe (0.4-0.6 μl) and a split ratio of 25. The first column is for volatility separation with DB-1 30 m length, 0.25 mm I.D., 1 μm film thickness (Agilent). The second column is for polarity separation Carbowax 1 m length, 0.1 mm I.D., 0.36 μm film thickness (Agilent). The primary oven is set at 35°C ramping to 275°C with a ramp rate of 2.2°C/min. The secondary oven is set at 75°C ramping to 325°C with a ramp rate of 2.27 °C/min. The GC images are processed and analyzed using GC Image (TM) Version 2.0b4, provided by Zoex. In this contribution, the total response of the MS detector was obtained from the evolution of specific compounds, and thus providing a semi-quantitative analysis. On some samples, an HP 6890A equipped with an FID detector was used to analyze the sample compositions to compare the relative area ratios of reacted compounds with and without catalytic treatment.

2.4 NMR-C^{13} Spectroscopy

NMR-C^{13} analysis was performed using Varian Mercury VX-300 MHz spectrometer with 4-nuclei auto-switchable probe PFG at 75MHz magnet. Approximately 7 drops of each sample was added to 0.75 ml d- THF NMR solvent in 5mm NMR tube. For each sample, 16384 scans were collected in consecutive 14 hours. We could identify acetic acid peaks at 20ppm and 179ppm, and ketone range from 180ppm to 210ppm.

2.5 Total Acid Number

The total acid number of the sample was measured using TitroLine alpha plus Free Fatty Acid Analyzer (TitroLine alpha plus, SCHOTT Instruments) by potentiometric titration. The total acid number is the quantity of base, expressed in milligrams of potassium hydroxide that is required to neutralize all acidic constituents present in 1 g of sample. 0.5 mg of each sample was collected and weighed in a 150 ml beaker and 80 ml of toluene and ethanol, 1:1, solvent mixture was added to the sample. 0.1M KOH titration solvent was added gradually to reach the equilibrium point of titration reaction. A calibration titration with 80ml of titration solvent was performed for each set of samples.

2.6 Water content

Water content analysis (% wt) of each sample was carried out by TitroLine Karl-Fischer analyzer (TitroLine KF, SCHOOT Instruments), in which 0.018g of sample was injected using syringe and titrated with KF reagent for volumetric titration. The water content percentage was determined upon the KF reagent consumed. For each set of samples perform a calibration titration with 0.18g DI-water.

3. RESULTS AND DISCUSSION

Liquid and Char yields were measured by % wt of the switchgrass reacted in the pyrolysis reactor, and gas yields were calculated by difference. Figure 2 compares the product yields from identical experiments with and without catalyst, showing a decrease in liquid yields by 28% and an increase in gas yields by 35% when reacted with catalyst. This result is consistent with aldol condensation reactions and ketonization reactions that reduce the number of moles of liquid compounds and produce more carbon dioxide.

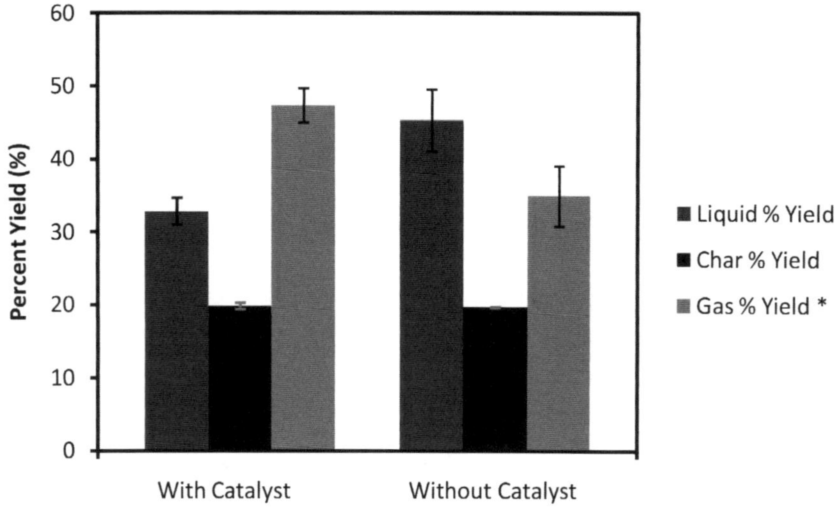

Figure 2: Product Yield Comparison
*Gas Yields were calculated by difference

The bubble graph in Figure 3 represents the peak volume response difference of the aqueous fraction of runs with and without catalyst. The positive responses are shown in the blue while the negatives are shown in red. A positive response means the volume/amount of a compound increases after catalyst treatment while the negative means the compound is consumed or reacted after treatment. The size of the bubble is equivalent to the magnitude of the difference. Note that a distinctively large red bubble at (12, 6) is acetic acid that is consumed after the catalyst treatment. Correspondingly, a large size blue bubble at (9.6, 7.4) is acetone being increased after the treatment. The acetone might be formed from ketonization reaction of acetic acid, shown as follows:

$$CH_3COOH + CH_3COOH \rightarrow CH_3COCH_3 + CO_2$$

However, the bubble size graph above is rather semi-quantitative due to the limitation of the mass spec detector. Thus, quantitative analysis is needed and can be carried out using GC with FID detector. Thus far, the main interest is focused on the conversion of the acid, relatively high concentration in pyrolysis oil, into a more stable product, in this case, acetone.

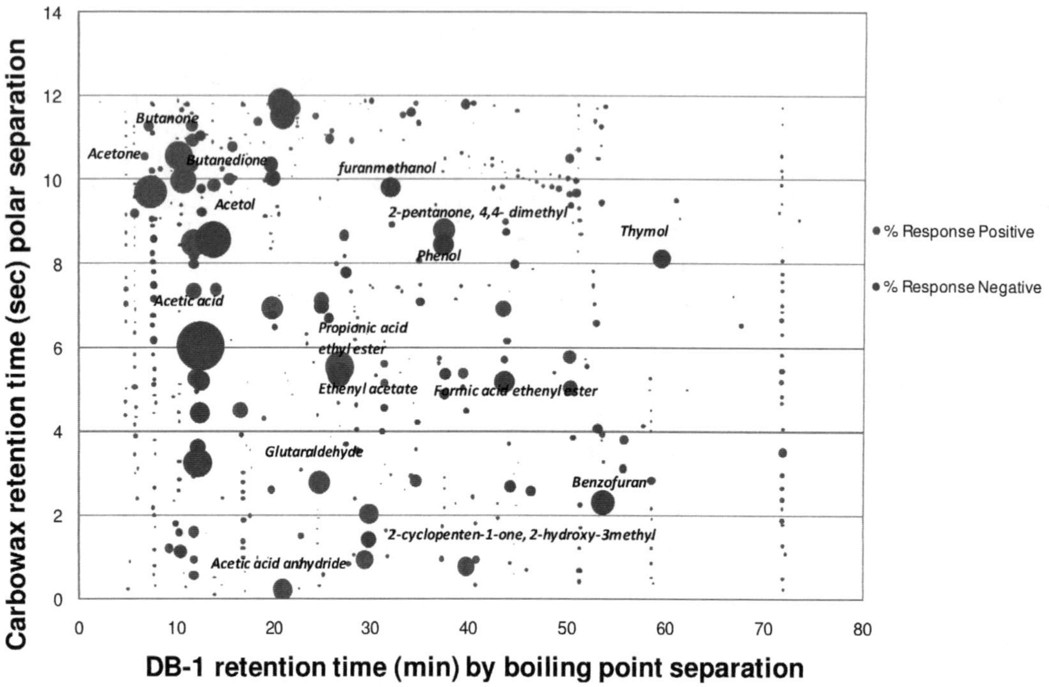

Figure 3: Positive and negative % Responses of peak volume with retention time from without catalyst 2D-GC and with catalyst 2D-GC samples.

Acetic acid and acetone standards were injected to confirm the retention times of these compounds in the aqueous fraction of the bio-oil samples. The injection of the samples with and without catalyst into GC FID showed that the area ratios of acetone/acetic acid are different, shown in Figure 4. Different amount of ceria and zirconia were coated onto inert silica alumina monoliths. The activity of these monoliths is much more stable compared to that of the pelletized catalysts, even for the same amount of loading.

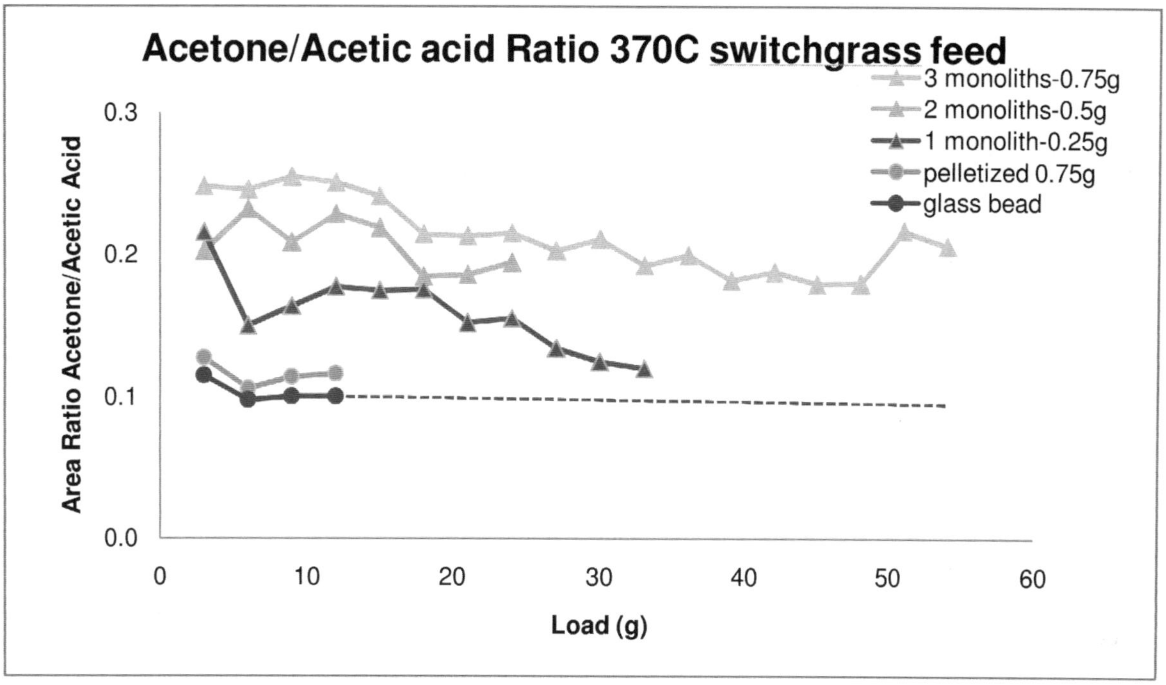

Figure 4: Acetone/Acetic Acid Area Ratio of feeding grass over coated monoliths vs. pelletized catalyst.

Another indicative parameter to investigate the activity of the catalyst is the CO_2/CO ratio by analyzing and quantifying outlet gas samples with various loads of fed grass. As ketonization reaction takes place, CO_2 is expected to be released more than when fed without catalysts. In Figure 5, for the 2 monoliths in series, the CO_2/CO molar ratio is about 1.4 to 1.8 and decreasing over time with gradual deactivation of the catalyst. However, without catalyst, this ratio stays low at about 1.1.

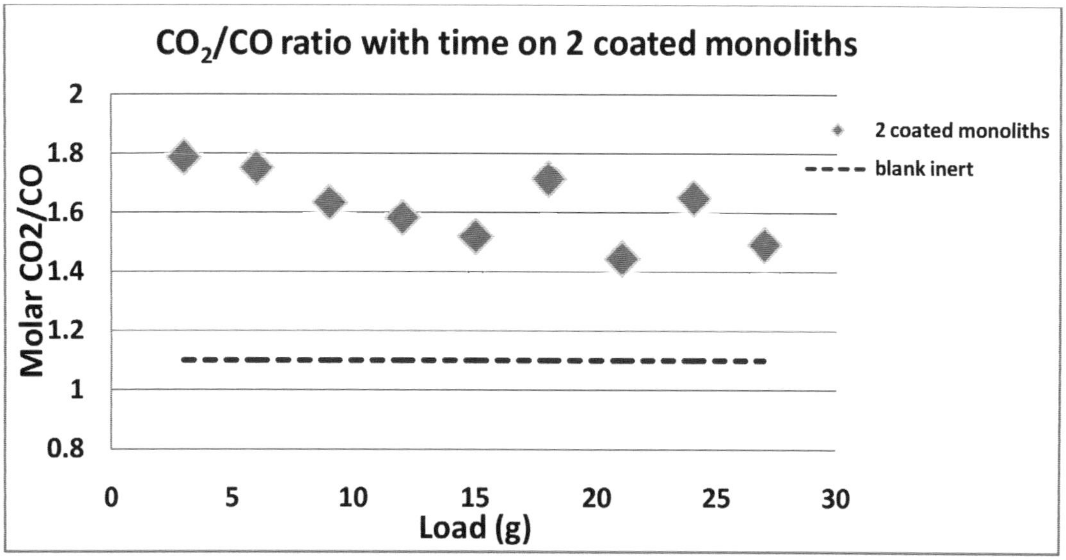

Figure 5: CO_2/CO molar ratio as indicator for catalyst activity. Herein shown ratio for 2 coated monoliths installed in series.

Besides acetic acid conversion, there are also other compounds being reacted with the presence of catalyst. In fact, propionic acid also converts in ketonization forming 3-pentanone, confirmed by GCMS. Cross ketonization of acetic acid and propionic acid results in the formation of butanone, that is also observed in the product spectrum change for with and without catalyst. Besides ketonization reactions, aldehydes in the bio-oil could also react in aldol-condensation to form large aldehydes. However, reactivity of the acids is much higher than that of the aldehydes [5]. There also exists competitive adsorption with other compounds, particularly the bulky molecules such as phenol, furfuraldehydes, etc., although these molecules might not react but rather take up sites on the surface of the catalyst.

After the catalyst upgrading, the total acid number drops approximately 10% due to the $Ce_{0.5}Zr_{0.5}O_2$ ketonization reaction which consumes acids to form ketones. The comparison of water content analysis between experiments with and without catalyst demonstrates that the percentage of water rises from 50% to 64% in average, which is also consistent with the ketonization reaction which produces water. Table 1 below summarizes the acid analysis results.

Table 1: Acid and Water Analysis of Aqueous Fraction of Bio-oil

Run#	Catalyst or Packing	Catalysis Temp. (°C)	TAN**	Water Content (%)
44c	Ce0.5Zr0.5O2	400	82.7	65.7%
48c	Ce0.5Zr0.5O2	400	94.2	63.6%
50c	Ce0.5Zr0.5O2	350	75.4	66.7%
53c1†	Ce0.5Zr0.5O2	400	76.3	53.4%
47	ground glass	400	97.5	53.4%
49	ground glass	400	89.2	50.3%
51	borosilicate	350	93.0	46.0%
52	borosilicate	350	104.6	50.1%

** Total Acid Number: the amount of KOH in mg needed to neutralize 1 g of the aqueous fraction.

†The catalyst in run 53c1 was used once before in run 44c

4. CONCLUSION

Upgrading switchgrass pyrolysis vapors over $Ce_{0.5}Zr_{0.5}O_2$ catalyst results in a decrease in liquid yields by 28% and an increase in gas yields by 35% under these conductions. The short aldehydes and carboxylic acids are converted to higher carbon chain aldehydes and ketones by way of aldol condensation and ketonization over the mixed oxides catalyst. This is marked by increasing the ratio of acetone to acetic acid in the aqueous products after reacting over the mixed oxides catalyst. While this only shows a part of the bigger picture, future studies will show more complete analysis of the total bio-oil.

REFERENCES

[1] Ragauskas, A. J. et al. *Science* **2006**, *311*, 484
[2] Huber, G. W.; Iborra, S.; Corma, A. *Chem. Rev.* **2006**, *106*, 4044
[3] Shafizadeh, F.; *J. Analytical Appl. Pyrolysis*, **1982**, *3*, 283
[4] Mohan, D.; Pittman, C. U. Jr.; Steele, P. H. *Energy & Fuels* **2006**, *20*, 848
[5] Gangadharan, A.; Resasco, D. E.; Mallinson, R. G. *Accepted* **2010**
[6]Hori, C. E.; Permana, H.; Ng, K. Y. S.; Brenner, A.; More, K.; Rahmoeller, K. M.; Belton, D. *Appl. Catal. B-Environ.* **1998**, *16*, 105
[7]Noronha, F. B.; Fendley, E. C.; Soares, R. R.; Alvarez, W. E.; Resasco, D. E. *Chem. Eng. J* **2001**, *82*, 21

(135h) Integrated Pyrolysis Quench Catalytic Cascade for Insitu Product Stabilization and Upgrading

Shaolong Wan[1], Xiaohan Zhang[1], Trung Pham[1], Roberto Galiasso[1], Lance Lobban[1], Daniel Resasco[1] and **Richard G. Mallinson**[2], (1)University of Oklahoma, Norman, OK, (2)School of Chemical, Biological and Materials Engineering, University of Oklahoma, Norman, OK

The operation and product distribution from switchgrass pyrolysis in an integrated kg/h plant are presented. Product quenching with aqueous and organic product streams segregate and stabilize the liquid products. An integrated catalytic cascade that converts fractions of the product vapors between quenches, before condensation, will be discussed.

See more of this Session: Biomass Pyrolysis
See more of this Group/Topical: Fuels and Petrochemicals Division

(140a) Progress in Nutrient Recovery From Anaerobically Digested Wastes

Shulin Chen, Quanbao Zhao and Craig Frear, Biological Systems Engineering, Washington State University, Pullman, WA

Recovering energy from solid wastes such as organic fraction of municipal wastes and animal manures has been recognized as having great potential for generating renewable energy and reducing greenhouse gas emissions. A more desirable concept, however, is bio-refining with which the waste materials are converted to a series of co-products in addition to biogas. Anaerobic digestion is the most suitable technology for such an application because of its capability in handling mixed waste feedstocks at a relatively low cost. A major technical challenge is to identify targeted products and develop related processes to produce these products economically. Due to the fact that animal manure and many types of municipal wastes such as food wastes are high in nutrient contents, recovering these nutrients as co-products not only benefits resource recycling but also contributes to environmental protection in terms of air and water quality. The Bioprocessing and Bioproduct Engineering Laboratory at Washington State University has been actively investigating the chemistry of the effluents from dairy manure and food waste anaerobic digestion and from that chemical research, developing technologies for recovering nutrients from the effluents. This paper provides a summary of the most recent progress. The information will be useful to the engineering community in selecting and designing systems to recover nutrients from waste streams.

See more of this Session: Conversion of Solid Wastes to Energy and/or Product
See more of this Group/Topical: Topical G: Innovations of Green Process Engineering for Sustainable Energy and Environment

(140b) A Spectrophotometric Method for Quantitative Determination of Xylose In Fermentation Medium

Patrisha J. Pham[1], Rafael Hernandez[1], Benjamin Estill[2], Andro H. Mondala[1] and William T. French[1], (1)Dave C. Swalm School of Chemical Engineering, Mississippi State University, Mississippi State, MS, (2)Department of Biological Sciences, Mississippi State University, Mississippi State, MS

The characterization of released sugars during fermentation is very important. Difficulty arises in the availability of a rapid, inexpensive and sensitive method for the detection of target sugars because fermentation broths are a complex mix of nutrients, waste products and cell debris. Since many of the ingredients are non-chromophoric and can't be detected by absorbance, target sugars can be identified from the rest of the fermentation medium by its conversion to a chromophoric substance that can be measured spectrophotometrically. A method involving UV spectrophotometry for the quantitative determination of xylose as the target sugar was developed. This widely known method is based on the dehydration of xylose to furfural in acidic medium with heat. The furfural produced condenses with orcinol in the presence of Fe^{3+} as catalyst to form a colored complex that can be measured by UV spectrophotometry. Factors affecting xylose concentration measurements such as acidic medium concentration, heating time and amount of Fe^{3+} catalyst were investigated. A continuous scan revealed the working wavelength to be 671 nm. The effect of other components in the fermentation broth was found to be negligible. Absorbance shows a linear relationship to xylose in the range of 100-500 mg/L. The xylose concentrations from actual fermentation were determined with the method and compared with YSI 2700, an enzyme electrode employed for xylose measurements and HPLC-ELSD method, a common technique for measuring xylose. Dilution is necessary for comparable xylose concentrations with YSI 2700 and HPLC.

See more of this Session: Conversion of Solid Wastes to Energy and/or Product
See more of this Group/Topical: Topical G: Innovations of Green Process Engineering for Sustainable Energy and Environment

(140c) Transesterification of Activated Sludge Oil by Ionic Liquid Supported Acid Catalysis

Patrisha J. Pham[1], Rafael Hernandez[1], Dyane Marie Acevedo Vélez[2] and William T. French[1], (1)Dave C. Swalm School of Chemical Engineering, Mississippi State University, Mississippi State, MS, (2) Chemical Engineering, University of Puerto Rico - Mayagüez, Mayagüez, PR

Progress in the transesterification of vegetable oils has been achieved using ionic liquids. Ionic liquids have been found to provide an ideal medium for increasing the biodiesel yield, stabilizing the inorganic acid and removing the glycerol by-product. In this present study, transesterification of triglycerides common in activated sludge oil with methanol to biodiesel was investigated using various imidazolium-based ionic liquids under acidic conditions. The influences of the nature of the anion and the alkyl groups on the imidazolium cation, ionic liquid dosage, reaction temperature and reaction time were also explored.

See more of this Session: Conversion of Solid Wastes to Energy and/or Product
See more of this Group/Topical: Topical G: Innovations of Green Process Engineering for Sustainable Energy and Environment

(140d) Pressure Cycles of Ozonation and Aeration for Activated Sludge Solubilization

Chia-Jung Cheng and Andy Hong, Civil & Environ. Eng., University of Utah, Salt Lake City, UT

Waste activated sludge from wastewater treatment plants is a universal problem. Traditional disposal methods such as composting, land application, and incineration post environmental and cost issues. Anaerobic digestion has been most common for excess waste sludge; however, retention time is often extensive because of kinetic limitation in the hydrolysis step. Ongoing researches have shown considerable benefits of solubilizing sludge prior to anaerobic digestion in solids reduction as well as biogas formation.

We tested a new sludge solubilization technique to obtain increased yields of soluble cell materials. The new technique incorporates rapid, successive compression and decompression cycles using ozone and air only. During compression, steepened gas concentration gradient across the cell membrane drives dissolved gas into the cell, increasing the gas concentration within the cell. When the pressure is rapidly reduced, gas formation and expansion occur within the cell, causing the formation of gas pockets in the cell (as transport across the membrane is slower than decompression). Gas accumulation in the cells during repeated pressure cycles leads to eventual rupture of the cells. Microscopic pictures of activated sludge were taken prior to pressure cycles, after 10 pressure cycles with air reaching 1000 kPa, and after 10 pressure cycles with 0.06% O3 reaching 1000 kPa. Prior to treatment, an abundance of filamentous bacteria and floc along with discernable microbes was present as in typical activated sludge. Pressure cycles with air show similar but diluted biomass with ciliated stalks broken into smaller pieces, while pressure cycles with ozone reveal little biomass but disperse nondescript flocs and seemingly gluey patches in the liquid. The application of pressure cycles resulted in disintegration of activated sludge. In its final form, this paper will present solubilization results under different rupturing conditions as characterized by TS, TSS, TDS, VDS, VSS, sCOD, tCOD, BOD, as well as confirmatory identifications of DNA and protein in the slurry before and after treatment.

See more of this Session: Conversion of Solid Wastes to Energy and/or Product
See more of this Group/Topical: Topical G: Innovations of Green Process Engineering for Sustainable Energy and Environment

(140e) An Evaluation of Production of Fuels From Sewage Sludge

Y. Madueke, Ramalingam Subramaniam, Cunwen Wang, Stephen Dufreche, Mark Zappi and **Rakesh Bajpai**, Chemical Engineering, University of Louisiana at Lafayette, Lafayette, LA

Efforts to produce biofuels as alternatives for petrofuels often runs into issues of renewable raw materials tat can be utilized. As a result, we have explored use of sewage sludge for biofuels. This paper will discuss the different constituents in sewage sludge that can be used for this purpose. Lipids and polymers present in the sludge will be identified and chemistries for their transformations will be presented.

See more of this Session: Conversion of Solid Wastes to Energy and/or Product
See more of this Group/Topical: Topical G: Innovations of Green Process Engineering for Sustainable Energy and Environment

APPLICATION OF COAL FLY ASH AS A SOLID BASIC CATALYST IN PRODUCING BIODIESEL

Omotola Babajide[1,*], Leslie Petrik[1], Nicholas Musyoka[1], Bamikole Amigun[1,2], Farouk Ameer[1]

[1]Environmental and Nano Sciences group, University of the Western Cape, South Africa
[2]Sustainable Energy Futures, Council for Scientific and Industrial Research (CSIR) Stellenbosch, South Africa

Keywords: Fly ash, Heterogeneous catalyst, Sunflower oil, Transesterification, Methyl ester

Abstract

South Africa is largely dependent on the combustion of coal for electricity production; Eskom's coal-fired power stations consume approximately 109 million tons of coal per annum, producing around 25 million tons of ash, to supply the bulk (93%) of South Africa's electricity. The management of this fly ash has been a concern with various approaches for its beneficial use being investigated. This work presents the results of transesterification reaction using sunflower oil as feedstock with methanol and class F fly ash catalyst derived from a coal fly ash dump in South Africa to produce methyl esters (biodiesel). The influence of various reactions parameters such as loading amount of active components, methanol: oil ratio, reaction time, temperature and catalyst deactivation was investigated and the product synthesized was tested. The important fuel properties of the methyl esters compared well with ASTM biodiesel standard.

Introduction

The recent energy crisis, largely driven by factors of economic growth and social integration, has lead to the construction of new coal power plants in South Africa. Specifically, South Africa generates 93 % of its electricity from coal combustion. Eskom's coal-fired power stations consume approximately 109 million tons of coal per annum, producing around 25 million tons of ash, to supply the bulk of South Africa's electricity [1]. Almost 90% of the ash produced in the generation process is called fly ash or pulverised fuel ash (a waste by-product of burning coal to generate electricity). The disposal and management of the huge quantities of fly ash generated has been a concern to the policy advisers, industrial and the scientific communities. The utilization of fly ash as a suitable feedstock for use as heterogeneous catalyst for the transesterification would allow beneficiation of fly ash in an environmentally friendly way which would make the process more economically viable.

Biofuels are derived from renewable sources such as naturally occurring fats and oils, which may be obtained from variety of plants and animals. Since, diesel-fuel figures as an essential function in the industrial economy, biodiesel-synthesis has constituted the essence of current research in the field of bio-fuels [2]. Biodiesel, a green fuel, is comprised of mono-alkyl esters of long chain fatty acids which are derived from fatty acid glycerides, and are major components of an oil or fat. Transesterification is the key and foremost important process to convert vegetable oil or animal fat into a diesel-like fuel [3]. This is illustrated in Figure 1.

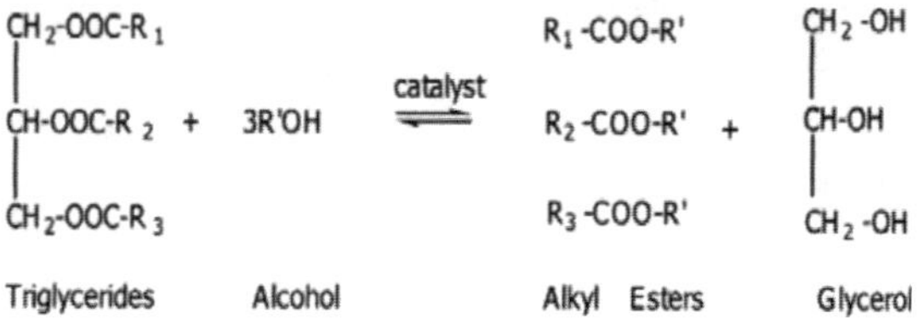

Fig. 1: Transesterification of Triglyceride with Alkyl Alcohol

While searching for promising solid base catalysts, several attempts have been made using basic zeolites, alkali and alkali earth oxides, alkali and alkali earth carbonates, supported guanidine and basic hydrotalcites. [4]. Fly ash, an inorganic residue arising from coal combustion processes consists essentially of SiO_2 and Al_2O_3 and being an environmental pollutant, constant generation of fly ash in huge quantities is a matter of concern. In view of its important economical and environmental implications, the disposal and utilization of fly ash represents a significant challenge. Development of environmentally safe applications for the utilization of coal fly ash is very important. Consequently, attempt was made to develop an efficient, inexpensive and environmental friendly fly ash-based heterogeneous catalyst for the esterification and/or transesterification reactions, for example for the production of biodiesel from sunflower oil.

Experimental

Table 1

Fatty Acid composition of Sunflower oil

Fatty Acid	Content (%)
Palmitic (16:0)	6.08
Stearic (18:0)	3.26
Oleic (18:1)	16.93
Linoleic (18:2)	71.73
Linolenic (18:3)	0.56

Pulverized fly ash collected from the ash collection systems at Arnot Thermal Power station in the Mpumalanga District of South Africa was used as a support. The elemental chemical composition of the fly ash (on dry basis) was SiO_2 50.91%, Al_2O_3 30.91%, Na_2O 0.10%, CaO 6.2%, Fe_2O_3 3.46%, MgO 1.48%, TiO_2 1.65%, MnO 0.02%, K_2O 0.60%, P_2O_5 0.56%. Before its use, material was ground to a fine powder. The average SiO_2/Al_2O_3 ratio of Arnot fly ash was found to be 1.65. The major elemental chemistry of the Arnot fly ash is consistent with $SiO_2+Al_2O_3+Fe_2O_3 \geq 70\%$ meaning that it is "class F" fly ash (ATSM method C 618). This is a typical composition of fly ash from combustion of bituminous coal from South Africa [5]. The alkali salt KNO_3 (99% purity from Merck Chemicals, South Africa) was used in as-received form to prepare its stock solution (aq. 5% wt/wt). Apart from fly ash from a pulverized coal power plant, another primary source of fly ash is from a municipal waste incineration plant.

Catalyst preparation and characterisation

The catalysts were prepared using established method of wet impregnation and were characterized to measure the properties such as pore size, surface area, structure to thus understand the performance of the catalyst using the following techniques. (X-ray powder diffraction (XRD), Scanning electron microscope (SEM), Fourier transformed infrared (FTIR) spectroscopy, BET surface area)

Results and discussion

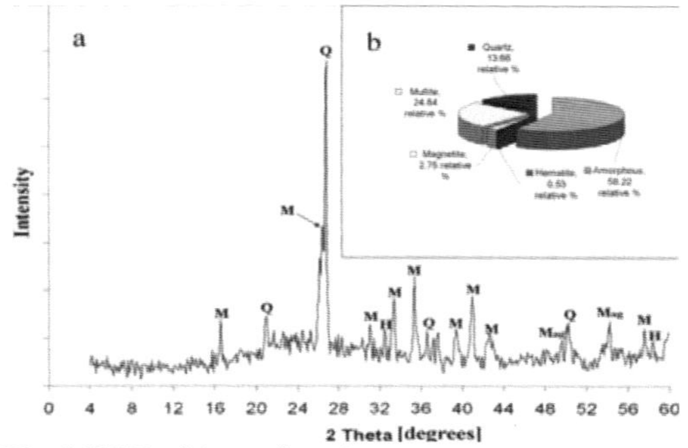

Fig. 1 XRD of Arnot fly ash

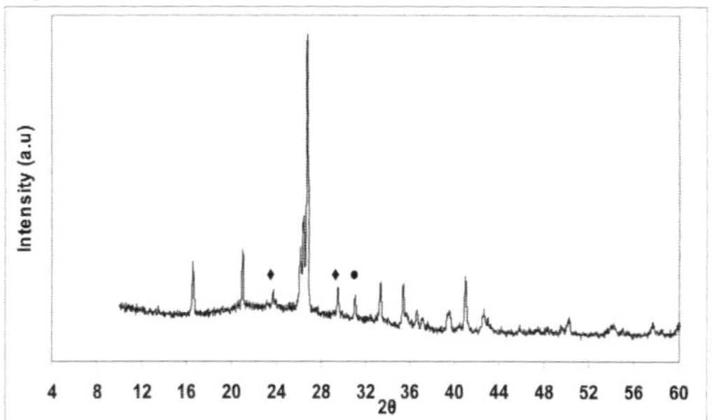

Fig. 2 XRD patterns of potassium impregnated fly ash catalyst with characteristic peaks due to KNO_3 • and K_2O♦.

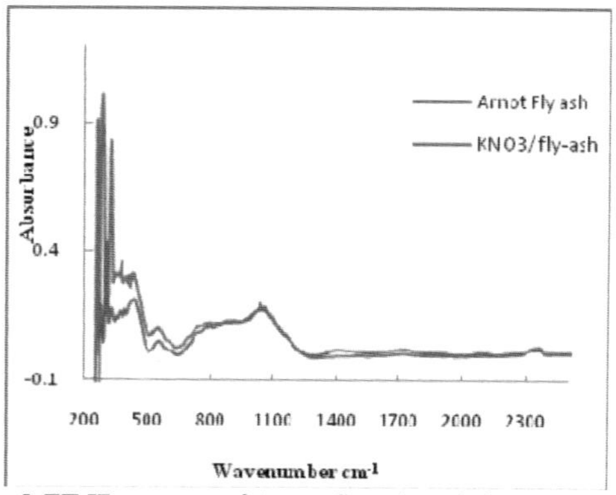

Fig. 3 FT-IR spectra of Arnot fly ash and the potassium impregnated fly ash catalyst

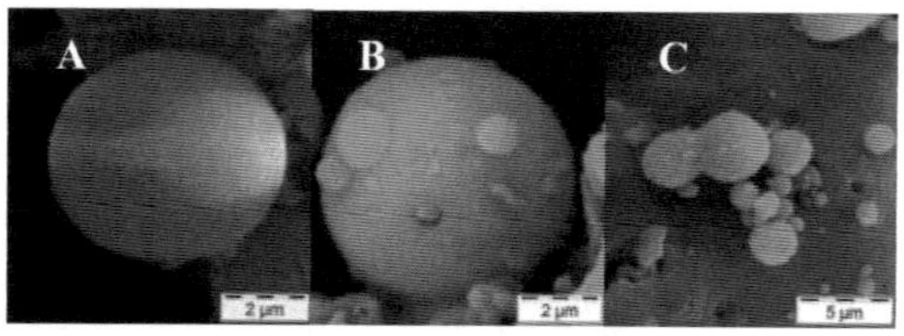

Fig. 4 SEM images of the (a) Arnot fly ash (b) and (c) potassium impregnated fly ash catalyst

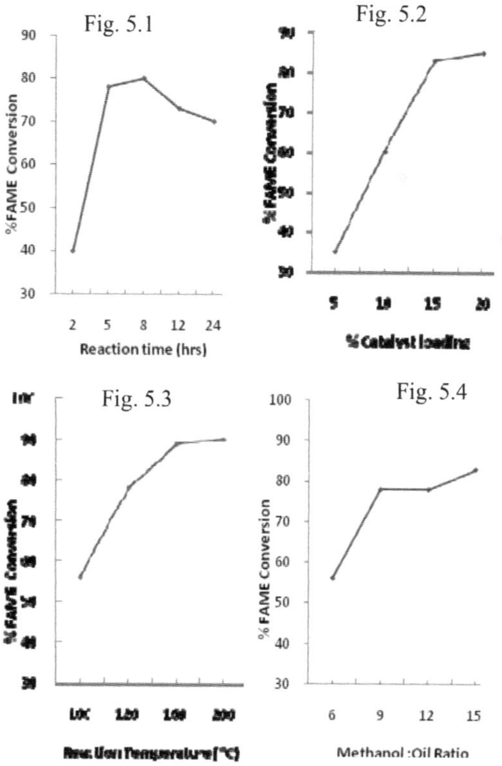

Figs. 5.1-5.4 Influence of process parameters on FAME conversion using the impregnated fly ash at different reaction conditions.

It can be observed from Fig 5.1, that the oil conversion was not greatly affected by an increase in the reaction time. This implies that the reaction equilibrium had been reached after 5hrs when the reaction temperature was 160°C as the experiment was conducted under reaction conditions of 160°C, 15% catalyst loading and a methanol: oil ratio of 15:1.

Fig 5.2 shows that the highest conversion of 87.45% was obtained at a catalyst loading of 20% wt/wt which was comparative with the conversion value of 86.56% at 15%wt/wt. As a consequence, the optimal amount of catalyst used in this pilot was 15wt % loading as it was sufficient in catalyzing the reaction despite the challenge posed by the three phase reaction system where initial mass transfer control is inhibited due to heterogeneous kinetics.

The results shown in Figure 5.3 and from the activity profile, it was observed that as reaction temperature increased, there was a corresponding increase in the conversion of the triglycerides to methyl esters, showing peaks at temperatures ≥ 160°C. The highest conversion was 89.34% at a reaction temperature of 200°C which compares to the conversion value of 87.12% at the reaction temperature of 160°C. Influence of reaction temperature on oil conversion under reaction time of 5hrs, a catalyst loading of 15%wt/wt and a methanol: oil ratio of 15:1.

From the Figure5.4, it can be seen that the highest methyl ester conversion of 86.1% was observed at a methanol: oil molar ratio of 15:1. It can be also be observed from the activity profile that as the molar ratio increased from 6:1 to 15:1, the conversion was found to increase from 56.23 to 86.1%. Further addition of methanol into the reaction phase did not show any significant increase in the conversion. Moreover, a limitation to the use of a greater amount of methanol needed to propel the forward reaction is that an occurrence of flooding of the active sites by the methanol molecules rather than triglyceride molecules which might hinder the completion of the triglyceride conversion to methyl esters as conversion at methanol oil ratios higher than 15 appeared to decline [6]

Conclusion

The use of fly ash loaded with KNO_3 has been found suitable for the conversion of sunflower oil to methyl esters and glycerol. The fly ash based catalyst loaded with 5% wt KNO_3 at a reaction temperature of 160°C exhibits maximum FAME conversion (86.13%). The highest conversion of oil was obtained at a catalyst loading of 20%wt/wt. Influence of process parameters such as reaction time, reaction temperature and molar ratio of methanol to oil were also carried out. The oil conversion was not greatly affected by an increase in the reaction time. However, reaction time between 5 to 8 hours will lead to better oil conversion. The highest conversion was observed at a temperature of 200°C. The activity profile shows that increase in the methanol to oil ratio from 6:1 to 15:1 resulted in increased conversion from 52.23% to 86.1%. A decline in the oil conversion was observed when the recovered catalyst was recycled for use. The FAME produced was tested and found to meet ASTM standard.

REFERENCES

[1] Eskom fact sheet (2009). Ash management in Eskom; Generation fact sheets Available online at http://www.eskom.co.za/live/content.php?Item ID=28. Accessed May 2010.

[2] Maa F, Hannab MA. Biodiesel production: a review1. Bioresource Technology 1999:70:1-15.

[3] Schuchardta U, Serchelia R, Matheus R. Transesterification of Vegetable Oils: a Review. J. Braz. Chem. Soc 1998; 9(1):199-210.

[4] Georgogianni KG, Katsoulidis AP, Pomonis PJ, Kontominas MG. Transesterification of soybean frying oil to biodiesel using heterogeneous catalysts. Fuel Processing Technology 2009; 90:671–76

[5] Gitari, M., Petrik Leslie, Key D., Etchebers O. and Okujeni C. Mineralogy and Trace Element Partitioning in Coal Fly Ash/Acid Mine Drainage Co-Disposed Solid Residues.http://whocares.caer.uky.edu/wasp/AshSymposium/AshLibraryAgenda.asp (2005). (Accessed on the 24th March, 2010).

[6] Shu Q, Zhang Q, Xu G. Wang J, Preparation of biodiesel using s-MWCNT catalysts and the coupling of reaction and separation. Food and Bioproducts Processing 2009; 87:164-70.

---

*To whom all correspondence should be addressed.

Email address: obabajide@uwc.ac.za (O. Babajide)
Tel./Fax: +27 21 959 3878.

(149a) Reducing the Impacts of Resource Temperature Degradation and Varying Ambient Temperature On Air-Cooled Binary Plant Performance

Daniel S. Wendt, Advanced Process & Decision Systems, Idaho National Laboratory, Idaho Falls, ID and Gregory L. Mines, Biofuels & Renewable Energy Technology, Idaho National Laboratory, Idaho Falls, ID

As the geothermal industry moves to use geothermal resources that are more expensive to develop, there will be increased incentive to use more efficient power plants. Because of increasing demand on finite supplies of water, this next generation of more efficient plants will likely need to reject heat sensibly to the ambient (air-cooling). Though they use no water, air-cooled systems have higher capital costs, reduced power output (rejection at a higher temperature), lower power sales due to higher parasitics (fan power), and have greater variability in power output (variation in the dry-bulb temperature). These effects are compounded as the resource temperature decreases.

Several currently available technologies have been studied to evaluate their potential to cost-effectively increase plant overall efficiency and total power generated during the plant operational life. These technologies include recuperation, in which turbine exhaust superheat is used to preheat the condensed working fluid prior to heat exchange with the geothermal fluid; reheat, where an intermediate heat exchange between the working fluid and geothermal fluid takes place during the expansion of the working fluid in the turbine; non-consumptive use of EGS makeup water to supplement the heat rejection; and substitution of working fluids at an intermediate point in the geothermal plant operational life.

Aspen Plus process models were developed to calculate optimal equipment sizing for maximizing the net power output of various process configurations in which geothermal resource temperature, geothermal fluid flow rate, ambient temperature, working fluid composition, heat exchanger pinch points, turbine and pump efficiencies, and process equipment and piping pressure drops were specified. A minimum geothermal fluid outlet temperature constraint was set to mitigate concerns of mineral precipitation on the heat exchanger surfaces. Two scenarios modeled were a baseline geothermal process employing the Organic Rankine Cycle (ORC) and a modified geothermal process design employing recuperation, reheat, and makeup water heat rejection in the Organic Rankine Cycle.

Following completion of the "plant design" modeling described above, a process modeling analysis was performed to evaluate the effect of hourly ambient temperature variation and declining resource temperature over the operational life of the two power plant designs. The equipment specifications from the baseline and modified geothermal plant process designs were fixed and correlations were incorporated into the process models to adjust heat transfer coefficients, pump and turbine efficiencies, and pressure drops in order to evaluate process performance as geothermal resource and ambient temperatures varied. The capability to vary turbine nozzle geometry was included in all simulations to provide an additional degree of freedom in establishing optimal process conditions. In all cases evaluated, the geothermal fluid flow rate was assumed constant.

A comparison of the baseline geothermal plant and modified geothermal plant capital costs and total power generated during the plant operational life was then performed for a scenario with a defined resource temperature decline and annual temperature profile. The methodologies used to select the plant design conditions, including ambient design temperature, working fluid selection, and turbine nozzle area for a declining resource temperature are discussed.

See more of this Session: Geothermal Engineering
See more of this Group/Topical: Topical D: Chemical Engineering in Oil and Gas Production and Other Complex Subsurface Processes

A Downhole Fluorimeter for Measuring Flow Processes in Geothermal and EGS Wellbores

Michael Mella[1], Steven Olsen[2], Peter Rose[1], Daniel Bour[3]Susan Petty[3]and Joel Harris[4]

[1]Energy and Geoscience Institute at the University of Utah · 423 Wakara Way Suite 300, Salt Lake City, UT 84108

[2]Red Rock Analog Design · 1218 Emerson Ave, Salt Lake City, UT 84108

[3]Alta Rock Energy Inc · 7900 E Green Lake Drive N Suite 202, Seattle, WA 98103

[4]Department of Chemistry, University of Utah

Keywords

Borehole Fluorimeter, Borehole Flowmeter, Fracture Characterization, EGS, Geothermal

Abstract

The conventional approach to measuring flow downhole in geothermal or EGS wellbores is through the use of a mechanical spinner tool, which makes use of a rotating paddle suspended and lowered (or raised) through the wellbore on a cable. But, such tools are notorious for failing-especially at high flow rates and at high temperatures. Likewise, non-uniform wellbore diameters greatly complicate spinner-tool data interpretation. An alternative approach to measuring flow rates in a geothermal or EGS wellbore is described. Making use of conventional tracer dilution methods, the tool is designed to work under both laminar and turbulent flow regimes. Volumetric flow rate measurement will allow the tool to accurately measure flow even through washout regions where borehole diameter varies. The tool is designed to be deployed on a single-conductor wireline and will be heat shielded. An optical fiber bundle will serve both to deliver the tool's excitation light signal and to return the emission signal from the fluorescent dye that will be introduced from a pump attached to the end of the tool. Temperature and pressure measurements will also be made using Resistance Temperature Detectors (RTD) and pressure transducers.

Introduction

Downhole temperature, pressure and flow rates are essential pieces of information for any geothermal field. Wire-line tools transmit data to the surface giving real-time readouts with depth. From the data logs fluid entry zones and temperature profiles of the wellbore can be obtained. Flow rates can be calculated and fluid entry zones can be interpreted in combination with temperature data. Temperature-Pressure-Spinner (TPS) tools are the current industry standard in obtaining this valuable information. TPS tools are fairly simple and straight forward instruments to understand and operate and can be manufactured to operate at high temperatures for extended periods of time (Davarzani 1988). When combined with caliper log data in combination with temperature and pressure data, TPS log data can be used to calculate volumetric flow rates.

As important as TPS tools are and how much valuable data they provide, there are limitations of TPS tools. TPS reliability relies on the velocity of water to turn a mechanical device on the tool. Flow rates that represent the two extremes in geothermal systems pose problems for TPS tools. Spinner tools typically need velocities in excess of 5 ft/min to turn the impeller element (Smollen 1996). Flow rates in the laminar flow range are not measurable. At very high flow velocities spinner tools may not be able to keep up with the velocity and the impeller itself may break off. The flow in between the two extremes may also require tools with differing impellers to accurately measure velocities. A second drawback to TPS tools is the inability to accurately measure flow in some open-hole or wash-out regions. If the tool is in a space of unknown diameter the velocity may not be used to accurately calculate the volumetric flow rate. The tool that is detailed in this paper is one that uses measured tracer concentration to determine flow rates.

Measuring flow through a pipe using tracing dyes dates to when Reynolds conducted his flow experiments in the late 1800's (de Nevers, 2005). Radioactive tracer tools have been employed for measuring flow in oil field injection wells (Smollen 1996). Radioactive tracers have the obvious drawback of using radioactive materials. For this reason radioactive tracing is used almost exclusively in injection wells. Most radioactive tracer tools do not measure concentration but rather are time of flight type tools that measure tracer pulses passing two different detectors.

The fluorescent tracers used for the application described in this paper need to be highly detectable and thermally stable. These compounds will additionally have to fluoresce at a wavelength where there is little interference coming from naturally occurring chemicals in the geothermal brine. Among the chief tracers being considered is fluorescein. Fluorescein has been proven to be highly detectable and thermally stable for use as a geothermal tracer (Adams et al. 1991). Other tracers that could be used include one in the family of naphthalene sulfonates (Rose et al. 2003), PTSA (Rose et al. 1998) or even quantum dot tracers (GTP 2010).
This paper describes a tool that will measure tracer concentration and calculate volumetric flow rates based on tracer dilution techniques. Using this approach, mass flow rate ($\dot{m}$) is calculated from the concentration of the tracer (X) and the rate at which the tracer is delivered (Q):

$$\dot{m}\left(\frac{m_{water}}{t}\right) = \frac{Q\left(\frac{m_{tracer}}{t}\right)}{X\left(\frac{m_{tracer}}{m_{water}}\right)} \tag{1}$$

A volumetric flow rate can then be calculated using mass flow rate. Tracer dilution techniques have been successfully used for measuring flow in geothermal production wells at the surface for several years (Hirtz et al., 2001 and Lovelock, 2001). The design process of such a tool to measure and deliver tracers at geothermal conditions is the subject of the remainder of this paper.

Tool Outline

The tool will be capable of high temperature operation for extended periods of time and be able to deliver and measure water soluble tracer in the wellbore and to send information back to the surface in real-time. The tool will consist of a fiber optic at one end for detecting tracer, a tracer delivery system at the other end, and heat shielded electronics. It will be connected to a single conductor wireline that will serve as both the structural tethering element and communication element to the surface. At the surface the tool will be controlled by a computer that will be capable of sending and receiving instructions. The tool will also contain pressure and temperature measuring devices and log this information at the same time it is logging the tracer signal. A cartoon sketch is shown in Figure 1 below.

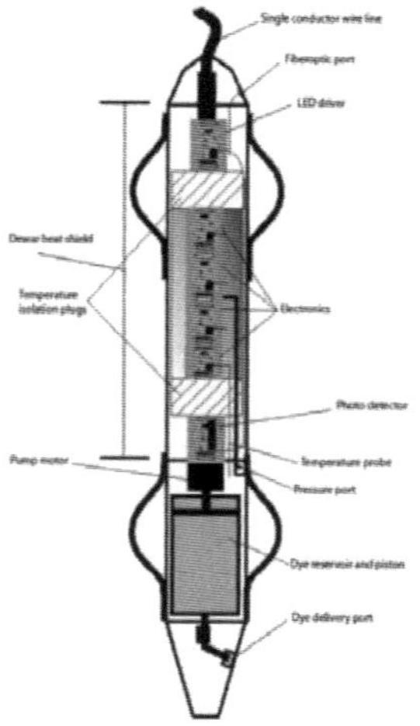

Figure 1. Sketch of the fluorimeter-flowmeter.

Flow Conditions and Tracer Mixing

Thorough mixing of the injected tracer within the wellbore is essential the accuracy of volumetric flow rates. Assuming an 8-inch well bore diameter calculations were made for several volumetric flow rates to determine the distance need to achieve 99% axial mixing in turbulent conditions. Calculation of mixing distance, L, was performed using Equation 2 listed below where f is the Fanning friction factor and is defined in equation

3.

$$L = \frac{0.56D}{\sqrt{f}} \tag{2}$$

$$f = 0.001375\left[1 + \left(20{,}000\frac{\varepsilon}{D} + \frac{10^6}{\mathrm{Re}}\right)^{\frac{1}{3}}\right] \tag{3}$$

Equation 3 is only valid for turbulent flow, or flows for which Re is greater than 2,300. In Equation 3, D is the pipe diameter, which was assumed to be 8 inches, and ε is the surface roughness assumed to be the roughness of commercial steel pipe. Re is the Reynolds number and is defined as

$$\mathrm{Re} = \frac{4Q}{\pi D \nu} \tag{4}$$

where Q is the volumetric flow rate, D is the diameter and ν is the kinematic viscosity. Kinematic viscosity is a function of temperature and the fluid itself. In this case the fluid viscosity was assumed to be that of water and the viscosity at different temperatures was calculated. From these calculations a plot was constructed showing mixing length vs. volumetric flow rate. The Reynolds numbers ranged from 41,000 to $1.7\text{x}10^7$. The plot is displayed below as Figure 2.

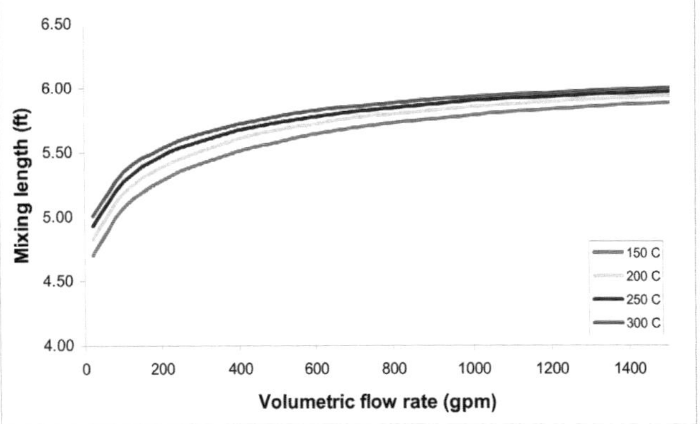

Figure 2. Mixing lengths required for complete axial mixing vs. flow rate.

From the plot one can see that the injection source and detection must be about 6 feet apart to approximate complete mixing even at very high flow rates (high velocities). The back mixing effect present in turbulent flow rates and present in most geothermal wells will ensure well mixed tracer provided the tool is longer than 6 feet.

The use of the tool at laminar flow conditions will be a challenge because of the lack of much back mixing inherent in the flow. Figure 3 below shows what tracer concentration across the wellbore might look like 8 feet from the injection source at different flow rates. As one can see from the plot, the tracer concentration would vary significantly from the borehole walls, at 1 and minus 1, to the center line.

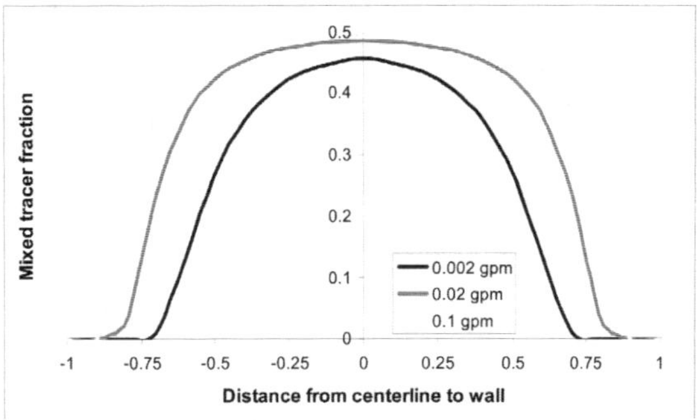

Figure 3. Tracer concentration profile across the wellbore at a distance of 8 feet from the injection point.

Operation at laminar flow regimes will require the addition of some sort of mixing device to provide for even mixing across the wellbore. Two of the ideas to help induce mixing include adding a piece that would have recirculation jets to induce back mixing or adding a mixing brush on the outside of the tool. Both of these options are depicted graphically in figure 4 below.

Figure 4. Cartoon representation of mixing devices, recirculation jets on the left and mixing brush on the right.

Temperature Effects

Perhaps the biggest challenge of deploying a tool in a geothermal wellbore is the high temperature. Using a Dewar-style heat shield in addition to a heat sink, estimates were made of the internal temperature of the tool as a function of time at geothermal conditions. With a stainless steel heat sink and an assumption of 5 watts of power dissipation within the tool, a plot of internal temperature versus time for several temperatures was produced and is shown below as Figure 5. Included in the plot are two lines that represent 85°C and 125°C. These two lines are significant because they show the temperatures at which traditional electronics (85°C) and automotive class electronics (125°C) will fail. The price increase to move from traditional to automotive grade electronics is not significant. Any electronics beyond these temperatures will need to be high temperature components that are available only at a significant increase in cost.

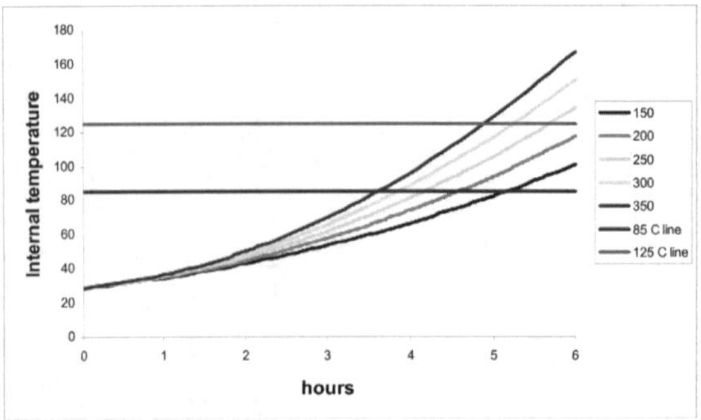

Figure 5. Expected internal temperature vs. time at various external temperatures.

From the plot it can be seen that use of automotive class components will allow for about 5 hours of operation at temperatures up to 300°C. To get to six hours of downhole operation at temperatures greater than 250°C it would appear that high temperature electronics would be called for.

In an attempt to extend the time that the tool can spend in the well, we investigated phase-change cooling. This entails the use of a low melting temperature alloy sealed inside the heat sink. Using this approach, thermal energy would be expended in the phase change of the alloy and delay temperature rise. For comparison, calculations were performed with a low melt alloy, AIM 70, and the results were plotted as seen in figure 6 below.

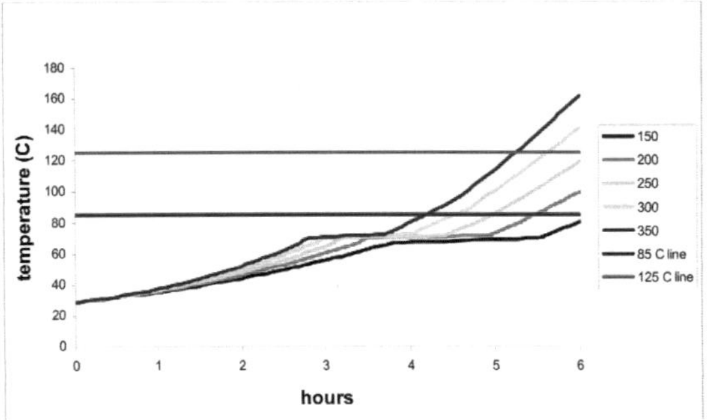

Figure 6. Expected internal temperature vs. time with heat sink that includes phase change material.

From the plot it can be seen that the phase change material would elongate the tool's time in the well by approximately 30-60 minutes depending on the temperature. It is unlikely that we would opt to use this strategy in a prototype tool but it may be part of a future tool. Overall it appears that the main challenges for temperature management include limiting internal heat generation and designing with higher grade automotive components.

Electrical Components

The tool will be controlled by internal electrical components that measure and report tracer signal, temperature and pressure, and control power. These components will be either in the heat-shielded portion of the tool or at the surface. The electronics will be manufactured in a module form allowing flexibility and ease of repair. The added benefit of manufacturing in a module method is that if one module is not performing well it can be swapped out with a better designed module in future iterations of the tool.

Power Supply and Depth Encoder

The power supply will consist of two components. The first is regulation of the power delivered at the surface and the second is the regulation of power at the tool. The powersupply at the surface will convert AC to DC power sent down the wireline. The power supply at the surface will regulate voltage based on the power consumption in the tool and the resistance that is encountered in the line. Figure 7 below shows the relationship of voltage to power consumption with different line resistances encountered. The tool side power supply will be able to give real time information to the surface power supply to determine how much voltage to send.

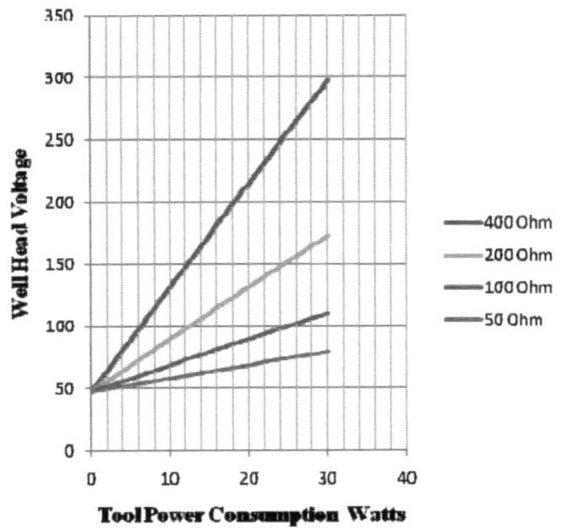

Figure 7. Well head voltage vs. power consumption with different line resistances.

The tool side power supply will involve input protection against over voltage and voltage converters to supply DC power at a variety of voltages to the various components. The power supply will also make use of real time supply optimization to request the right amount of power from moment from the surface supply. This should help to reduce the amount of heat that would be dissipated in the tool. A second strategy that will be used will be turning unused components off when they are not needed. For instance, when the tool is on the way to the zone of interest only intermittent temperature and pressure readings may be desired, the other components can be turned off to reduce internal heat generation.

In addition to supplying power along the wireline, the amount of line paid out will be used to calculate the depth of the tool at any given time. The controls at the surface will contain a depth-encoder device that will be able to accurately calculate the depth of the tool by counting the revolutions of the wireline spool. The depth information will be added to the pressure, temperature, and flow rate data.

Light Source and Detector

The light source is provisionally a high efficiency blue LED that will provide excitation at 465 nm and will operate at 125°C. The LED will be coupled to a collimator which will be connected to the fiber optic source. The LED operation strategy will include modulation to effectively subtract dark current as well as sampling the light to get an accurate indication of intensity out.

The detector will be a photo multiplier tube (PMT). Another detector that was considered was an avalanche photodetectors (APD). The advantages of the APD are low cost, ruggedness and a small electrical footprint. The disadvantages include low sensitivity and inability to operate above 50°C. In contrast the PMT has several orders of magnitude better sensitivity and can operate at temperatures up to 175°C. The disadvantages of PMT's include larger electrical footprint, high voltage (low current) operation and increased cost.

Pressure, Temperature and Inclination

The temperatures within the borehole and internal to the tool will be closely monitored throughout the test. A platinum RTD will be placed through one end of the tools sub-assembly and a second RTD will record temperatures within the tool body. Because some of the components could be destroyed by heat internal temperature will be measured as well. Pressure will be measured using a high temperature piezoelectric pressure transducer using a standard pressure filter port through the tools bottom sub-assembly. An accelerometer will be included in the tool to measure inclination and motion.

Tool Control and Data Transmission

In general much of the tool control will be located within the software being run at the surface monitoring station. Things that will be controlled will be the on/off of individual components, the LED intensity, the PMT gain and the flow rate of the dye pump.

Field Programmable Gate Arrays (FPGA) will be utilized in the communication through the wireline. The FPGA's will allow for flexible signals, in this case data will be packetized and sent with internal validation. Signal prioritization will be controlled by a CAN bus within the tool.

Test and Deployment Plan

The module design of the electronics will afford the ability to test each of the parts independently before attempting to test the tool as an entire unit. The electronics, fiber optics, pump and detector will be tested in autoclave reactors that will simulate geothermal pressures and temperatures. Once the individual components are tested a test fixture will be constructed to test the tool under flowing conditions. A long pipe with inlets at intervals along the length will serve as a test fixture. Water at a specified flow rate can be flowed through the pipe and then water will either be drawn off or added at a known flow rate through the inlets along the length of the pipe. The test fixture will be used to test the tool under flow conditions and serve as a calibration tool for the fluorimeter. Tests in the lab should reveal problems prior to a field test.

Conclusion

The design of a single-conductor wireline fluorimeter-flowmeter for geothermal applications is described. The tool will be useful under both turbulent and laminar flow conditions. Using high temperature electronics, the tool will be able to log well bores in excess of 300°C for periods of greater than 6 hours. It will provide the user with temperature, pressure and volumetric flow rates at each point along the wellbore.

References

Adams, M.C.; Davis J. Kinetics of Fluorescein Decay and its Application as a Geothermal Tracer. *Geothermics* **1991**, *20*, 53-66.

Davarzani, J.; Sloan, M. Simultaneous TPS Logging Systems. *GRC Transactions* **1988**, *12*, 441-448.

de Nevers, N. *Fluid Mechanics for Chemical Engineers*, Third Edition; McGraw Hill: New York, 2005.

U.S. Department of Energy Geothermal Technologies Program (GTP). http://apps1.eere.energy.gov/geothermal/projects/projects.cfm/ProjectID=176 (accessed 27 April 2010).

Henfling, J.A.; Norman, R.A. *Precision Pressure/Temperature Logging Tool*; SAND98-0165; Sandia National Laboratories: Albuquerque, NM, January 1998, 1-21.

Hirtz, P.N.; Kunzman, R.J.; Broaddus, M.L.; Barbitta, J.A. Developments in Tracer Flow Testing for Geothermal Production Engineering. *Geothermics* **2001,** *30*, 727-745.

Lovelock B.G. Steam Flow Measurements Using Alcohol Tracers. *Geothermics* **2001,** *30*, 641-654.

Rose, P.E.; Benoit, W.R.; Adams, M.C. Tracer Testing at Dixie Valley, Nevada Using Pyrene Tetrasulfonate, Amino G, and Fluorescein. *GRC Transactions* **1998**, *22*, 583-587.

Rose, P.E.; Mella, M.; Kasteller, C. A New Tracer for Use in Liquid- Dominated, High Temperature Geothermal Reservoirs. *GRC Transactions* **2003**, *27*, pp 403-406.

Smollen, J.D. *Cased Hole and Production Log Evaluation*; PennWell Books: Tulsa, OK, 1996.

An Integrated Model to Compare Net Electricity Generation for CO_2- and Water-Based Geothermal Systems

Vikas Agarwal[1] and Brian J. Anderson[1]
[1]Department of Chemical Engineering
West Virginia University
Morgantown, WV 26506

Keywords: Wellbore flow, Reservoir simulation, Power cycle, Fluid flow, CO_2, Modeling

ABSTRACT:

Utilization of supercritical CO_2 as a geothermal fluid instead of water has been proposed by Brown[1] and its advantages have been discussed by him and researchers[4,5]. This work assesses the net electricity which could be generated by using supercritical CO_2 as a geothermal working fluid and compares it with water under same reservoir conditions of temperature and pressure. This procedure provides a method of direct comparison of water and CO_2 as geothermal working fluids, in terms of net electricity generation over time.

An integrated three-part model has been developed to determine net electricity generation for CO_2- and water-based geothermal reservoirs. This model consists of a wellbore model, reservoir simulation and surface plant simulation. To determine the bottomhole pressure and temperature of the geothermal fluid (either water or CO_2) in the injection well, a wellbore model was developed using fluid-phase thermodynamic equations of state, fluid dynamics, and heat transfer models. A computer program was developed that solves for the temperature and pressure of the working fluid (either water or CO_2) down the wellbore by simultaneously solving for the fluid thermophysical properties, heat transfer, and frictional losses.

For the reservoir simulation TOUGH2 has been used to model the temperature and pressure characteristics of the working fluid in the reservoir. The EOS1 module of TOUGH2 has been used for the water system and the EOS2 module of the TOUGH2 code has been employed for the CO_2 case.

The surface plant is simulated using CHEMCAD, a chemical process simulator, to determine the net electricity generated. A binary organic Rankine cycle is simulated. The calculated net electricity generated for the optimized water and CO_2 systems are compared over the working time of the reservoir.

---

1. Introduction

There is enough heat in the crust of the earth to satisfy a large part of the world's probable energy needs for a very long time. This heat could be tapped by Enhanced Geothermal Systems (EGS) which aims to extract the heat beneath the earth crust by injecting a fluid through a wellbore, which flows through a hot artificially fractured reservoir and then is forced out from the production well, thus heat is carried to the surface of the earth from beneath the surface. Conventional geothermal power plants are based on extracting the heat from the existing geothermal reservoirs having naturally occurring porosity, permeability which allows the flow of water through them. However, there is vast majority of hot dry rocks that could be found in western U.S.[2] which are assessable through current drilling practice. The heat from these hot dry rocks could be tapped by enhancing the permeability of the rock through hydraulic stimulation and thereafter allowing the fluid to flow through them. One of the most common options is to use water as a heat transfer fluid. However, some of the researchers like Brown[1], Pruess[3] etc. have considered and studied the use of CO_2 as a heat transfer fluid.

Properties that motivate the use of CO_2 as a geothermal fluid emphasized are:

1. The density of cold CO_2 and that of hot CO_2 obtained from production well have large differences, which could provide a significant buoyant force[4] and thus circulation power consumption would be less and more net electricity could be generated.

2. CO_2 is a poor solvent may not dissolve minerals from reservoir thus reducing scaling problems caused by silica dissolution[4]. Conversely water is a powerful solvent, makes difficult to operate EGS reservoir in a stable manner.

3. At the elevated temperature of reservoir, CO_2-mineral reactions may provide an additional advantage for the sequestration of CO_2[5].

This work primarily concerns the comparison of CO_2 and water as a geothermal fluid in terms of net electricity generation over a period of time. The aim of the comparison is to understand under identical set of reservoir conditions which choice is better if one fluid has to be picked over other. A comparison is being made using the same flow rate of both the fluids operating under same surface power cycles and same reservoir conditions compared to work by Pruess et al[3] that compared those under constant delta P.

Any geothermal power plant comprises of three components: the wellbore (injection/production), the geothermal reservoir (from which heat is extracted), and the surface plant (where heat extracted is transformed into electricity for human use). These three components need to be studied separately yet need to be coupled to understand the overall impact on net electricity generation.

In this work, to determine the temperature and pressure profiles of two fluids (water/CO_2) a wellbore model has been developed. TOUGH2, a numerical simulator has been employed for reservoir simulations using the EOS1 module of TOUGH2 for simulating flow of water and EOS2 for CO_2. Finally to determine net electricity generation the chemical process simulation software, Chemcad, has been used for surface plant calculations.

2. Wellbore model

Geothermal fluids either CO_2 or water are being injected into the reservoir through injection wells and the heated fluid are being taken out from the reservoir through production wells. As the fluid travels through the wellbore it either loses enthalpy to or gains enthalpy from the wellbore formation and the rate of loss or gain of enthalpy is a function of time and depth. Therefore geothermal wells should be considered nonisothermal. It is important to know the bottom-hole pressure and temperature in the case of an injection well (to know the enthalpy of the injected fluid to the reservoir) and the wellhead pressure and temperature for the production well. It is the temperature and pressure of the fluid obtained at the wellhead of the production well that determines the amount of electricity that could be generated from the geothermal fluid. Figure 1 shows the basic well configuration used for the modeling. Although the model is developed for any depth and any radius but in this work reservoir is considered to be at a depth of 3 km. Direction of fluid flow in the well could be either upwards or downwards.

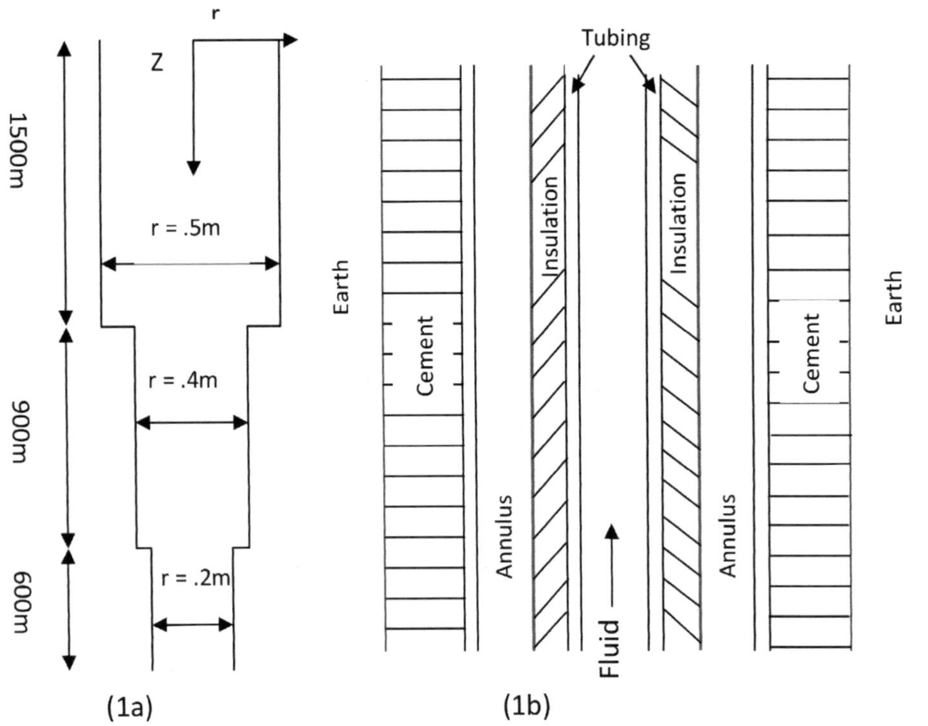

Figure 1: Basic wellbore configuration (1a) Tube diameter variation with depth (1b) Different layers of materials in the radial direction.

Basic flow equations:

We shall now discuss the basic equations used and the methodology in determining the temperature and pressure distribution of the geothermal fluid along the well. The assumptions which are incorporated in the development of this model are:

1) As the diameter of the wellbore is very small when compared to its length, no variation of temperature, pressure and fluid velocity in the radial direction of well is assumed. All properties changes along vertical direction only.
2) For designing flow of CO_2 in the well, the well is taken to be completely dry with respect to water.
3) The geothermal gradient of the earth (which is defined as the rate of change of temperature with depth) is taken to be constant in this work. However, the code could easily be modified to give geothermal gradient variation with depth.
4) The acceleration term of the pressure gradient is neglected. This can be a reasonable assumption, considering that velocity of the fluid doesn't change as fast as the pressure changes with depth. Velocity is a function of density and density is a function of pressure and temperature. As geothermal fluid goes down the wellbore, its pressure increases and hence density tends to increase but due to thermal gradient of earth, temperature also increase which causes the density to decrease and hence the net effect on density and velocity is not that large as in the case of pressure. So, change in velocity is sluggish in comparison to change in pressure. Thus, velocity gradient and hence acceleration term can be neglected in comparison with pressure gradient. It can be justified when the velocity profile is obtained based on this assumption. Based on the similar argument, acceleration term can be dropped when the fluid is ascending the production wellbore.
5) Vertical heat diffusion is neglected as a first effort of model development. It can also be a reasonable assumption because velocity of the fluid is high and convective heat transfer dominates the diffusive heat transfer.

The equation for the pressure drop in a well is given by:

$$\frac{dp}{dZ} + \rho v \frac{dv}{dZ} + \rho g + f \frac{\rho v^2}{2d} = 0 \qquad (1)$$

where,

p = pressure, N/m^2

Z = distance in vertical direction from the wellhead, m

ρ = density of fluid, kg/m^3

v = velocity in z direction, m/s

g = acceleration due to gravity, m/s^2

f = Moody friction factor

d = wellbore diameter, m

All units are taken in SI system.

In Equation (1) the $\rho v \frac{dv}{dZ}$ term represents the acceleration term, ρg represents the gravitational term and $f\frac{\rho v^2}{2d}$ represents the frictional term of the pressure gradient along the well.

Neglecting the acceleration term of the pressure gradient as discussed above, one is left with

$$\frac{dp}{dz} + \rho g + f\frac{\rho v^2}{2d} = 0 \quad (2)$$

The friction factor, f, can be calculated as[11]

$$f = [-2\log\{\frac{\epsilon/d}{3.7065} - \frac{5.0452}{Re}\log(\frac{(\epsilon/d)^{1.1098}}{2.8257} + (\frac{7.149}{Re})^{0.8981})\}]^{-2} \quad (3)$$

Where,

ϵ = Pipe roughness factor

Re = Reynolds number

$$Re = \frac{dv\rho}{\mu} \quad (4)$$

Temperature calculations:

The variation of the formation temperature along the radial direction as a function of temperature can be given by diffusivity equation:

$$\frac{\partial^2 T_e}{\partial r^2} + \frac{1}{r}\frac{\partial T_e}{\partial r} = \frac{C_e \rho_e}{K_e}\frac{\partial T_e}{\partial t} \quad (5)$$

where,

T_e = Temperature of earth (°C) at time t

r = Distance from center of well, m

K_e = Thermal conductivity of earth, W/m/ K

C_e = Specific heat of earth, J/kg

ρ_e = Density of earth, kg/m^3

Hasan and Kabir[6] solved the above diffusivity equation using Laplace transform method involving dimensionless variable and concluded the following solution:

$$T_D = \ln\left[e^{-0.2t_D} + (1.5 - 0.3719e^{-t_D})\sqrt{t_D}\right] \tag{6}$$

where,

T_D= Dimensionless temperature which is defined as

$$T_D \equiv -\frac{2\pi K_e}{Q}(T_{wb} - T_{ei}) \tag{7}$$

t_D = Dimensionless time

$$t_D = \frac{\alpha t}{r_{wb}^2} \tag{8}$$

where,

R_{wb} = Outer radius of cement, m

T_{wb}= temperature at R_{wb}, °C

T_{ei} = Initial temperature of the earth as a function of depth, °C

A = Thermal diffusivity, $\alpha = \frac{K_e}{\rho_e C_e}$

T_{ei} is a function of depth and geothermal gradient and can be defined as

$$T_{ei} = T_{eibh} - (L - Z)g_G \tag{9}$$

where,

L = Length of the well, m

T_{eibh} = Bottomhole temperature of the production well, °C

g_G = Geothermal gradient, °C/km

T_{ei} can also be defined on the basis of wellhead temperature in the injection well and geothermal gradient.

To relate the dimensionless temperature (T_D) to the temperature of fluid (T_F) energy balance is carried out as depicted in Figure 2, heat transfers takes place from the earth which is at higher temperature to the cold fluid in the tube of the injection well and vice versa for production well. All resistances to heat transfer across the well, i.e, resistance offered by cement, annulus fluid (air), insulation and tube thickness can be assumed in series.

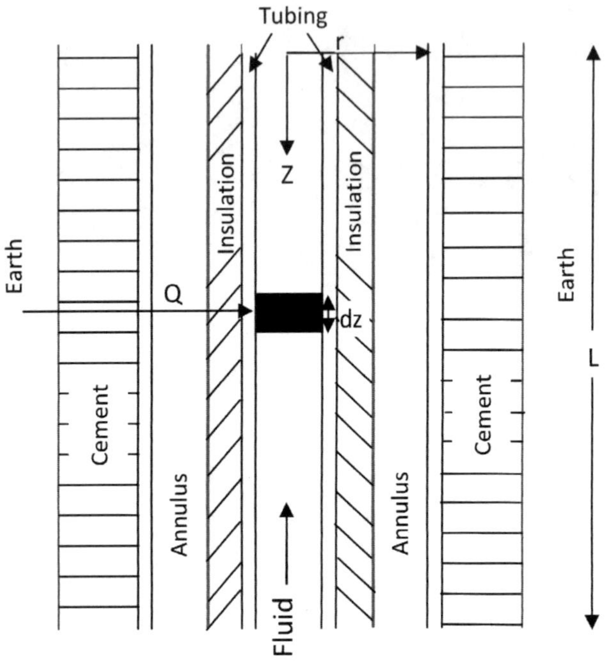

Figure 2: Energy balance on fluid.

Consider an elemental fluid at a depth of z and of control volume Adz, if A is the cross section area. Assuming flow rate of fluid to be W kg/s and the specific enthalpy of fluid to be H kJ/kg.

Conserving energy on the control volume gives:

$$(WH)_{Z+dz} + Wg(Z+dz) + \left(\frac{WV^2}{2}\right)_{Z+dZ} + Qdz = (WH)_Z + wg(Z) + \left(\frac{WV^2}{2}\right)_Z \tag{10}$$

In Equation (10), the sign of Q changes if the heat is transferred from the fluid to the formation which is the case of injection well.

Rearranging Equation (10) gives

$$\frac{dH}{dz} + g + V\frac{dV}{dZ} = \pm\frac{Q}{W} \tag{11}$$

Where a negative sign implies production well and positive sign implies injection well.

Substituting the definition of enthalpy as a function of pressure and temperature, Hasan and Kabir[6] obtained the following Equation (12).

$$\frac{dT_f}{dZ} = C_J \frac{dP}{dZ} + \frac{1}{C_P}\left[\mp \frac{Q}{W} - g - V\frac{dV}{dZ}\right] \tag{12}$$

where,

T_f = Temperature of fluid, °C

C_J = Joule- Thompson coefficient, °C/ Pa

C_p = Specific heat of the fluid, J/kg/°C

Heat flow per unit length can be defined as

$$Q = 2\pi r_{to} U_{t0}(T_f - T_{wb}) \tag{13}$$

where,

r_{t0} = Outer radius of tubing, m

U_{t0} = Overall heat transfer coefficient, J/s-m^2-K

For single phase fluid, the Joule Thompson coefficient C_J can be approximated as[6]

$$C_J = \frac{1}{\rho C_P} \tag{14}$$

Since the CO_2 is under supercritical conditions throughout the operation of the geothermal energy generation process, it has been assumed that the Joule Thompson coefficient for CO_2 is also given by Equation (14).

Combining Equations 7, 9, 12, 13 & 14 and eliminating T_{wb}, a final equation obtained for production well by Hasan and Kabir[6] used in this model is

$$T_F = T_{eibh} - g_G\left[(L - Z) - (1 - e^{(Z-L)L_r})/L_R\right] \tag{15}$$

Above Equation (15) can be modified for the injection well as follows

$$T_F = T_{eiwh} + g_G\left[(Z) - (1 - e^{(-Z)L_r})/L_R\right] \tag{16}$$

where,

T_{eiwh} = Temperature at the wellhead of the injection well (°C)

L_R is called relaxation parameter and is defined by Ramey[7] as

$$L_R = \frac{2\pi}{C_P W}\left[\frac{r_{to}U_{to}K_e}{K_e+(r_{to}U_{to}T_D)}\right] \quad (17)$$

The correlation for overall heat transfer coefficient is obtained from Hasan and Kabir.[8]

Thermophysical properties of water such as density, viscosity, enthalpy, specific heat etc. are calculated using ASME steam table obtained as an Excel Add-In from Bernhard Spang[9]. For CO_2, Altunin[12] correlations are used, calculated using an Excel Add-In developed by our group. A computer program was developed using Microsoft Visual Basic that solves for the temperature and pressure of the working fluid (either water or CO_2) down the wellbore by simultaneously solving for the fluid thermophysical properties, heat transfer, and frictional losses. Input variables were specified for both water and CO_2 as shown in Table 1 although the code can be used for any flow rate, well radius, geothermal gradient, or injection conditions. A very high geothermal gradient of 80 °C/km is employed in this study to compare H_2O and CO_2 under commercial conditions[2]. The properties of the earth surrounding the well specified in this work are shown in Table 2.

Table 1. Specification of input variables

Well head pressure of the fluid at injection well	8 MPa
Well head temperature of fluid at injection well	35°C
Surface temperature of earth	30°C
Flow rate	15 kg/sec
Geothermal gradient	80°C/km
Well depth	3 km
Inner radius r1	0.5 km
Inner radius r2	0.3 km
Inner radius r3	0.2 km

Table 2. Specification of earth properties

Thermal conductivity of earth	2.42 W/m•K
Diffusivity of earth	10^{-6} m^2/sec
Thermal conductivity of cement	1.50 W/m•K
Annulus fluid	Air

Results obtained after running the code for various times are shown in Figure 3a, 3b and 3c.

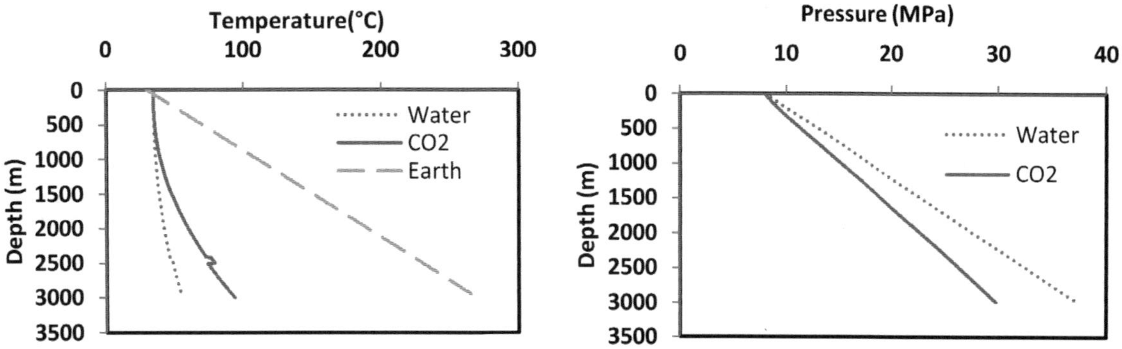

Figure 3a: Temperature and pressure profiles of CO_2 and water in the injection well after 0.001 years.

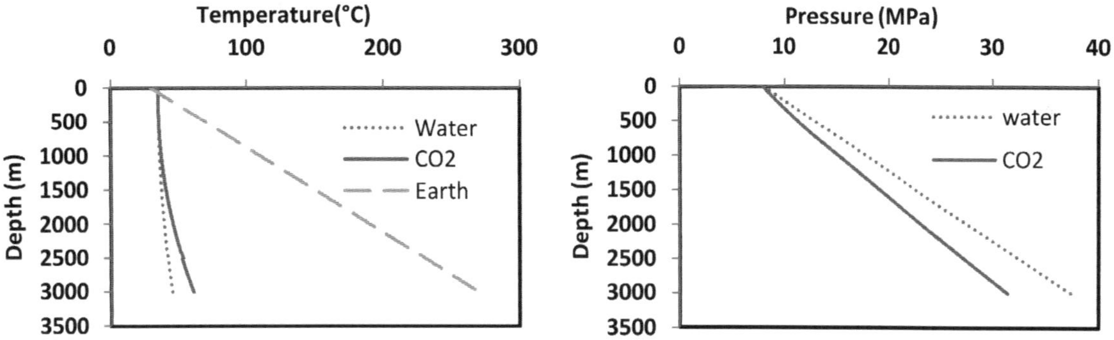

Figure 3b: Temperature and pressure profiles of CO_2 and water in the injection well after 10 years.

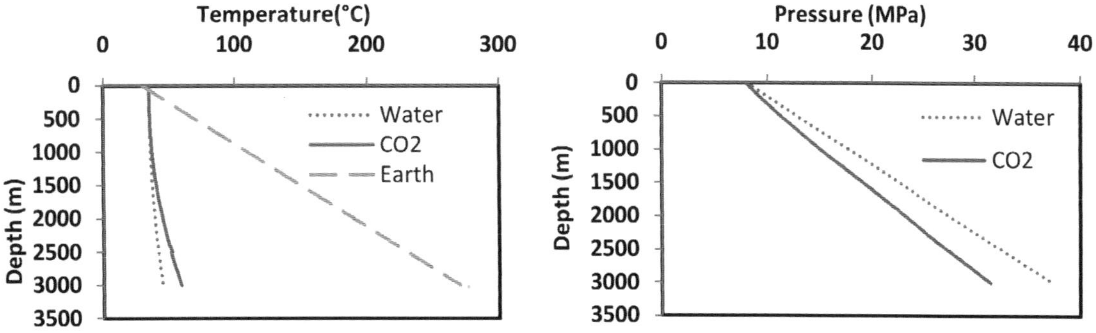

Figure 3c: Temperature and pressure profiles of CO_2 and water in the injection well after 30 years.

As shown in the Figure 3 and also expected theoretically, CO_2 gains more temperature from the surrounding formation than water due to the virtue of its low specific heat capacity. On the same basis, it loses more temperature than water in the production well. As time progress temperature of earth adjacent to the injection well (see Figure 1) starts to decrease and thus the rate of heat gained by the fluids (both water and CO_2) from the earth starts to decrease.

The increase in pressure for water is more than that of CO_2 because of higher density of water than supercritical CO_2. The net gain in pressure decreases over time for both the fluids. The bottom-hole temperature and pressure of the injection well obtained from the wellbore model is subsequently used in the reservoir simulation.

3. Reservoir simulation

A five spot pattern is considered for reservoir simulation as shown in Figure 4. Due to the high level of symmetry only 1/8 of the whole reservoir needs to be modeled. As mentioned before, a flow rate of 15 kg/sec is being injected on the basis of 1/8 of the reservoir, thus on the full well basis a flow rate of 120 kg/sec is being injected. All results shown in this work are on the basis of 15 kg/sec.

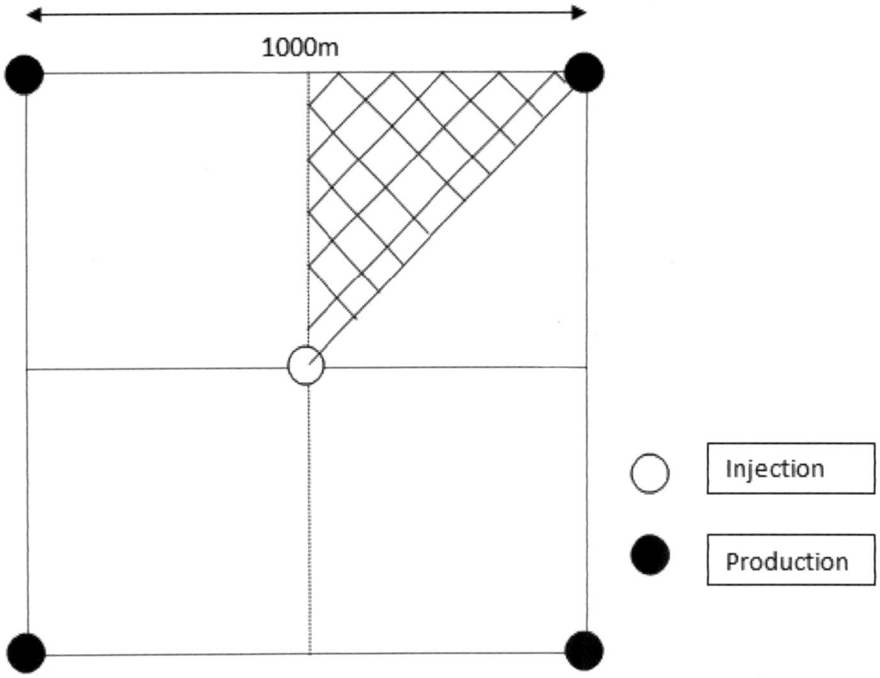

Figure 4: Five spot pattern reservoir

Parameters used for reservoir simulation using TOUGH2-MP are shown in Table 3. These are based on the parameters used by Karsten Pruess[3].

Table 3. Parameters of the reservoir	
Thickness	305 m
Grid block dimensions	70.71 m ×70.71 m
Permeability	60×10^{-15} m^2
Rock Density	2650 kg/m^3
Rock specific heat	1000 J/kg/°C
Rock thermal conductivity	2.1 W/m/°C
Initial reservoir temperature	270°C
Initial reservoir pressure	32 MPa

Results obtained from reservoir simulation are shown in Figure 5. It is assumed that the temperature of the fluid coming out of the production block is same as the production block.

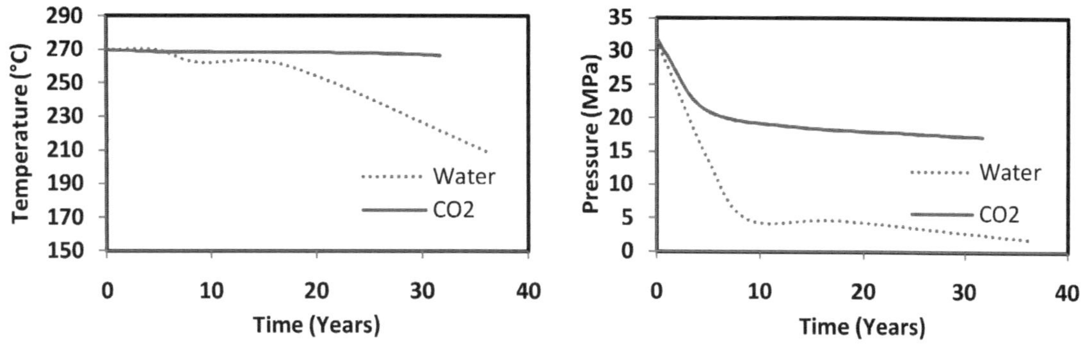

Figure 5: Temperature (left) and pressure (right) of the production block with time.

Results showed that more pressure and temperature drop is observed in the reservoir using water as a flowing fluid than CO_2 having exactly same flow rate. Thus, it can be concluded that there is more heat sweep of the reservoir having water as a heat transmission fluid.

It is required to know the temperature and pressure of water and CO_2 at the top of the production well. For this wellbore model is being used. The bottomhole temperature and pressure obtained from reservoir simulation results at the production block served as the input parameter for the wellbore model. Results obtained from wellbore model are shown in Figure 6.

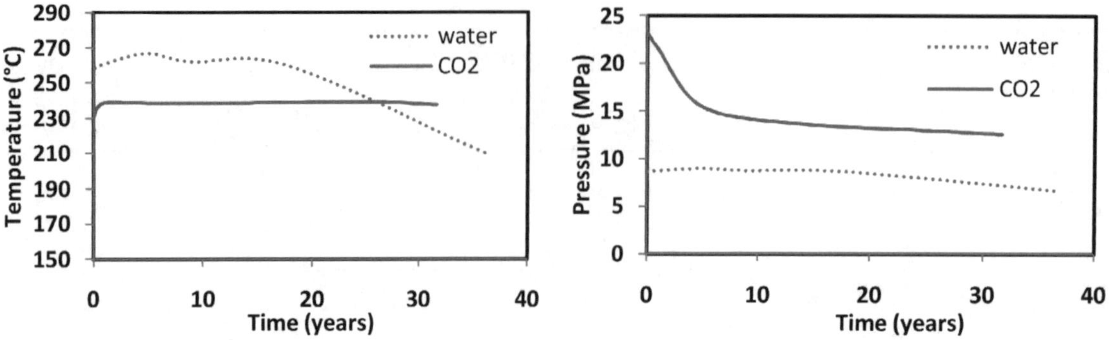

Figure 6: Temperature (left) and pressure (right) at wellhead of the production well with time.

It is important to note that a pump has been placed for water in order to avoid two phase flow in the well as the surface plant is designed for only hot pressurized water and not for wet steam. The role of pump is to increase the pressure of the water from the production block back to 32 MPa. However, amount of work consumed by the pump has been accounted in the net electricity generation calculations.

4. Surface plant

The temperature and pressure of the geothermal fluid over time at the wellhead of the production well determines the amount of net electricity generation. A surface plant is designed based on the Rankine cycle to determine electricity production. In this cycle, an organic fluid such as isobutene, iso-pentane, R-32 (refrigerant) etc. takes heat from the geothermal fluid (water or CO_2) and is then get vaporized because of low boiling point. This high-pressure, vaporized organic fluid is then passed through the turbine/generation unit and work (electricity) is obtained. After coming out of the turbine it is then condensed in the condenser and then again pumped back to the heat exchanger (vaporizer) where it gets vaporized and thus the cycle is completed. An extra heat exchanger has been used in the present work in order to extract more heat from the geothermal fluid coming out of the first heat exchanger. The amount of work obtained through this cycle will also depend upon the choice of organic fluid. As mentioned by the Augustine et al.[10], considering the temperature of the geothermal fluid at the wellhead of the production well, the maximum efficiency would be obtained using iso-pentane as the secondary fluid. Hence, iso-pentane is the secondary fluid used in the Rankine cycle. Figure 7 shows the schematic diagram of the surface power plant being simulated.

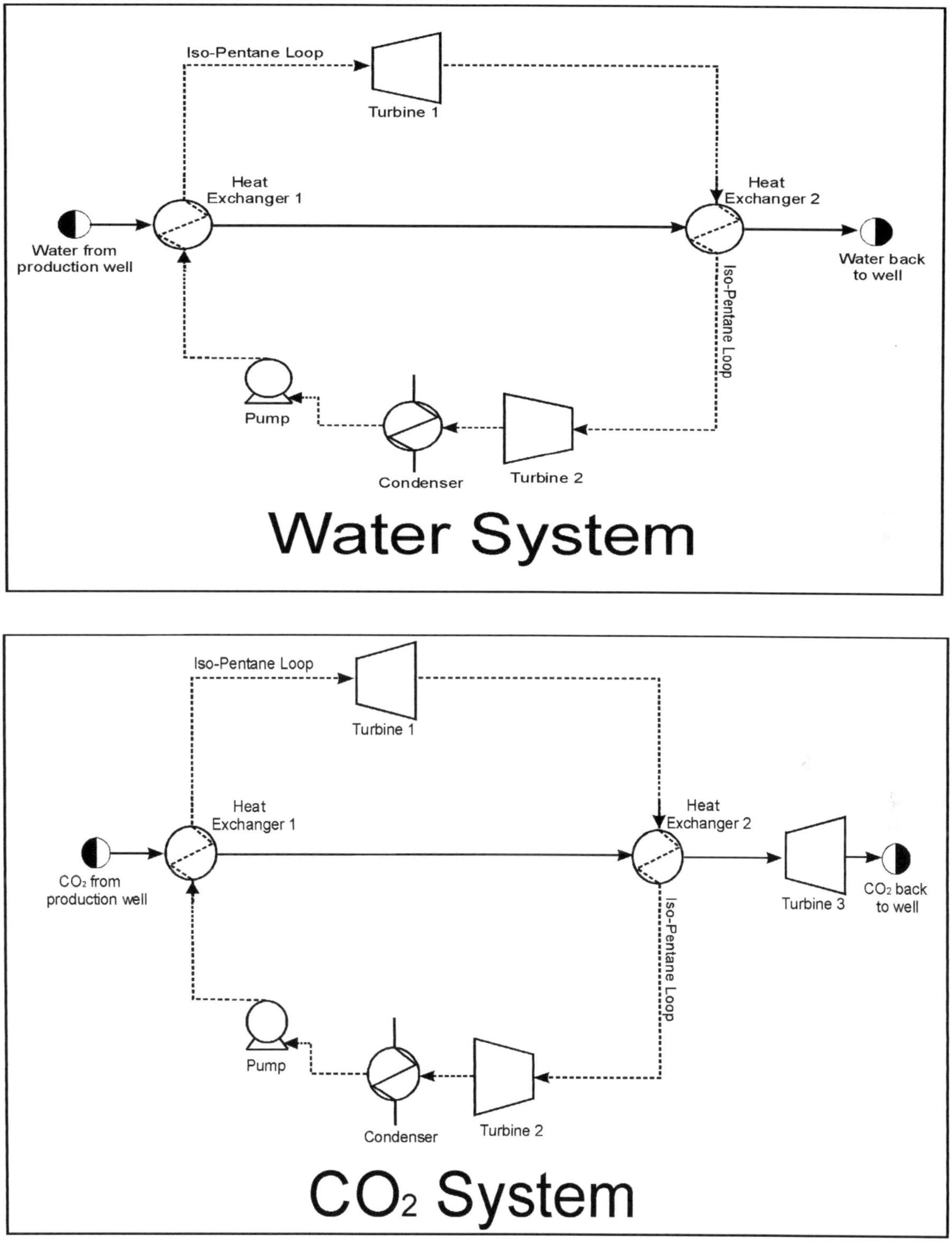

Figure 7: Schematic diagram of power plant for water (Top) and CO_2 (Bottom) as the heat transmission fluid.

It is worth noting that an extra turbine could be employed for CO_2 as it coming out of the production well at a higher pressure and would add more to the net electricity generation. The cycle is being operated in such a way that the heat transmission fluid either water/ CO_2 coming out of the cycle nearly equals the injection conditions at the injection well, hence overall cycle is completed. Chemcad 6 has been used to calculate net electricity obtained through the cycle. The overall results are shown in Figure 8. Theoretically, water produces more electricity than CO_2, but the rate of the decline in net power generation is found to be much more rapid for water than CO_2 as shown in Figure 8.

Although CO_2 has lower pressure drop in the geothermal well and reservoir, but due to its lower heat carrying capacity, less heat is being extracted from the reservoir and thus less power is being produced.

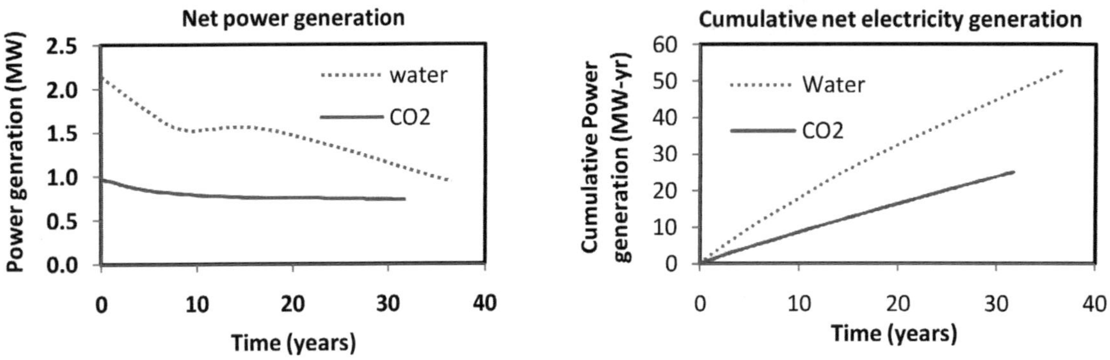

Figure 8: Net power (MW) and net cumulative electricity (MW-yr) could be generated over time.

5. Conclusion

Based on theoretical calculations in this work and other specifications assumed, water is found to produce nearly 70-80% more net power in the initial 10 years of power generation than CO_2 keeping all other parameters exactly same. However, power generation employing water reduces considerably over time giving a disadvantage over CO_2 which seems to produce nearly consistent power generation. After 30 year of plant operation water produces 80% more cumulative electricity than CO_2.

A important conclusion about CO_2 could be drawn is that it may not require external power circulation as found in this work and the flow of CO_2 through the complete geothermal cycle is possible just on the basis of density difference of cold and hot supercritical CO_2. So, it could be used as a geothermal fluid in the areas having higher geothermal gradient but facing water scarcity problems.

Loss of water in the reservoir would cost money but loss of CO_2 could be a mean of CO_2 sequestration and hence could earn carbon credits. More frequent fouling in heat exchangers and other units of power plant could be seen employing water because of high minerals dissolved in it, however with CO_2 it might not be the case.

Karsten Pruess[3] in his work has shown that approximately 3.7 times flow rate of CO_2 could be achieved over water keeping the pressure drop of the reservoir to be same and further concluded that CO_2 has more heat extraction rate over water under constant bottom-hole conditions of injection and production well. However, in this study, flow rate of both the fluids are kept same and the more pressure of CO_2 obtained over water (at the production block of the reservoir) is being harnessed at the surface power plant by providing an extra turbine. Further studies are being conducted on the effect of CO_2 flow rates on net electricity production.

References:

1. Brown, D. A Hot Dry Rock geothermal energy concept utilizing supercritical CO2 instead of water. In: Proceedings of the Twenty-Fifth workshop on Geothermal Reservoir Engineering, Stanford University, Pages 233-238, 2000
2. U.S. Department of Energy, Geothermal Technologies Program-Energy Efficiency and Renewable Energy.
3. Pruess, K., Enhanced geothermal systems(EGS) using CO_2 as working fluid- A novel approach for generating energy with simultaneous sequestration of carbon, Geothermics, Volume 35, Pages 351-367, 2006.
4. Fouillac, C., Sanjuan, B.,Gentier, S., Czernichowski-Lauriol, I., Could sequestration of CO2 be combined with the development of Enhanced Geothermal Systems? Third Annual Conference on carbon capture and Sequestration, Alexandria, VA, 2004.
5. Metz, B., Davidson, O., De Coninck, H., Loos, M., Meyer, L., IPCC Special Report on Carbon Dioxide Capture and Storage, Intergovernmental Panel on Climate Change; Geneva (Switzerland); Working Group Iii, Cambridge University Press, 2005.
6. Hasan, A.R. and Kabir, C.S. – Fluid Flow and Heat Transfer in wellbores, Chapter 5, Pages 65-69.
7. Ramey, H.J., Wellbore Heat Transmission, Trans., AIME, 225, 1962.
8. Hasan, A.R. and Kabir, C.S. – Fluid Flow and Heat Transfer in wellbores, Appendix-A, Pages 151-153

9. Bernhard Spang , Water97-v13.xla-Excel Excel Add-In for properties of Water and Steam in SI-Units, Hamburg, Germany, February 2002 http://www.cheresources.com/iapwsif97.shtml
10. Augustine, C., Filed, R., DiPippo, R., Gigliucci, G., Fastelli, I., Tester, J., Modeling and Analysis of Sub- and Supercritical Binary Rankine Cycles for Low- to Mid- temperature Geothermal Resources, GRC Transactions, Vol.33, 2009.
11. Chen, N.H., An Explicit Equation for Friction Factor in Pipe, Ind. Eng. Chem. Fund., 18, 296, 1979.
12. Altunin, V.V., Thermophysical Properties of Carbon dioxide, publishing house of standards, Moscow, 1975.

Assessment of Effective Heat Transfer Area with Sorbing Tracers

Kevin Leecaster[1]
Pete Rose[1]
Susan Petty[2]
[1]Energy & Geosciences Institute/Univ. of Utah
423 Wakara Way, Suite 300
Salt Lake City, Utah 84108
801 585 0793
kleecaster@egi.utah.edu

[2]AltaRock Energy, Inc.
7900 E. Greenlake Way North Suite 202
Seattle, WA 98103

Assessing the surface area from which the working fluid draws heat in geothermal energy extraction systems could help with the management of existing resources and with the design and construction of engineered geothermal systems in hot dry rocks. In otherwise identical aquifers hosting coupled injection & production wells, a difference in the amount of surface area available to heat the injected fluid will influence the amount of energy that can be extracted without a loss in fluid enthalpy at the production well occurring. Thermal drawdown can occur when the heat extraction rate through fluid convection exceeds the conduction rate from the host rocks to the fracture surfaces where most heat transfer occurs. Predicting the host rock thermal response to the induced fluid convection requires an estimation of, among other things, the effective surface area available to transfer heat to the fluid. Operating geothermal reservoirs often have flow paths plugged by mineral precipitation near the injection wells. This decreases the available heat transfer area and could lead to premature thermal drawdown if not identified and addressed. Creation of new fractures within hot dry rocks will open up unconventional geothermal reservoirs. A tool to assess the exposed surface area within these new reservoirs would help to support this advancing technology. Since adsorption of solutes is a function (in part) of how much surface area/fluid interaction occurs in a system, sequential tracer breakthrough curves through the same system with only a change in its surface area is a potential surface area measurement tool.

We are working on developing thermally stable tracers that will help geothermal energy producers to assess the surface area available between injection and production wells. Research results are presented that show how the use of tracers that interact (i.e. sorb) with the surface of the geomedia can help to refine the estimation of the available effective heat transfer area. A series of high temperature miscible displacement experiments through media with different size fractions of Ottawa sand demonstrates the relationship between the retardation of a reactive tracer relative to a conservative tracer and the amount of surface area that the fluid encounters. Thermal characteristics of the tracer and temperature effects on its retardation through the media are presented, as well as a discussion on analysis, detectability, and availability of potential reactive tracer candidates for field applications.

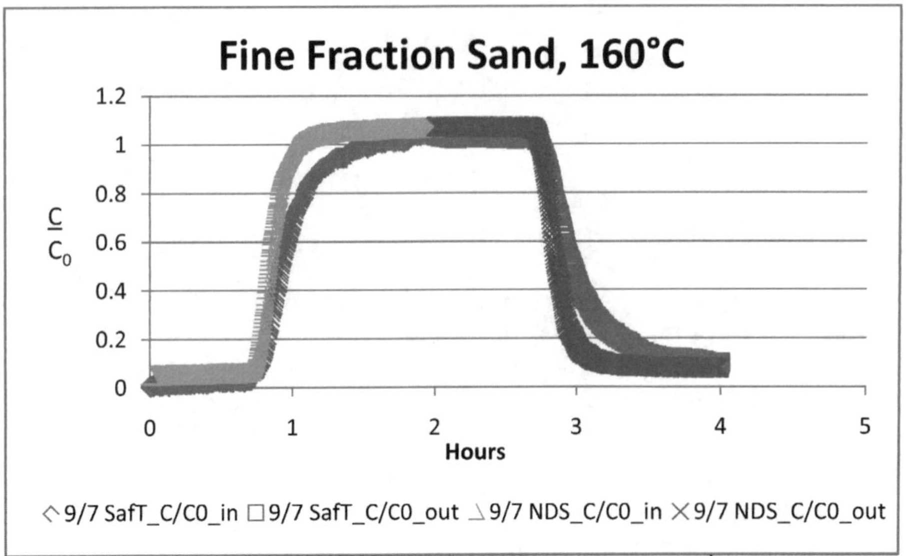

Figure 1 Breakthrough curves from step input of 1.4 mg L^{-1} safranin T and 100 mg L^{-1} 1,5 naphthalenedisulfonate in 200 mg L^{-1} NaCl, 175 mg L^{-1} $CaCl_2$, and 10 mg L^{-1} tetraethylorthosilicate flowing at 2 mL min^{-1} through Ottawa sand fraction of 180 to 425 μm at 160°C.

Two food colorings, yellow #5 & blue #1, and three safranines were examined for thermal stability. Our tests indicate that the blue #1 is more themally stable than the yellow dye, but only up to a working temperature of around 150°C. Of the tested molecules trimethyl safranin proved the most thermally stable (table 1). Our results indicate that it could be a viable option in reservoirs hotter than 220°C. The miscible displacement experiments show that safranin is retarded relative to a conservative tracer at temperatures up to 160°C (figure 1). The safranin T in that experiment had a retardation factor of 1.12 with a standard deviation of 0.008 from two experiments. Additionally, results shown in figure 2 demonstrate a difference in the retardation factor derived from breakthrough curves from media of the same mineralogy but different particle sizes over three different temperatures. The retardation of the reactive tracer decreases at higher temperatures, but distinctly shows greater retardation with smaller particle sizes (i.e. greater surface area).

Table 1 Thermal Stabilities

Tracer	Temperature (°C)	Hours at Temperature	Degradation Rate (percent loss/day)
Yellow #5	150	41	57%(±0.1%)
Blue #1	150	41	11%(±0.9%)
Phenosafranin	140	26	21%(±5.6%)
Safranin O	150	72	14%(±1.8%)
Safranin T	220	136	6% (±2.3%)

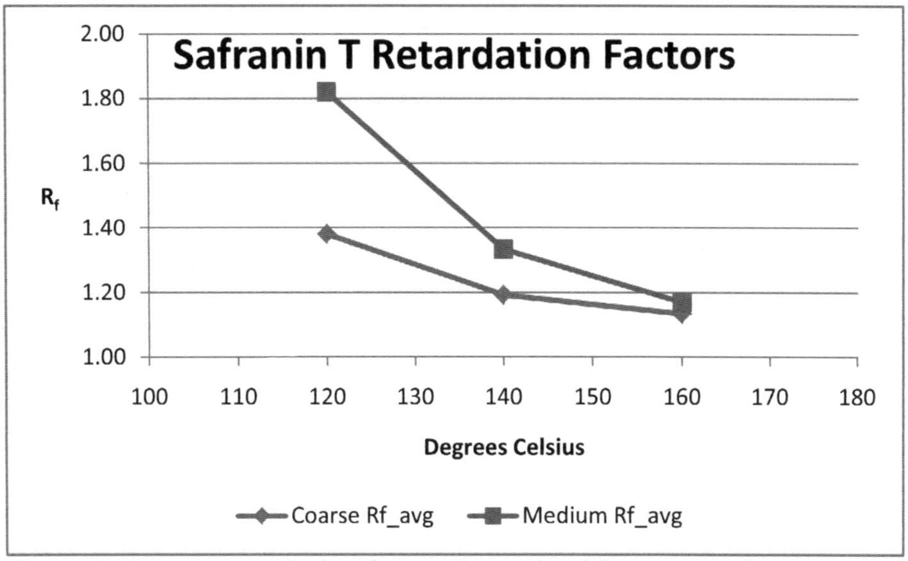

Figure 2 Average retardation factors determined from ratio of conservative and reactive tracers' median retention times. Coarse fraction sand is between 600 and 500 μm and medium fraction is between 500 and 425 μm.

THE CHEMICAL AND THERMAL STABILITY OF PROPPANTS UNDER GEOTHERMAL CONDITIONS

Kristie McLin[1], Daniel Brinton[1,2], Prashanth Mandalaparty[2], Clay Jones[1], and Joseph Moore[1]

1. Energy & Geoscience Institute, University of Utah, Salt Lake City, UT 84108
2. Dept. of Chemical Engineering, University of Utah, Salt Lake City, UT 84108

Keywords: EGS, fracture, proppant, reservoir engineering

Abstract

Enhanced geothermal system (EGS) reservoir fracture creation and management may require the use of proppants to maintain fracture conductivity. Hydraulic fracturing accompanied by the addition of proppant is a common procedure in the oil and gas industry. In contrast, the use of proppants by the geothermal industry has been limited. A variety of proppants are available, and the most commonly used consist of silica sand, ceramic, resin coated sands, and bauxite. These solids remain in newly formed fractures to keep them open. In geothermal systems, proppant will need to withstand high temperatures, acidified fluids, acid treatments, and cleanouts while maintaining the porosity and permeability of the fracture. However, proppant particles may act as nucleation sites and promote precipitation, or they may dissolve, consequently affecting fracture performance. Thus, the performance of proppants in geothermal reservoirs must be understood to avoid costly mistakes in the creation and maintenance of EGS reservoirs.

A series of static experiments have been conducted at 200°C using 30/60 Sinterball bauxite proppant provided by Sintex Minerals. The proppant consists of corundum with minor quartz, hematite and chlorite. De-ionized water has been used for the fluid phase. The experiments were run for periods of one to two months.

X-ray diffraction, scanning electron microscope and QEMScan analyses were performed on the proppant before and after testing to evaluate textural, mineralogical and compositional changes that occurred as a result of water rock interactions. Only minor changes were observed. These changes include the presence of surface pitting, resulting from dissolution of impurities in the proppant grains and deposition of minor amorphous silicates and aluminum oxides.

Introduction

In the oil and gas industry, hydraulic fracturing is often accompanied by the addition of proppants. Proppants are sized particles that are added to fracturing fluids and remain in the fracture to physically hold the newly created fracture open while maintaining conductivity within the fracture. A variety of proppants are available, and the most commonly used proppants consist of silica sand, ceramic and resin coated sands, and bauxite. In enhanced geothermal systems (EGS), proppant will need to withstand high temperatures, acidified fluids, acid treatments, and cleanouts while maintaining the porosity and permeability of the fracture. However, proppant particles may act as nucleation sites and promote precipitation, or they may dissolve, consequently affecting fracture performance. Thus, the performance of proppants in geothermal reservoirs must be understood to avoid costly mistakes in the creation and maintenance of EGS reservoirs. The goal of this investigation is to develop improved methods for maintaining permeable fracture volumes in EGS reservoirs by experimentally evaluating the performance of proppants (natural and manufactured) under conditions found within geothermal environments and determining the effects of mineral deposition and dissolution on propped fractures.

Previous Studies

A study in the Northeast German Basin used high-strength ceramic proppant in a well at 4080-4190m depth and 140°C (Legarth et al, 2003). Propped fractures were created, and inflow performance of well was significantly enhanced. Even so, post-frac productivity was not as good as had been predicted. The possible reasons given for this discrepancy are chemical and mechanical processes during shut-in or undersized and poorly connected fractures.

Previous studies from the oil and gas industry have reported that sintered bauxite proppant is resilient under high stress environments (Cooke, 1977). When formation stresses are high, these proppants deform, rather than crush. This deformation allows for the increase of contact area between proppant grains or between proppant and the wall rock. Sand has been shown to crush, decreasing porosity and permeability in the fracture (Cooke, 1977).

Experimental Setup

A variety of proppants are currently available on the market. These include silica sands, ceramic proppants and bauxite proppants. Our current series of tests have utilized Sintex Minerals and Services, Inc. Sinterball Bauxite proppant. SinterBall Bauxite 30/60 is a high strength bauxite proppant used for high stress, temperature, and flow rate hydraulic fractures. The high alumina content of this bauxite is what provides the strength, chemical resistance and thermal stability it needs to ensure optimum production. Mechanical properties are shown in Table 1a-c, and the relationship between closure stress and permeability is shown in Figure 1.

Table 1a-c. Mechanical properties of the SinterBall Bauxite 30/60 proppant as provided by Sintex Minerals and Services, Inc.

a.

Technical Data	
Specific Graviy	3.63
Bulk Density	130 lb/ft^3
Crush @ 12,500 psi	2.1%
Absolute Volume	.033 gal/lb
Mean Particle Diameter	.501 mm

b.

Sieve	Retained
30	3%
40	77%
50	18%
60	2%

c.

Closure Stress (psi)	Permeability (darcies)	Conductivity (md-ft)	Beta Factor (atm-sec^2/kg)
6,000	169	2,342	0.376
8,000	137	1,854	0.422
10,000	120	1,580	0.470
12,000	103	1,321	0.581
14,000	86	1,073	0.741

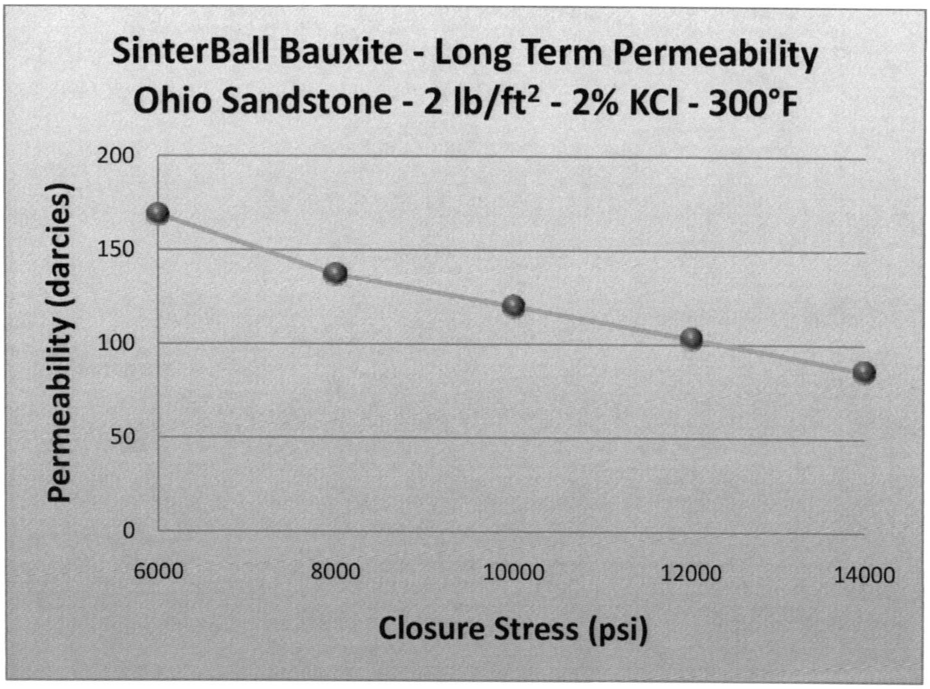

Figure 1. Relationship of closure stress to permeability of the SinterBall Bauxite 30/60 proppant as provided by Sintex Minerals and Services, Inc.

The experimental apparatus is shown in Figure 2. It consists of a stainless steel reactor made of 316-grade stainless steel, which was rated for 4000 psi at 600°C. The reactor is heated by means of electrical heat tape. The temperature is controlled via a k-type thermocouple using a temperature controller and a SPECVIEW interface. The reactor was designed with removable top and bottom caps which allow for solids to be easily added and removed. The experimental apparatus is modified from the previous experimental setup described in Seyfried et al. (1987), which was adopted by Kaszuba et al. (2003, 2005) and Rosenbauer et al. (2005). Experiments were conducted at 200°C. The first two experiments used de-ionized water as the liquid phase.

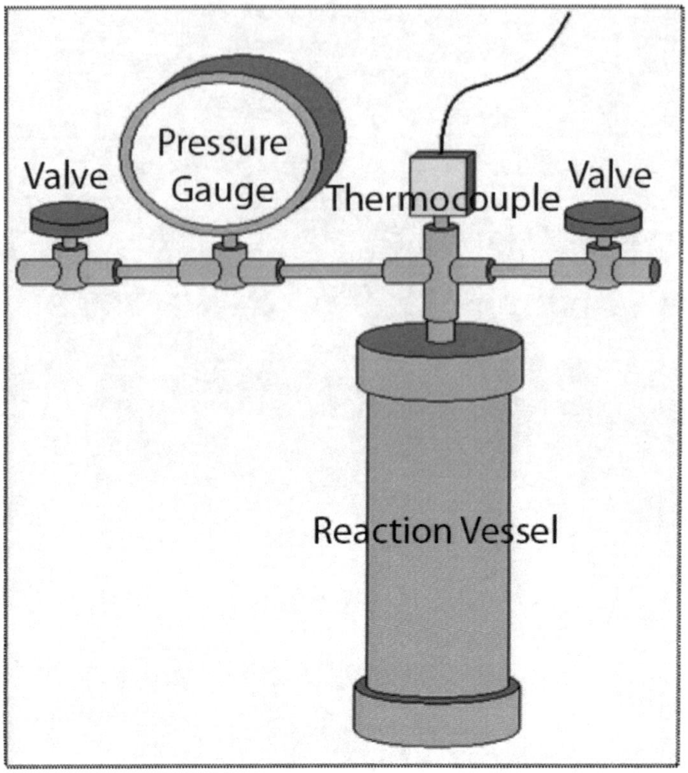

Figure 2. Reactor vessel setup.

Batch experiments on Sintex 30/60 proppant were conducted using de-ionized water at 200°C. Experiments were conducted in identical reactors. The initial samples were prepared by weighing an appropriate amount of proppant and placing it into the reactor. DI water was slowly added to the reactor to allow the water to fill the pore spaces until the water level was just below the top of the reactor. The reactors were then sealed and held at a constant temperature of 200°C for one to two months.

After the reaction period, the samples were retrieved from the reactor. The proppant and other newly precipitated phases were retrieved by removing the end caps and removing the sample, being careful to minimize disturbance to mineral precipitates and crystals. Reactor liquids were separated from the solids by vacuum filtration with Whatman 40 filter paper. The solids were air dried and analyzed using SEM/EDS, XRD and QEMScan. Reactor liquids were acidified with sulfuric acid to prevent any further precipitation and were analyzed using ICPMS to determine the concentrations of ions. Preliminary SEM and XRD analyses were conducted on the dried reactor solids to identify the precipitated phases and dissolution patterns in the mineral matrix after the experiments. The SEM analysis was done at different resolutions to examine the textural features and chemical composition of the proppant prior to and after reaction. In these experiments, the reactor systems were allowed to cool over night before liquids and solids were removed. This was done to avoid the flashing of reactor liquids resulting in precipitation of dissolved solids. However, the extended cooling time was likely sufficient for significant precipitation to occur even in the absence of flashing.

Results

A gray clay mineral was observed in reaction vessels at the end of the first two experiments. Aggregates of this mineral were found near the base of the first reactor. In contrast, this mineral was dispersed throughout the proppant in the second reactor. This mineral was not observed in the un-reacted proppant. In addition to the differences in the distribution of the clay mineral, the proppant at the base of the second reactor had formed a dense hard mass that was cemented to the walls of the vessel. Efforts to dislodge the proppant were largely unsuccessful. X-ray diffraction (XRD), scanning electron microscope (SEM), and QEMScan analyses were performed on the proppant before and after reaction to evaluate textural, mineralogical and compositional changes that occurred as a result of water rock interactions.

Scanning Electron Microscope Analyses

An FEI SEM was used to image the un-reacted and reacted proppant. The proppant was left uncoated to increase the accuracy in chemical analyses using an energy dispersive X-ray unit (EDX). The un-reacted proppant is shown in Figures 3a-c. XRD analysis shows that the proppant is mostly corundum with minor quartz and hematite. Figure 3c shows a close up of the un-reacted proppant and the corundum plates. There are occasional pits in the un-reacted proppant, and these pits are more iron rich than the rest of the proppant surface.

Figure 3a-c. (a). Magnified untested proppant. (b). SEM image of untested proppant, showing minor surface pitting. (c). SEM image of untested proppant showing corundum plates and close up of pitted surface.

The proppant that was reacted in experiment 1 is shown in Figures 4a, b. It can be seen that the surface texture has changed and that dissolution of the corundum plates has occurred (Fig. 4b).

Clay minerals (Fig. 5) were found in the bottom of the reaction vessel. X-ray diffraction analyses indicate these clays are Mg and Fe rich chlorites.

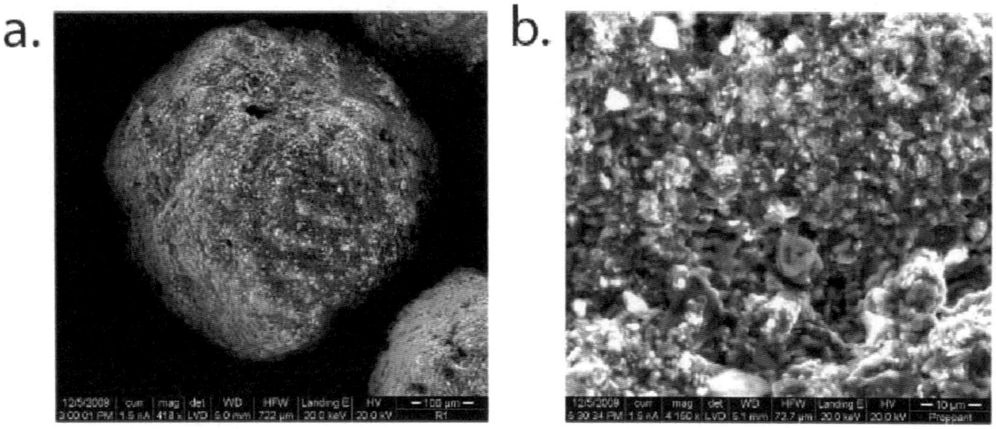

Figure 4a, b. (a). SEM image of proppant reacted in experiment 1. (b). Close up SEM image of proppant reacted in experiment 1 showing dissolution textures on the corundum plates.

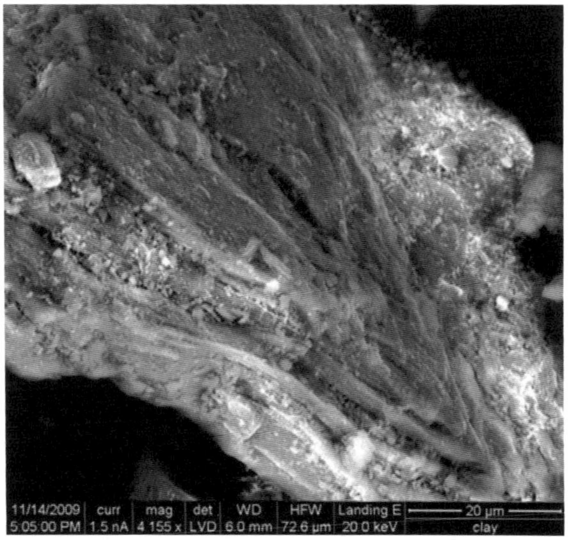

Figure 5. SEM image of clay found in the bottom of the reaction vessel from experiment 1.

The proppant from experiment 2 shows similar dissolution, but it has also been coated with a variety of other mineral phases. Figure 6a shows the texture of the "gibbsite". Gibbsite is a hydrate of alumina and is common in bauxites with diaspore and boehmite. We use the term "gibbsite" to represent these phases, as they have not been positively identified by X-ray diffractometry. Figures 6b,c show a silicate coating on proppant from the reacted sample. The texture of the silica phase suggests that it amorphous.

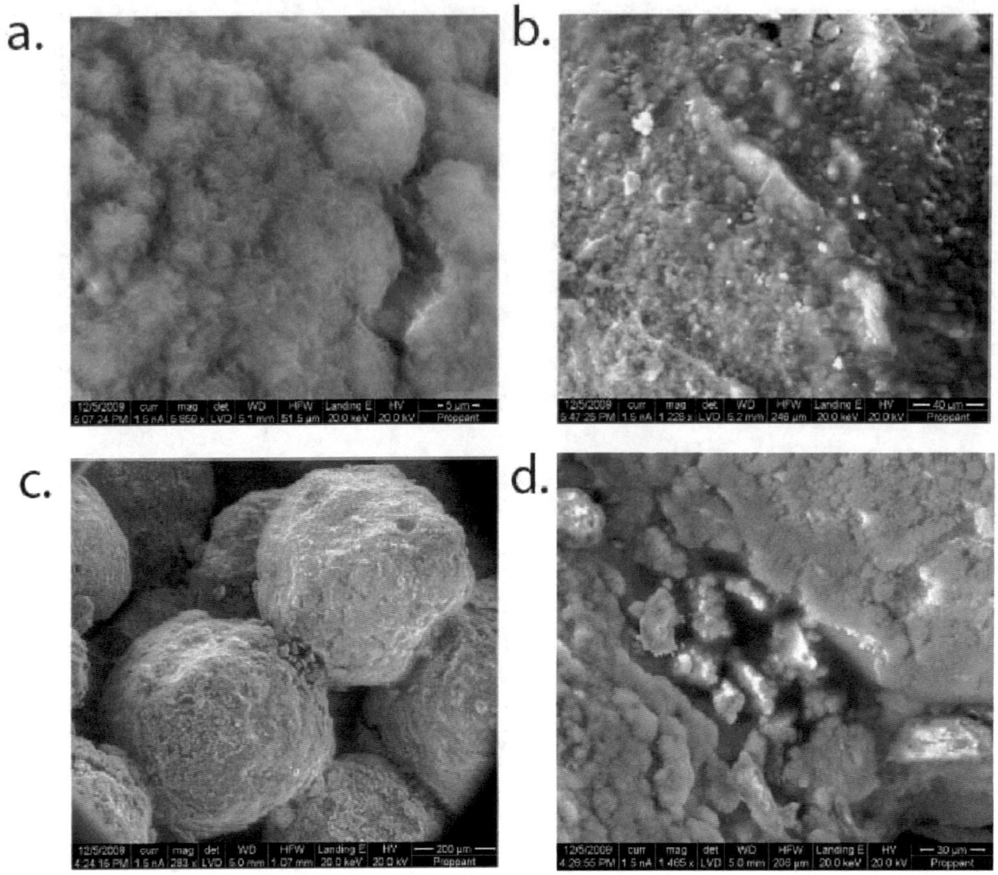

Figure 6a-d. SEM images of proppant from experiment 2. (a). Aluminum oxide coating on proppant. (b). Layered coatings of aluminum oxide and amorphous silicate on proppant. (c). Amorphous silicate cementing two proppant grains together. (d). Close up image of the amorphous silicate cement from (c).

X-Ray Diffraction Analyses

The XRD analyses provide additional detail on the mineralogy of the samples. Figure 7 shows the XRD scans and mineralogy of the un-reacted and reacted proppant samples from experiment 1 and experiment 2 (top and bottom of reaction vessel). There is no significant change observed in the bulk mineralogy of the samples. It is possible that other minerals are present in the sample with concentrations of less than a few percent, but are not visible because: (1) they do not have a strong primary peak making them indistinguishable from the background; (2) their peaks overlap with another mineral of higher concentration; or, (3) they are iron rich and therefore are prone to absorb X-rays. Figure 7a shows the glycolated clay sized fraction of the un-reacted and reacted samples from experiment 1 and experiment 2 (top and bottom of reaction vessel). There are no prominent peaks at low angles indicting well formed sheet silicates are of low abundance. Chlorite, illite, and aluminum hydroxide (bauxite) minerals occur in trace amounts. The broad hump at low angles suggests that a poorly formed "clay-like" mineral may be present. Several subtle differences can be observed in the glycolated clay sized X-ray diffraction patterns. A secondary quartz peak at about 21 2 theta is not present in the 'Experiment 2 Bottom' sample indicating a lower abundance of quartz; and primary feldspar peaks (K-feldspar and plagioclase) are seen between 27.5 and 28 2 theta in all but the un-reacted proppant. Particular attention was paid to the white, fine-grained material produced by experiment 1. This material was concentrated and analyzed separately. Figure 7b is an X-ray diffraction pattern of the white mineral aggregate products in experiment 1. Figure 7c shows diffraction patterns of the clay sized fraction (< 5 micrometers) of the same material. The dominant sheet silicate identified in the X-ray pattern is chlorite, which is accompanied by a trace of illite.

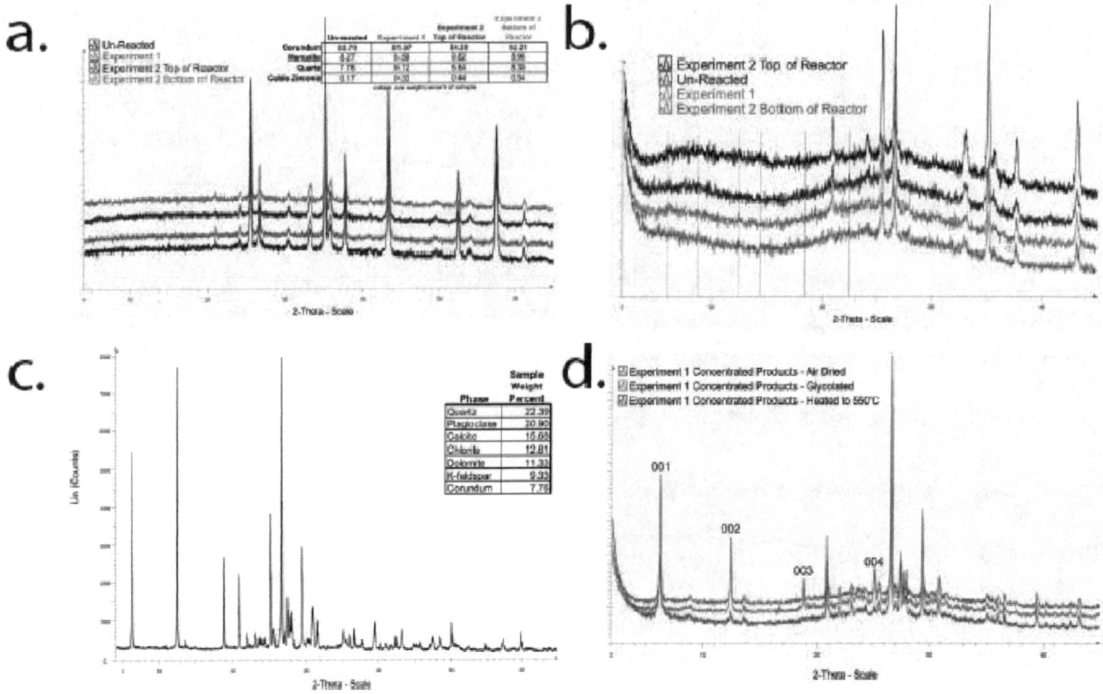

Figure 7a-d. (a). X-ray diffraction patterns of the un-reacted and reacted bulk proppant. (b). X-ray diffraction patterns of the glycolated clay sized fraction (<5 micrometers) of the un-reacted and reacted proppants. Vertical solid lines have been drawn at the primary peak locations of the bauxite minerals, boehmite (14.5 2 theta), gibbsite (18.3 2 theta), and diaspore (22.3 2 theta). A trace of boehmite is present in the sample 'Experiment 2 Top of Reactor', likewise a trace of gibbsite is present in the 'Un-Reacted' sample. Vertical dotted lines show the peak locations of the chlorite 001 and 002 peaks, a trace of which can be seen in sample 'Experiment 1'. The vertical dashed line shows the location of the 001 illite peak, which is present in trace amounts in sample 'Experiment 2 Bottom of Reactor'. (c). X-ray diffraction pattern and composition of the fine-grained mineral aggregate produced in experiment 1. It should be noted that some proppant is present in the sample, as indicated by the presence of corundum. (d). X-ray diffraction patterns of the clay sized fraction (<5 micrometers) of the material produced in experiment 1. Chlorite is the dominant sheet silicate ("clay" mineral) in the sample, with a trace of illite (001 at 8.8 2 theta, shown by vertical black line). The chlorite peaks are labeled 001, 002, etc. The large peak at 26.6° 2 theta is quartz; the broad hump centered at this angle is due to amorphous silica. There are no shifts between the air dried and glycolated patterns indicating there is no swelling clay in the sample. After heating to 550°C for one hour, the intensity of 001 peak increases greatly and shifts slightly (a few tenths of a degree 2 theta) and the 002, 003, & 004 reflections are significantly weakened; this is characteristic of chlorite.

QEMScan Analyses

QEMScan analysis of proppant from experiment 2 (Fig. 8) shows the mineralogical variation in the composition of the proppant itself, as well as the deposition of various minerals on the surface of the proppant. The proppant is mainly composed of Al and Fe oxides. The surface mineral deposits are identified as aluminum oxides and various silicates, including plagioclase, muscovite, quartz, and undifferentiated amorphous silicates. These mineral deposits appear to be layered (Fig. 8), likely as a function of their solubilities.

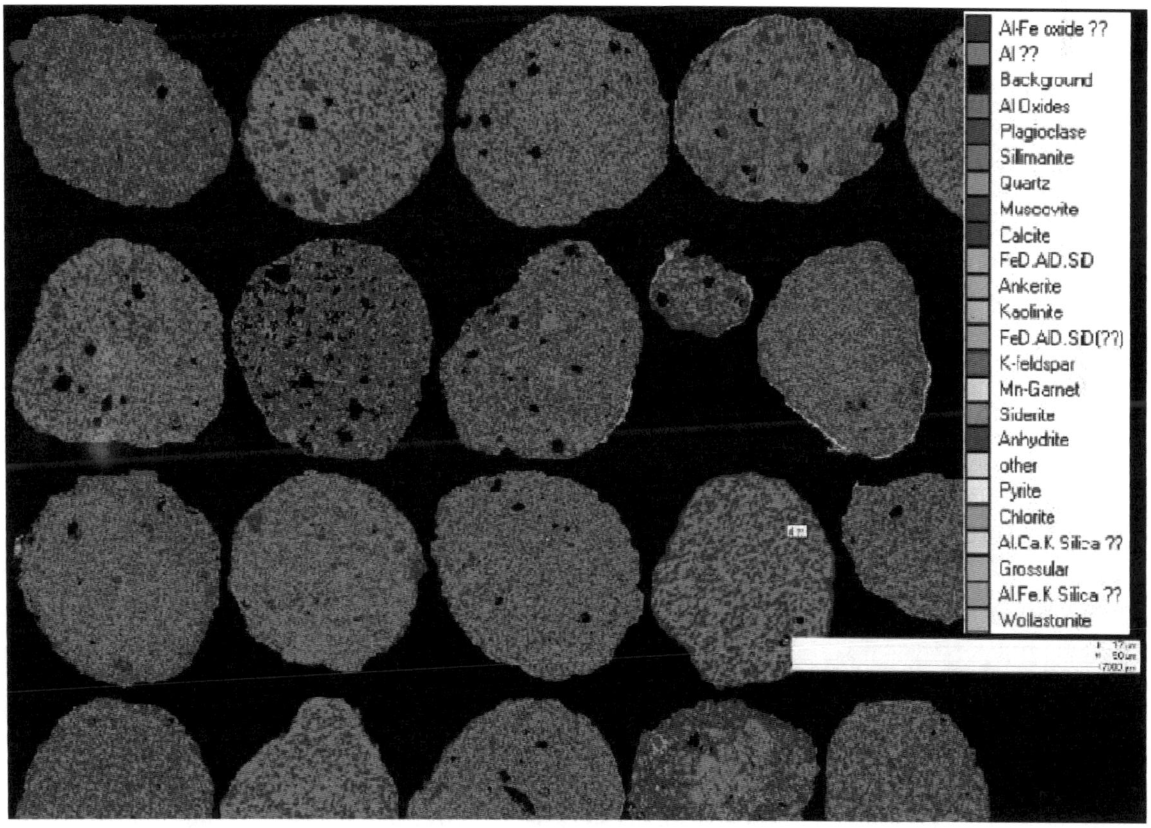

Figure 8. QEMScan analysis of proppant reacted in experiment 2.

Discussion

Mineral deposition on the proppant during experiment 2 is likely the result of the experimental methodology in removing the fluid from the apparatus. By allowing the fluid to cool within the reactor, minerals that were in equilibrium with the fluid at 200°C would have been at concentrations above their saturation limits as the fluid cooled. Due to this effect, the apparatus used for future experiments was modified to allow extraction of reactor liquids while preventing flashing and minimizing cooling time and associated precipitation. An additional valve at the bottom of the reactor will be used to remove reactor liquids. High pressure nitrogen will force liquids out through the bottom of the reactor. Fluids will flow down a length of tubing and through an ice bath to quickly cool the fluids. The fluids will then be collected and diluted to prevent mineral precipitation.

In order to more closely simulate in situ geothermal conditions, de-ionized water spiked with silica will be used for on-going experiments. This should minimize the dissolution of minerals from the proppant. The silica-spiked solution will be prepared using tetraethyl orthosilicate (TEOS) as the silica source. TEOS will be added to de-ionized water in an attempt to achieve a silica concentration of 265 ppm (the quartz saturation limit at 200°C). Granite will also be added to understand the chemical interactions between the proppant, rock, and fluid to more closely represent a geothermal system. Granite will be used as the rock type for on-going testing because it is an appropriate proxy for typical reservoir rocks at EGS sites. Several pieces of granitic core from Glacier Peak, Canada, a fossil geothermal system, will be used.

Conclusions

When using DI water at 200°C, the proppant has shown some signs of dissolution in these experiments. Changes to the chemistry of the fluid, as well as to the methodology of the experiments, should yield more realistic results as to the durability of bauxite proppants under geothermal conditions. The results of these experiments will be used as constraints in geochemical models to further understand these reactions and predict the chemical behavior of proppants in a reservoir environment.

Acknowledgements

This study is funded by the U.S. Department of Energy under grant **DE-FG36-08GO18189.**

References

Cooke Jr., C.E., 1977. "Fracturing with a high-strength proppant." Journal of Petroleum Technology, p. 1222-1226.

Kaszuba, J. P., Janecky, D. R., Snow, M. G., 2005. "Experimental evaluation of mixed fluid reactions between supercritical carbon dioxide and brine: Relevance to the integrity of a geologic carbon repository". *Chemical Geology*, v. 217, p. 277-293.

Kaszuba, J. P., Janecky, D. R., Snow, M. G., 2003. "Carbon dioxide reaction processes in a model brine aquifer at 2000 C and 200 bar: Implications for geologic sequestration of carbon." *Applied Geochemistry*, v. 18, p. 1065–1080.

Legarth, B., Tischner, T., Huenges, E., 2003. "Simulation experiments in sedimentary, low-enthalpy reservoirs for geothermal power generation, Germany." *Geothermics*, v. 32, p. 487-495.

Rosenbauer, R. J., Koksalan, T., Palandri, J.L., 2005. "Experimental investigation of CO_2-brine-rock interactions at elevated temperature and pressure: Implications for CO_2 sequestration in deep-saline aquifers." *Fuel Processing Technology*, v. 86, p. 1581–1597.

Seyfried Jr., Janecky, W.E., Berndt, D.R., 1987. "Rocking autoclaves for hydrothermal experiments, II. The flexible reaction cell system." In: Ulmer, G.C.; Barnes, H.L. (Eds.), Hydrothermal Experimental Techniques. John Wiley & Sons, p. 216-239.

Predicting Thermal Conductivity of Geothermal Reservoir Rocks

Dan Brinton, Joseph Moore, John McLennan, Clay Jones

Abstract

Understanding rock temperatures and stress fields will be important in predicting fracture behavior during enhanced geothermal reservoir development. For many geothermal reservoirs, thermophysical property data (coefficients of thermal expansion, thermal conductivity and specific heat) are limited. Thermal conductivity, diffusivity and specific heat are relevant for quantifying the power potential and forecast for prospective and operating geothermal plants and govern the way temperature gradients move through the reservoir. The coefficient of thermal expansion and certain other mechanical parameters of the reservoir such as Young's modulus and Poisson's ratio intrinsically contribute to thermal alteration of in-situ stresses in response to temperature perturbations during reservoir development by hydraulic fracturing and during heat production.

Recognizing the lack of a substantial database of thermophysical properties, standard geophysical well logs and mineralogic analysis of core and cuttings from the Raft River Geothermal Field have been used to predict thermal conductivity. Predictions are compared with the experimentally-determined values from core samples collected from Raft River Geothermal well RRG-3C.

The work will support ongoing investigations to thermally and hydraulically stimulate well RRG-9 under a US DOE sponsored grant. RRG-9 was drilled to 6089 feet measured depth (MD) in the Raft River Geothermal Field located in southern Idaho, near the Idaho-Utah state line. The field lies within the Basin and Range province. In the 1970s to early 1980s, the field was the site of intense study by the US Department of Energy. During that period a significant amount of geologic and reservoir data were collected. The reservoir is developed in Precambrian metamorphic and granitic rocks beneath approximately 5000 ft of Tertiary volcaniclastic deposits. U.S. Geothermal, Inc. operates a 13-MW plant at the site.

Introduction

The Raft River valley is located in southern Idaho, just north of the Utah border. Geothermal wells have been drilled in the valley to depths of up to 6500 feet and have intersected three distinct geologic formations. The Precambrian basement is comprised of quartz monzonite which is overlain by a series of four metasedimentary units: The Older Schist, the Elba Quartzite, the Upper Narrows Schist and the Quartzite of Yost (Blackett & Kolesar, 1983). Rock units from the Precambrian basement have been correlated with rocks in the Raft River Range (Covington, 1980). Above these basement rocks lies a thick (up to 5250 feet) series of highly fractured tuffs, sandstones and siltstones known as the Salt Lake Formation. The Salt Lake Formation is overlain by the Raft River Formation, comprised of sand, silt and gravel; the boundary between the Salt Lake and Raft River formations is gradational (Blackett & Kolesar, 1983).

Two major north-south trending fault zones—the Bridge Fault and the Horse Well Fault—have been identified on the east side of the valley in the Tertiary sediments (Blackett & Kolesar, 1983). These fault zones are responsible for the upward transport of the geothermal water from greater depths. In addition to the Bridge and Horse Well faults, there is some evidence, including lateral displacement of volcanic rocks at the surface (Blackett & Kolesar, 1983), for a shear fault in the Precambrian basement known as the Narrows structure.

Meteoric recharge of the hydrothermal system likely occurs in the Raft River and Albion mountains (Covington, 1980). Water is circulated at depth in the fractured basement, flows along the Narrows structure and then flows up the Bridge and Horse Well faults to the surface (Covington, 1980). The majority of the hydrothermal activity in the valley coincides with the proposed intersection of the Bridge and Horse Well fault zones with the Narrows structure.

Nine wells have been drilled to exploit the Raft River geothermal resource. Raft River Geothermal well 9 (RRG-9) is currently used as an injection well. A thermo-hydraulic fracture treatment campaign is scheduled to be conducted in late 2010 or early 2011. Because of the temperature difference between cold injection fluids and hot reservoir rocks, thermal fracturing, in addition to hydraulic fracturing, is expected to occur over the life of the reservoir. This thermal fracturing is anticipated to further increase the total fracture surface area beyond the surface area resulting from hydraulic fracturing alone. Perkins and Gonzalez suggested that these thermal fractures may extend perpendicular to the main fracture face creating a fracture network (1985).

The extent to which thermal fracturing occurs and makes a significant contribution to the total fracture surface area of the geothermal reservoir will be controlled by the thermophysical properties of the rock, existing fractures and the in-situ stress regime. The thermal conductivity, specific heat and density of the reservoir rocks will determine the rate of cooling in the reservoir. The coefficient of thermal expansion of the reservoir rock will control stress changes in the cooled region. Thermophysical properties such as thermal conductivity, specific heat and coefficient of thermal expansion are important in modeling and predicting fracture shearing and propagation in geothermal reservoirs. Laboratory-obtained values of these key properties have been limited by the scarcity of geothermal core. Standard geophysical well logs and mineralogic composition data from analyses of core and cuttings can be used to predict thermal conductivity.

Thermal Fracturing

During exploitation of a geothermal reservoir, thermal energy is removed from the reservoir by cool water flowing through the rock in the case of high permeability hydrothermal reservoirs, or along fracture faces in the case of enhanced geothermal systems where matrix permeability is assumed to be small. Heat extraction results in an ever-growing cooled region. As the cooled region continues to grow, the temperature gradients in the reservoir become smaller until it becomes uneconomical to operate the well.

In an enhanced geothermal system, the surface area available for heat transfer is limited by the surface area of the relatively few natural and engineered fractures in the reservoir formation. The rate of thermal drawdown in an enhanced geothermal system would be significantly greater than in a traditional permeable hydrothermal reservoir. Bažant et al. suggested that thermal fracturing associated with reservoir cooling may increase the effective size of the reservoir by opening up new cracks for fluid flow and heat transfer (1979). As the surfaces of a fluid channels cool, shrinkage cracks will begin to form normal to the main fluid channel. Bažant et al. predicted that water flux through these shrinkage cracks could be expected to be on the same order of magnitude as water flux through the main cracks (1979). Perkins and Gonzalez addressed the same phenomenon in hydraulic fracture applications in hot, permeable, convection dominated oil reservoirs from the perspective of modification of in- situ stresses due to rock shrinkage (1985).

It is evident that in low matrix permeability fractured geothermal reservoirs that heat conduction controls thermal output and depletion. The thermal conductivity, heat capacity and density of a substance govern the way heat flows through the substance. As applied to reservoir management and thermal fracturing, it will be important to be able to quantify these key properties in order to predict the way transient temperature profiles will move through the reservoir.

Predicting Thermal Conductivity

Recognizing the importance of knowing the reservoir properties such as thermal conductivity, as well as the difficulty and expense associated with an extensive coring and laboratory testing campaign, the present work outlines a method for obtaining the reservoir thermal conductivity from geophysical well logs and phonon conduction theory. This method was first described by Williams and Anderson (1990). Various methods of predicting rock thermal conductivity have been outlined, including prediction directly from geophysical logs or from rock compositional data; however these methods are strictly empirical and largely ignore the physics of heat transfer in solids. Williams and Anderson, however, combined easily-obtained well log data with the theory of phonon conduction in order to produce a model which can be applied generally (1990).

Heat is conducted in non-metallic solids primarily by crystal lattice vibration (Berman, 1976). This is known as phonon conduction. Energy can be stored and transferred in solids in the form of atoms vibrating relative to their equilibrium position (Yang, 2004). In the presence of a temperature gradient, lattice waves known as phonons travel down the gradient from regions of greater vibration, associated with elevated temperature, to regions of lesser vibration (Berman, 1976; Yang, 2004). The thermal conductivity of a crystal is a function of the heat capacity of the crystal, the phonon velocity and the mean free path (Yang, 2004). In non-porous rocks, then, thermal conductivity is a complex function of composition, crystal geometry and temperature. In porous rocks, prediction of thermal conductivity is expected to be further complicated by pore geometry, the presence of pore fluids and rock-fluid interactions.

Williams and Anderson presented a semi-empirical model of thermal conductivity in low-porosity rocks. Their model was based on geophysical well log responses, phonon conduction theory and laboratory testing of aggregate minerals. The equation for phonon conduction presented by Williams and Anderson is given as Equation 1

$$\lambda_{phonon} = 0.7531 + 0.1005\left(\frac{aV_m\mu^2}{3K_ST}\right) \quad 1$$

where a is the mean interatomic distance, V_m is the mean phonon velocity, μ is the shear modulus, K_s is the bulk modulus and T is the temperature (Williams & Anderson, 1990). The temperature in Equation 1 is the temperature at which the experimental data reported by Williams and Anderson was obtained: 298 K. When this relationship was first presented, Williams and Anderson derived the shear and bulk moduli from compressional and shear wave sonic velocities and the mean interatomic distance was inferred from an approximation which used the photoelectric factor obtained from the photoelectric log. In the current study of Raft River geothermal wells, neither shear wave velocity nor photoelectric factor logs were available. Alternate methods for approximating shear wave sonic velocity and photoelectric factor have been used.

Brocher presented an empirical relationship between compression and shear wave velocities based on a large data set which included data from Quaternary alluvium and Salinian granites (2005). "Brocher's regression fit" is given as Equation 2.

$$V_S = 0.7858 - 1.2344V_P + 0.7949V_P^2 - 0.1238V_P^3 + 0.0064V_P^4 \quad 2$$

The mineralogy of core and cuttings samples from well RRG-3C and RRG-3D were determined using X-ray diffraction spectroscopy (XRD). Approximate elemental compositions were determined from the mineralogy. The photoelectric factor for each mineral was calculated from the approximate elemental composition as the sum of the products of the atomic weights, photoelectric cross sections and numbers of each atom in the mineral, all divided by the molecular weight. Calculated photoelectric factors were compared with photoelectric factors tabulated by Hashmy and Alberty (1992). Table 1 compares the photoelectric factors obtained by the two methods. The photoelectric factors for each sample were then calculated as the average of the photoelectric factors of the minerals weighted by their mass fractions.

A principal component analysis (PCA) was conducted on well log data from Raft River well RRG-3 over the interval from approximately 4250 to 5800 feet MD. The PCA was performed on neutron porosity, gamma, formation density and compressional wave slowness logs along with a log of the difference between the neutron porosity and density porosity logs. PCA is a multivariate data analysis method which allows multivariate data sets to be reduced to a smaller number of variables or principal components. This is accomplished by performing an eigenvector decomposition of the variance-covariance matrix. The resulting eigenvectors and eigenvalues define the principal components of the data set.

Any data point can be projected onto the principal component just as, in two dimensions, any point can be projected onto the x or y axes. The projection of a data point onto the principal component is called the *score* of the data point. Principal component scores have been correlated with the photoelectric factor from analysis of core and cuttings. A linear relationship (Figure 1; Equation 3) emerged between the principal component scores and the photoelectric factor. While the goodness of fit is not ideal, an analysis of variance indicates that the linear relationship is statistically significant. The calculated value of the F statistic is 8.66 which is greater than the critical value of the F statistic (F=5.12) at the 95% confidence level.

Table 1 - Photoelectric factors of minerals present in core and cuttings from well RRG-3

Mineral	Photoelectric Factor	
	Calculated	Tabulated[a]
Biotite	12.39	6.27
Calcite	5.09	5.08
Chlorite	3.51[b]	6.3
Chlorite/Smectite	5.61[c]	
Epidote	6.87	
Fluorite	6.56	6.67
Hematite	21.94	21.48
Illite	2.44[d]	3.45
K-feldspar	2.89	2.86
Kaolinite	1.52	1.83
Microcline	2.89	2.86
Muscovite	2.44	2.4
Orthoclase	2.89	2.86
Plagioclase[e]	1.76[f]	1.68[f]
Albite	1.76	1.68
Anorthite	3.06	3.13
Pyrite	17.42	16.97
Quartz	1.81	1.81
Titanite	22	
Zeolite[e]	1.55[g]	
Analcime	1.55	
Chabazite	2.43	

a. from Hashmy and Alberty (1992)
b. as clinochlore
c. as corrensite
d. as muscovite
e. undifferentiated
f. as albite
g. as analcime

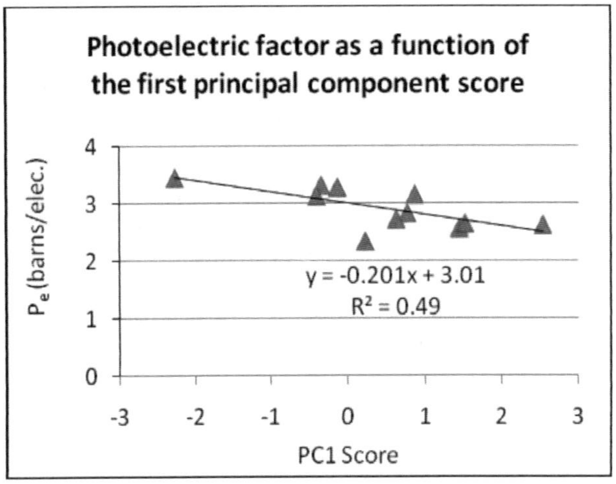

Figure 1 - A linear correlation between photoelectric factor (y axis) and principal component score (x axis) was found. This relationship allows for the determination of photoelectric factor from well logs.

$$P_e = 3.01 - 0.201 \cdot (PC1Score); \quad R^2 = 0.49 \qquad 3$$

The mean interatomic distance has been calculated as prescribed by Williams and Anderson: The effective atomic number has been calculated as (Ellis, 1987)

$$Z = 10P_e^{0.2778} \qquad 4$$

The mean atomic mass has been calculated as described by Williams (1989)

$$\overline{M} = \frac{(Z + 26.0)}{1.88} \qquad 5$$

Finally, the mean interatomic distance has been calculated as

$$a = \left(\frac{\overline{M}}{\rho}\right)^{1/3} \qquad 6$$

Thermal conductivity has been calculated for the rock solids over the entire depth range by the method of Williams and Anderson (Equation 1).

It is worth noting that the equation put forth by Williams and Anderson is a strong function of sonic wave velocities through both the mean phonon velocity and the bulk moduli. Thermal conductivity determined using this method is affected by the presence of alteration and microfracturing (Williams & Anderson, 1990). This was reemphasized by Pribnow and Sass (1995). They also pointed out that the Williams/Anderson model does not account for the effect of shear wave splitting associated with anisotropic rocks. The model then is most likely to be accurate in isotropic or flat-lying anisotropic rocks (Pribnow & Sass, 1995). Additionally, because the sonic tool analyzes waves traveling vertically, the thermal conductivity from the Williams/Anderson method is the *vertical* thermal conductivity (Williams & Anderson, 1990). While still valuable, for the present application the quantity of most interest is the *horizontal* thermal conductivity.

Experimental

Laboratory testing has been conducted on two core plugs from Raft River well RRG-3C to determine thermal conductivity. The core sample plugs were saturated and loaded to approximate in-situ conditions. The thermal conductivity of the two samples at various temperatures is reported in Table 2.

To allow comparison between the model of Williams and Anderson and the results of laboratory experiments, the experimental thermal conductivities have been extrapolated to 25°C, the temperature at which the Williams/Anderson experiments were conducted. Table 3 compares extrapolated experimental thermal conductivities with those predicted by the model.

Conclusions

The photoelectric factor of reservoir rocks shows a linear relationship with the principal component score derived from a principal component analysis of the neutron porosity, gamma, formation density, compression wave slowness and porosity difference logs of Raft River well

RRG-3. Results of the laboratory thermal conductivity testing of plugs from core taken from Raft River well RRG-3 show good correlation with conductivity determined using principal component analysis and the method of Williams and Anderson for the sample taken from 5272 feet. The discrepancy observed for the sample taken from 5270 feet may be due to anisotropy. Additional experimentally-determined results are necessary to draw further conclusions.

Table 2 - Thermal conductivity of core from Raft River well RRG-3C. Samples were saturated with slightly saline water and loaded to approximate in situ conditions.

Sample ID	Depth (ft)	Saturant	Confining Pressure (psi)	Pore Pressure (psi)	Average Test Temperature (°C)	Average Thermal Conductivity (W/m-K)
EGM1-1	5270.1	0.4% NaCl	2120	100	34.0	1.775
					68.0	1.850
					138.0	1.929
					Average over Temperatures	1.815
EGM2-5	5272.15	0.4% NaCl	2120	100	33.4	1.210
					69.6	1.307
					133.5	1.315
					Average over Temperatures	1.277

Table 3 - Comparison between thermal conductivity determined by laboratory analysis and by the method of Williams and Anderson

Depth (ft)	Sample Number	Extrapolated Laboratory Thermal Conductivity (W/m-K)	Thermal Conductivity from method of Williams and Anderson (W/m-K)
5270	EGM1-1	1.755	1.293
5272	EGM2-5	1.187	1.194

Acknowledgements

The authors wish to thank the management of U.S. Geothermal, Inc. for providing drill cuttings for analysis. This work has been supported by the United States Department of Energy grant DE-EE0000215.

References

Bažant, Z. P., Ohtsubo, H., and Aoh, K. (1979). Stability and Post-critical Growth of a System of Cooling or Shrinkage Cracks. *International Journal of Fracture , 15* (5), 443-456.

Berman, R. (1976). *Thermal Conduction in Solids.* Oxford: Clarendon Press.

Blackett, R. E., and Kolesar, P. T. (1983). Geology and Alteration of the Raft River Geothermal System, Idaho. *Transactions of the Geothermal Resources Council,* (pp. 123-127).

Brocher, T. M. (2005). Empirical Relations between Elastic Wavespeeds and Density in the Earth's Crust. *Bulletin of the Seismological Society of America , 95* (6), 2081-2092.

Covington, H. R. (1980). Subsurface Geology of the Raft River Geotheremal Area, Idaho. *Transactions of the Geothermal Resources Council,* (pp. 113-115).

Ellis, D. V. (1987). *Well Logging for Earth Scientists.* New York: Elsevier.

Hashmy, K., and Alberty, M. (1992). Difficult Lithologies. In D. Morton-Thompson, & A. M. Woods, *Development Geology Reference Manual.*

Perkins, T. K., and Gonzalez, J. A. (1985). The Effect of Thermoelastic Stresses on Injection Well Fracturing. *Society of Petroleum Engineers Journal* , 78-88.

Pribnow, D. F., and Sass, J. H. (1995). Determination of Thermal Conductivity for Deep Boreholes. *Journal of Geophysical Research , 100* (B6), 9981-9994.

Williams, C. F. (1989). Heat and Fluid Flow Within the Earth's Crust: In Situ Measurements and Mathematical Models. New York: Columbia University.

Williams, C. F., and Anderson, R. N. (1990). Thermophysical Properties of the Earth's Crust: In Situ Measurements From Continental and Ocean Drilling. *Journal of Geophysical Research , 95* (B6), 9209-9236.

Yang, J. (2004). Theory of Thermal Conductivity. In T. M. Tritt, *Thermal Conductivity: Theory, Properties, and Applications* (pp. 1-20). New York: Kluwer Academic/Plenum Publishers.

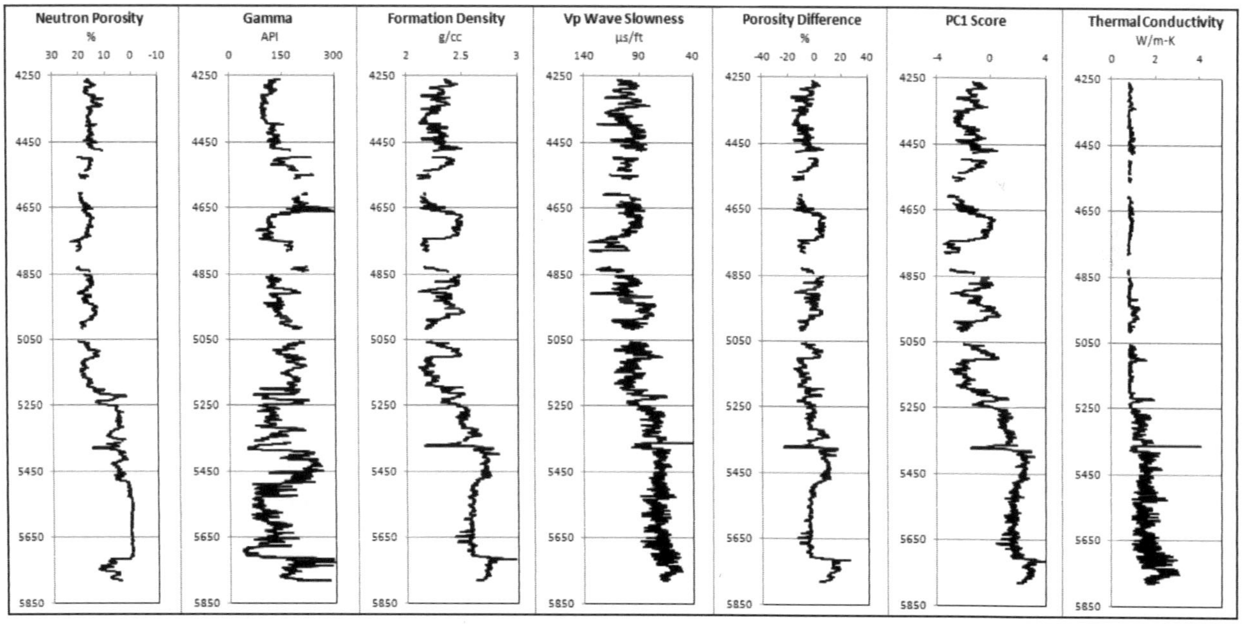

Figure 2 - Neutron porosity, gamma, formation density, and compressional wave slowness geophysical well logs. Also a log of the difference between neutron porosity and density porosity logs. These logs were used in a principal component analysis. Logs of Principal Component 1 Score and Calculated Thermal Conductivity are also shown. Thermal conductivity was calculated using the Williams/Anderson model.

(246a) Modeling and Simulation of Subsurface Processes in Oil Recovery and Carbon Storage

Angelo Lucia, University of Rhode Island, Kingston, RI and George A. Zyvoloski, Earth & Environmental Sciences Division, Los Alamos National Laboratory, Los Alamos, NM

Subsurface flow processes are important in oil recovery from shales and tar sands, carbon storage, and the movement of contaminants in ground water basins and typically involve complex physical phenomena that span several time and length scales. These processes are often modeled using coupled mass and energy balance equations along with phase equilibrium relationships to describe multi-phase flows through porous media. Simulation of steady or unsteady-state behavior of subsurface processes, reservoir simulation, includes a wide class of applications where both accurate phase densities and accurate phase equilibrium are essential. In addition, phase stability and phase equilibrium (or multi-phase flash computations) need to be performed quickly and reliably because they often consume a large fraction of the total reservoir simulation time (anywhere from 40 to 50%).

In this talk we describe a novel multi-scale modeling and simulation approach to subsurface processes. The proposed approach combines information from three time and length scales – the molecular scale, the bulk fluid phase scale, and the reservoir scale. At the molecular length scale we use coarse-grained NTP Monte Carlo simulations to determine internal energies of departure, UD, and create look-up tables of UD as a function of temperature, pressure and composition. This molecular scale information is communicated to the bulk phase length scale via look-up tables and interpolating formulae in order to predict the energy parameter in the new, multi-scale Gibbs-Helmholtz constrained (GHC) cubic equation of state (EOS) developed by Lucia (2009). The GHC EOS is used to compute fluid phase densities and component fugacity coefficients and solution fugacities which, in turn, are used to determine the number and types of phases as well as their corresponding compositions using Gibbs free energy minimization with full analytical partial derivatives. These minimum Gibbs free energy-based multi-phase flash computations are coupled to the reservoir simulator Finite Element Heat & Mass Transfer (FEHM) primarily developed and evolved by Zyvoloski (1975-present) to determine spatial and temporal distributions of vapor and liquid phases and multi-phase flows throughout the reservoir. FEHM uses a control volume finite element formulation of coupled heat and mass balance equations, Newton's method with analytical partial derivatives to solve the nonlinear algebraic equations that come from a discrete representations of the governing partial differential equations, and a pre-conditioned Krylov subspace method (partial LU factorization with GMRES) to solve the linear equations at each Newton step.

A number of interesting numerical examples are presented to demonstrate the efficacy of the proposed multi-scale modeling and simulation approach. We present comparisons of the GHC and SRK equations in the context of reservoir simulations. We also show that the multi-phase flash computations can be streamlined without loss of accuracy to provide very fast and reliable determination of phase behavior, can be readily interfaced with FEHM, and that coupling the GHC EOS with FEHM permits larger integration time steps and reliable simulations covering time horizons of 20 to 100 years.

See more of this Session: Simulation of Complex Subsurface Processes I
See more of this Group/Topical: Topical D: Chemical Engineering in Oil and Gas Production and Other Complex Subsurface Processes

New research simulation framework for flow modeling of complex physical system

Denis Voskov

Introduction

One of the most challenging type of a modern oil recovery processes is a compositional simulations. The main complexity of it is due to the nonlinear coupling of a multi-component multiphase flow and a thermodynamical phase behavior problem. An Equation of State (EoS) model usually employ to describe the phase partitioning for a mixture [6, 3]. For a given temperature, pressure, and overall composition of each component in the mixture, EoS computations define the phase state and compositions for each phase. That usually leads to the significant grow of computational time since EoS computations (that is an iterative process itself) needs to be performed for each computational cell and on each nonlinear iteration.

For a general purpose modeling, the different time-approximation scheme can be unify in a single simulation framework [4, 15]. Since governing equations, described multi-component flow coupled with thermodynamical equilibrium, are highly nonlinear, a Newton-Raphson method is used for a linearization. Base variables are usually divided into primary and secondary set based on alignment with equations. However, there is no single opinion which variables should be used as a base variables for such kind of linearization.

One of the most used formulations is the natural variables, when pressure, saturations, and phase mole fractions are used as the base set of unknowns (see [6]). One of the essential part of the natural-variables formulation is the variable substitution technique when we should use different combination of variables and equations (conservation equations and thermodynamical constraints) for a different phase state. Another widely used formulation is a different variations of molar variables [14, 5]. Here pressure and overall mole fractions used as a primary variables. Secondary variable set can be different for different formulations. This formulations look attractive since it is possible to avoid variable substitution when phase status is changing.

The phase behavior of multi-component mixture with arbitrary number of phases is another challenging problem. Several methods have been developed to improve the correctness of the phase stability results (see [18, 19]). However, all of these methods are based on global optimization techniques, computationally expensive and not guaranteed the convergence to the right solution. In the proposed simulation framework we use stability test based on adaptive tie-simplex parameterization (see [11], [20]) that improve both robustness and efficiency of phase behavior computations for multiphase multi-component simulation.

After a brief overview of mathematical statement for the problem, we are going to show the behavior of several nonlinear formulations and solution strategies for two-phase isothermal compositional simulation. Next the extension of two-phase isothermal framework to the flow of an arbitrary number of phases is presented. This extension helps to model a very complex processes such as thermal steam injection, low temperature CO2 injection to hydrocarbon reservoirs, flow in the gas-hydrates systems etc. Several representative examples are shown.

Conservation equations and constraints

In this section a general description of the governing equations for a multiphase flow in porous media is given. The transport equations for a system containing n_c components and n_p phases can be written as:

$$F_i = \frac{\partial}{\partial t}\left(\phi \sum_j x_{ij}\rho_j S_j\right) \quad (1)$$

$$+ \operatorname{div} \sum_j x_{ij}\rho_j \mathbf{u}_j + \sum_j x_{ij}\rho_j q_j = 0, \quad i = 1, \ldots, n_c. \quad (2)$$

Here ϕ denotes porosity, ρ_j is the density of phase j, S_j is a phase saturation and x_{ij} is the composition of component, i, in phase, j. All relations are functions depended on base (nonlinear) unknowns. These unknowns can be almost arbitrary chosen based on the nonlinear formulation. Below we present several formulations for isothermal compositional problem.

For description of flow for any phase we use Darcy's law:

$$\mathbf{u}_j = \left(\mathbf{k}\frac{k_{rj}}{\mu_j}(\nabla p - \gamma_j \nabla D)\right), \quad j = 1, \ldots, n_p, \quad (3)$$

where k_{rj} is the relative permeability of phase j, μ_j is the viscosity and k is the absolute permeability tensor.

Another assumption usually used for modeling of large scale flow in porous media is the thermodynamical equilibrium for components. In order to satisfy this condition we need to add $n_c \times (n_p - 1)$ equations define the equilibrium of chemical potential or fugacities for each component in each phase:

$$\psi_{ij}(p, T, x_{ij}) - \psi_{i,k}(p, T, x_{ik}) = 0, \quad \forall j \neq k \in \{1, \ldots, n_p\}, \quad i \in [1, \ldots n_c]. \quad (4)$$

Here p and T denote pressure and temperature, $\psi_{ij}(p, T, x_{ij})$ is the fugacity of component i in phase j.

Depends on the choice of formulation, different constraints can be added to the system in order to provide a closed mathematical statement. The phase behavior is an important type of computation needs to complete numerical framework and we describe it in the next section.

Phase behavior

For the phase behavior of a multi-component multiphase flow in large scale porous media we use the assumption of spontaneous thermodynamic equilibrium for each component in each phase, described by (4). This equilibrium condition can be used in combination with other equations to form a closed mathematical statement of problem.

Consistently with two-phase compositional framework we are going to use a two-stage procedure for a computation of phase equilibrium. First, the number of existing phases in each cell is determined and number of independent variables is defined based on the cell status. Next, if the status is changed, we need to run flash calculations to determine the compositions of the existing phases, x_{ij}.

Various methods can be used for that type of computations. First stage, called stability test, is the challenging procedure for system with arbitrary number of phases since a lot of local minima in solution of equilibrium problem exist. The second stage can be done using multiphase flash procedure, described in [17]. Since both stages of phase behavior computation is the second order high dimensional optimization it is usually a very computationally expensive procedure.

Here we propose a different approach for a phase behavior computation. We are going to use the fact that phase behavior is a continuous function of pressure, temperature and overall composition. (see [23]). That give us an opportunity to use Compositional Space Adaptive Tabulation, developed for two-phase systems ([10], [11]), in order to parameterize tie-simplex associated with phase behavior for discrete points in compositional space and for discreet values of pressure and temperature.

The CSAT framework allows us to use a generalized negative-flash computations in order to define the number of existing phases and corresponding phase composition. In [24] we demonstrate that single multi-phase negative flash could be not enough for accurate phase state identification and suggested multi-stage negative flash in order to cover this complexity.

We can use pressure and temperature defined for a given grid cell and build interpolation for each tie-simplex (e.g. tie-triangle for three-phase system) in the table. Then we can use closest tie-simplex to identify phase state for a given grid cell. There is strong theoretical and practical evidence ([20]) that we need only limited number of such tie-simplex tables to cover the whole range of composition associated with multi-component multiphase displacement process.

Numerical framework

The simulation framework for the research purposes should be highly configurable in order to check different scientific conceptions and ideas. One of the important part of any simulation software is the nonlinear formulation that includes a set of equations described the problem and base unknowns used for linearization of problem. Here we use numerical implementation based on Automatic Differentiation Expression Templates Library ADETL described in [13].

In the framework we specified different phase states for the simulation region (grid-block) that is usually closely related on thermodynamical state (number of

Table 1: Phase status table for three-phase system

Status	Number of phases	Array of existing phases		
		Phase 1	Phase 2	Phase 3
1	1	x	-	-
2	1	-	x	-
3	2	x	x	-
4	1	-	-	x
5	2	x	-	x
6	2	-	x	x
7	3	x	x	x

existing phases). Depending on the state we can use a different set of governing equations and nonlinear unknowns. The resulting Jacobian is constructed based on the specified variable set automatically since all the procedures used in residual construction are based on AD-type variables that allow to compute derivatives simultaneously with construction of analytical relation.

One of the important things is the possibility to construct and differentiate the all possible thermodynamical states corresponding to the system with specific number of co-existing phases. As we mentioned before, the negative-flash procedure estimate the number of existing phases based on the sign of phase fraction ν_j. Than we can define the status of phases based on the following relation:

$$N_s = \sum_{j=1}^{n_p} 2^{j-1}\delta_{(\nu_j>0)} \tag{5}$$

Here $\delta_{(\nu_j>0)} = 1$ if $\nu_j > 0$ and 0 otherwise. The example of all possible states corresponding to three-phase system is presented in following table. It is possible to build a similar table for arbitrary number of phases. The phase status can be evaluated based on (5) and for the known number of status we can have the list of existing phases based on the table. That helps to unify the most of the computations based on a phase status for the particular grid-block.

Solution Methods

The problem we need to solve is highly nonlinear, and we use the Newton-Raphson method to linearize it. For each nonlinear iteration linear system contained Jacobian should be solved. In order to reduce the size of linear system, the complete set of equation is split into two parts. First part (primary equations) are the mass and energy conservation equations since they account for fluxes across cell boundaries and depend on the unknowns (variables) in neighboring gridblocks. The second part is the rest of the equations which are local for any discrete gridblock. The unknowns aligned with both parts are called primary and secondary variables. There is no special rules for alignment of equations and variables except the condition that both the primary and the secondary diagonal matrices should be invertible.

After the linear system is solved, we can update all the nonlinear variables. This update provides a nonlinear solution based on Newton-Raphson method. However, if we use this method blindly that can lead to inconsistency in nonlinear behavior since any Newton update is mathematical abstraction that could not fully preserve all the physics. In order to avoid nonphysical conditions and improve the nonlinear convergence, we introduce several corrections to the standard Newton update.

The time-stepping strategy based on convergence of nonlinear process in the L_∞ norm. Each successful timestep follow by increasing of the next timestep until its length reaches a limit. If the number of nonlinear iterations breaks a maximal number of iterations, current timestep should be reduced. Another reason for timestep to be reduced is the negative value of pressure on any Newton iteration or convergence problem of a linear solver.

Nonlinear compositional formulations

In this section we describe different nonlinear formulations for the isothermal compositional problem. In each formulation we are using different set of unknowns to solve the system described by (1) and (4) nonlinear equations. For each unknown we specified its dimension in square brackets. We also outline the sequence of steps used to compute the properties and the Jacobian matrix for each formulation.

Natural variables

For this formulation we are using following variables:

- p – pressure [1],
- S_1 – phase saturation [1],
- x_{ij} – phase composition of each component $[2n_c - 2]$.

In order to detect the phase change we use EoS computations. For single-phase gridblocks we need to run stability test to detect is the second phase can appear. In two-phase blocks we use S_1 to detect the phase disappearance.

In order to complete computational framework, we also need to express the molar phase fraction, ν_j, for two-phase cells as a function of saturation, namely,

$$\nu_j = \frac{S_j \rho_j}{\sum_k S_k \rho_k}. \tag{6}$$

Then the overall mole fraction of component i can be written as

$$z_i = \sum_j x_{ij} \nu_j. \tag{7}$$

For a single-phase mixtures, $\nu_j = S_j$ and $z_i \equiv x_{ij}$.

Molar variables

([5]) This formulation is a variant of the molar-variables formulation, where instead of the phase fraction $x_{i,1}$, equilibrium ratios, K_i, are used. In [12], use of $\ln K_i$ was suggested in order to make (4) more linear. The following unknowns are used:

- p – pressure [1],
- ν_1 – phase fraction [1],
- z_i – overall phase composition of each component $[n_c - 1]$,
- lnK_i – phase equilibrium ratio $[n_c - 1]$.

The phase appearance and disappearance is detected based on the similar to natural formulation rules except that to detect the phase disappearance ν_1 is used instead of S_1.

In order to complete computational framework, we need to define phase composition for two-phase state as

$$x_{i,2} = z_i / \left(1 + \nu_1 (K_i - 1)\right) \tag{8}$$

$$x_{i,1} = K_i x_{i,2} \tag{9}$$

Then density of phase can be found from EoS and saturation computed from modification of (6).

Overall-composition variables

In this formulation, the following variables are used:

- p – pressure [1],
- z_i – overall composition of each component $[n_c]$.

For a given Newton iteration, a phase stability test is performed to determine if the overall composition, z_i, will split into two phases. If the gridblock is single phase, no additional computations are necessary. If a two-phase state is detected, flash calculations are performed to obtain the phase compositions $x_{i,j}$. The system $\mathbf{F} = 0$ is then solved for the given p, T, and z_i where:

$$F_i = f_{i,1}(p, T, x_{i,1}) - f_{i,2}(p, T, x_{i,2}), \quad i = 1, \ldots, n_c, \tag{10}$$

$$F_{i+n_c} = z_i - V x_{i,1} - (1 - V) x_{i,2}, \quad i = 1, \ldots, n_c, \tag{11}$$

$$F_{2n_c+1} = \sum_{i=1}^{n_c} (x_{i,1} - x_{i,2}), \tag{12}$$

Note that derivatives of the phase fractions as a function of the base unknowns are required, which can be obtained as follows

$$\frac{\partial \mathbf{x}}{\partial \mathbf{z}} = \left(\frac{\partial \mathbf{F}}{\partial \mathbf{x}}\right)^{-1} \frac{\partial \mathbf{F}}{\partial \mathbf{z}}. \tag{13}$$

The rest of the computations can be done in a manner similar to the other mass (mole) formulations.

Volume-balance formulation

([1]) The main difference of volume-balance formulation for a fully-implicit simulation is the use of extensive molar variables, N_i, instead of intensive overall compositions, z_i. Usually N_i is defined as

$$N_i = \phi V \sum_{p=1}^{n_p} \rho_p S_p x_{i,p}. \tag{14}$$

In order to complete computational framework, we can compute $z_i = N_i / \sum_{i=1}^{n_c} N_i$. Following computations are similar to the molar formulation.

Isothermal two-phase examples

Here we present the numerical results for comparison of different nonlinear formulation described in previous section. For test we choose small 2D (20x20) reservoir with four high permeable channels: in the 3rd and 12th gridblock in the x-direction and in the 2nd and 8th gridblocks in the y-direction. The other parameters for the reservoir are

- the size of block is 100x100x1 meters
- the reservoir-matrix permeability is 1 D, the permeability of the channels is 20 D.
- porosity is 0.2
- relative permeability is a Corey function with gas exponents equal to three and oil exponents equal to two.
- production and injection wells located on opposite corners (1,1) and (20,20)

In this reservoir we tested three different compositional models

- 4 component, immiscible: initial oil is made of $\{C_1$ (20%), CO_2 (1%), C_4 (29%), C_{10} (50%)$\}$ at temperature $T = 100^oC$ and initial pressure of 80 bars (above the bubble point), injection gas is a mixture of two components $\{CO_2$ (90%),C_1 (10%)$\}$. The injection pressure is 140 bars, which is sufficiently above the dew point of the injected mixture; the production pressure is 50 bars, which is below the bubble-point pressure of the reservoir mixture.
- 4 component, miscible: the same initial mixture with injection gas $\{CO_2$ (90%),C_4 (10%)$\}$ and the same injection, production, and initial pressures. Note that injection pressure is higher than Minima Miscible Pressure (MMP) for this system that makes it miscible.
- 6 component, SPE5: system used for SPE 5 comparative study [7]. Gas is composed of $\{C_1$ (77%), C_3 (20%), C_6 (3%)$\}$. The oil composition is $\{C_1$ (50%), C_3 (3%), C_6 (7%), C_{10} (20%), C_{15} (15%)$\}$, C_{20} (5%)$\}$. Temperature of

Table 2: Test results for formulations comparison

N	Formulations	4 comp., immiscible		4 comp., miscible		6 comp. (SPE5)	
		Newton	EOS	Newton	EOS	Newton	EOS
1	Natural	373	5.02	379	6.17	247	8.10
	Molar	386	5.07	441	6.43	455 (45)	10.77
	Molar VB	403	5.07	484 (6)	6.92	403 (30)	10.12
	Overall	444 (62)	8.27	435 (47)	9.30	252	10.58
2	Natural	207 (29)	7.24	174	6.77	197 (43)	12.30
	Molar	213 (20)	6.79	268 (9)	7.99	535 (153)	13.49
	Molar VB	201 (2)	5.78	378 (43)	9.18	395 (67)	11.29
	Overall	323 (120)	12.01	271 (73)	12.27	259 (100)	20.77
3	Natural	206 (28)	7.69	174	8.36	181 (34)	11.55
	Molar	198 (20)	7.35	179	8.33	138	10.35
	Molar VB	203 (20)	7.31	220 (21)	8.73	136	10.71

reservoir is $T = 70^o C$, initial pressure is 210 bars, injection pressure is 340 bars and production pressure is 100 bars. In described condition the displacement process is multi-contact miscible.

Below you can see the table with results. Three types of simulation was done for all systems and nonlinear formulations:

1. Normal simulation with not-aggressive time step, corresponding CFL numbers for three cases are $CFL_{max} = 9.2$, 9.0 and 14.8 respectively;

2. Normal simulation with aggressive time step, corresponding CFL numbers for three cases are $CFL_{max} = 35.2$, 35.5 and 29.1 respectively;

3. Simulation with the flash reduction for aggressive time step, the CFL numbers are similar to previous.

In the last type of simulation we apply exact flash procedure and correct corresponding variables for any Newton iteration. The maximal number of Newton iterations allowed for each time step was $N_{max} = 20$.

For each test run you can see two numbers corresponding with an accumulative number of Newton iteration for each run, including number of wasted iteration in brackets and an average number of EoS iteration (either stability or flash) done for each Newton iteration in each grid block. For all the simulations number of linear iterations was quite similar and that's why we don't report it in the table. Since the amount of computations for each Newton iteration is almost constant, the number of Newton iterations really corresponds to the simulation performance. Of course, the number of EoS iterations affects the total time as well, but it can sufficiently reduces to almost nothing using CSAT approach [10, 11]. There is no results reported for the overall-composition formulation in last type of run because these numbers are identical to the second case since this formulation always requires exact flash for any Newton iteration.

As you can see, for the first type of simulation (lower time step limit), the number of Newton iterations is quite different. The formulation based on natural variables looks the most attractive among the rest for the full range of simulated problems, but the difference is not really big. The number of EOS iterations is almost constant for all formulations except overall-composition, since it needs more flash iterations for two-phase cells. This picture changes when aggressive time-step is applied. Now the natural formulation is a way more advanced. That happened since the nonlinear chopping of base variables in case of natural formulation deals with phase saturation directly. That helps to avoid oscillations along a fractional flow curve that seems to be a dominant nonlinearity for compositional simulation with an intensive phase change. At the same time, the formulation based on an overall-composition run exact flash for any nonlinear iteration in order to detect phase disappearance. For the non-fully-converged solution that can introduce fake phase change and leads to oscillation that grow the number of wasted iterations. The rest of a molar-variable-based formulation just suffers from a worse nonlinear convergence.

The performance of molar variable set change when exact flash is added for the Newton iteration. In this case, the difference between molar formulations and overall-composition formulation is that for both molar formulations we use phase fraction ν_1 to detect phase disappearance for two-phase gridblocks before the flash correction. If a two-phase state is still valid, we do an exact flash in order to correct phase composition x_{cp}. That avoids oscillation based on phase detection, that is typical for overall-composition formulation, and improves convergence of molar formulations. As you can see from the table, both molar formulations demonstrate the large improvement and become almost as efficient as the natural formulation. At the same time, the flash correction improves the natural formulation very slightly.

Examples of three-phase flow

Here we present several numerical examples of the displacement in the system with three EOS-based phase. The nonlinear formulation is based on an extension of natural variables. The nonlinear solver is the same simple chopping strategy. In all examples, the Stone 1 model is used to construct the oil relative permeability.

For a phase behavior computations we use the combination of the multiphase flash [17] with the stability analysis based on the Compositional Space Adaptive Tabulation (CSAT). For a three-phase system this algorithm includes parameterization of a three-phase region as well as all corresponding two-phase regions in the plane of the tie-triangle, corresponding to the three-phase region.

In [21] it was shown that it is not possible to enter the three-phase region from the one-phase region without a shock in the composition and this shock belongs to the extension of the tie-triangle plane. In [20] we show that the similar rule is valid for the system with an arbitrary number of phases. This fact helps to explain why CSAT approach is supposed to be effective even when a number of phases exceed two. We demonstrate this efficiency for two-phase isothermal [10, 11] and thermal [22] displacement problems and here we are going to demonstrate it for three-phase problems here.

All simulations are done based on 1D and 2D reservoirs models with following parameters:

- the size of block is 100x100x1 meters;
- the porosity is 0.2;
- the relative permeability is a Corey function with gas exponents equal to three and oil exponents equal to two;
- production and injection wells located on opposite sides of model (the first and the last blocks);
- the reservoir permeability is 1 D.

We run either 1D (100 grid blocks) or 2D (10×10 and 20×20 grid blocks) models. In 2D models we add two high-permeable (20 D) channels in order to provide a heterogeneity. We use the three-parameter Peng-Robinson Equation of State for a phase behavior model.

First type of the simulation is an injection of a steam-gas mixture in the heavy-oil reservoir represented by $\{C_1$ (5%), C_8 (10%), C_{20} (70%), H_2O (10%)$\}$. The injected steam-gas mixture is $\{C_1$ (20%), H_2O (80%)$\}$. Initial conditions are $p = 140$ and $T = 130^oC$, injection conditions are $p = 60$ bars.

As you can see in Fig. 1,a, in this case only one tie-triangle-plane is needed to parameterize an entire simulation composition that requires a phase identification. That happened because both the injection and initial compositions belongs to the almost single plane and all shocks between compositions in different phases belongs in this plane as well. Next we change injection composition and add the C_8 component, so the steam-gas mixture will become $\{C_1$ (15%), $\{C_8$ (5%), H_2O (80%)$\}$. In Fig. 1,b you can see that two tie-triangle-planes are parameterized by the CSAT method. Now two shocks that connects the injection and initial composition belong to the different tie-triangle-plane and we need both of them for a phase identification.

If we change the reservoir description and look at the displacement curve for the 2D homogenous reservoir (10×10 gridblocks), in Fig. 2,a you can see that the phase behavior for this case is quite similar. The solution is obviously more dispersive but we need only one more tie-triangle-plane to parameterize solution in this case. If we use better resolution (20×20 gridblocks), the number of parameterized tie-triangle-planes again drops to only two (see Fig. 2,b).

Next example is the injection of CO_2 into oil at cold conditions. The oil is modeling by $\{C_1$ (20%), C_2 (30%), C_6 (40%), CO_2 (10%)$\}$ and initial conditions are $p = 4$ bars and $T = -110^oC$. At this condition the second liquid hydrocarbon (CO_2-rich) phase appears in the system. We inject pure CO_2 at $p = 10$ bars and produce at $p = 2$ bars. In Fig. 3,a you can see solution of CO_2 displacement in 1D reservoir. Again, in order to describe phase behavior for the entire solution, CSAT parameterized only 5 tie-triangles-planes. And this number changes insignificantly

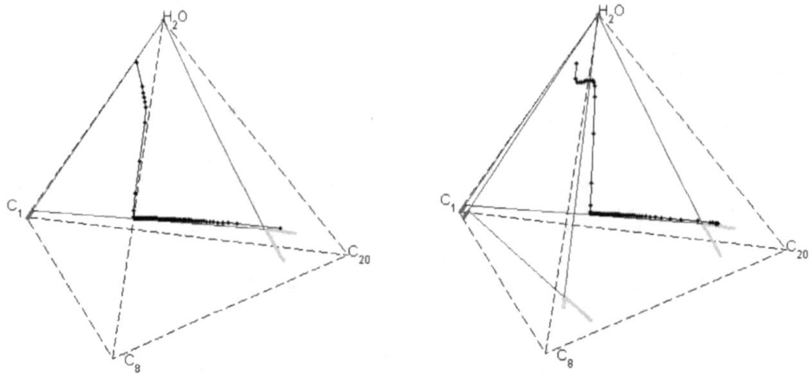

(a) (b)

Figure 1: Solution of oil displacement problem with steam-gas mixture for 1D homogenous reservoir. Tie-triangles with corresponding two-phase regions are presented in color.

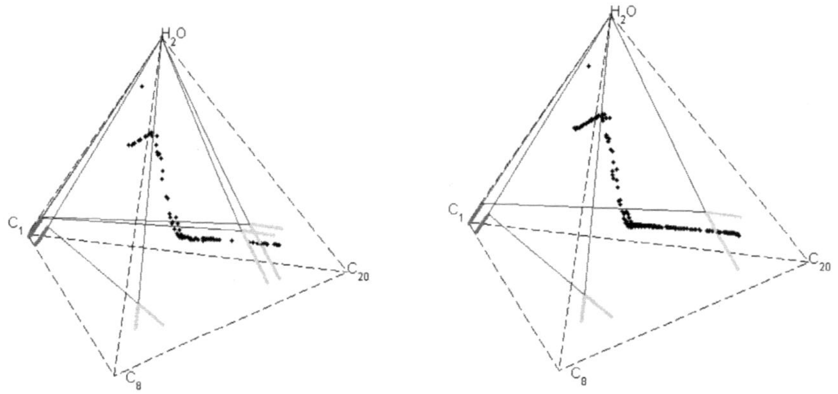

(a) (b)

Figure 2: Solution of oil displacement problem with steam-gas mixture for 2D heterogenous reservoir. Tie-triangles with corresponding two-phase regions are presented in color.

if we move from homogenous 1D model to heterogenous 2D (20×20 gridblocks) as seen in Fig. 3

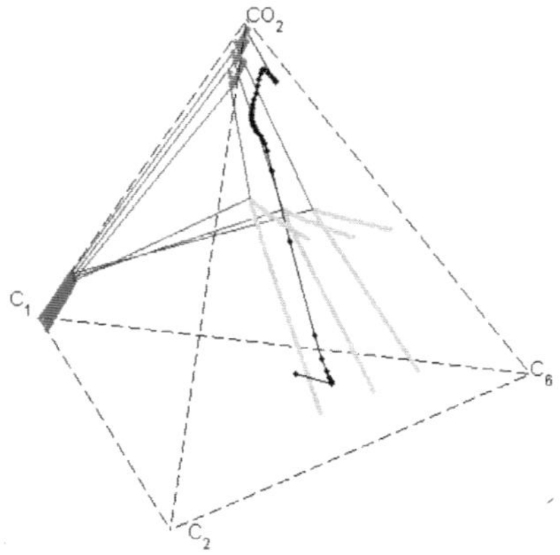

(a)

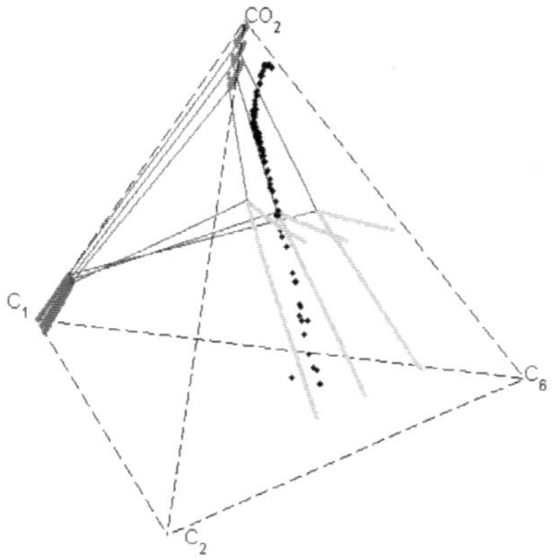

(b)

Figure 3: Solution of oil displacement problem with pure CO_2 at cold conditions for 1D (a) and 2D (b) reservoirs. Tie-triangles with corresponding two-phase regions are presented in color.

Summary and Conclusions

The new framework for General Purpose Research Simulator (GPRS) based on an Automatic Differentiation with Expression Templates Library (ADETL) library is presented. This framework is used for various research related to the simulation of a complex physical problems associated with modern Enhancement Oil Recovery (EOR) processes.

The comparison of several widely used nonlinear formulations for isothermal compositional simulation is performed. The aim of this investigation was the nonlinear behavior associated with gas displacement processes in heterogenous reservoirs.

Several results of flow in porous media simulation for EoS-based three-phase system are also reported.

Acknowledgement

Thanks to Alireza Iranshahr, Rami Younis and Yifan Zhou for contribution to the simulation framework. Acknowledgements to the Chevron, Schlumberger and Total companies and the SUPRI-B Industrial Affiliates program on reservoir simulation at Stanford University for financial support.

References

[1] Act G., Doleshall S., Farkas E., General Purpose Compositional Model. - SPEJ, 1985, vol. 25, N 4.

[2] Aziz, K. and Settari, A., Petroleum Reservoir Simulation, Applied Science Publishers, London, England, 1979

[3] Aziz, K. and Wong, T.W., Considerations in the Development of Multipurpose Reservoir Simulation Models, proceedings of the 1st and 2nd International Forum on Reservoir Simulation, Alpbach, Austria, September 12-16, 1988 and September 4-8, 1989.

[4] Cao H., Development of Techniques for General Purpose Simulators, Ph.D. Thesis, Stanford University, 2002

[5] Chien M.C.H., Lee S.T., Chen W.H., A New Fully Implicit Compositional Simulator. - SPE 13385, 1985.

[6] Coats K.H., An Equation of State Compostional Model. - SPEJ, 1980, v.20, N 5.

[7] Killough J.E., Kossack C.A., Fifth comparative solution project: evaluation of miscible flood simulators, SPE 16000, 1987.

[8] Michelsen M.L., The isothermal flash problem: Part I. Stability. Fluid Phase Equilibria, 9: 1-19, 1982.

[9] Michelsen M.L., The isothermal flash problem: Part II. Phase-split calculation, Fluid Phase Equilibria, 9: 21-40, 1982.

[10] Voskov D., Tchelepi H., Compositional Space Parameterization: Theoretical Background and Application for Immiscible Displacements, SPE Journal 14 (3), 2009, (paper SPE 106029 presented at 2007 SPE Reservoir Simulation Symposium, Woodlands, TX, February 26-28).

[11] Voskov D., Tchelepi H., Compositional Space Parameterization: Multi-Contact Miscible Displacements and Extension to Multiple Phases, SPE Journal 14 (3), 2009, (paper SPE 113492 presented at 2008 SPE/DOE Symposium on Improved Oil Recovery, Tulsa, Oklahoma, April 20-23).

[12] Wang P., Yotov I., Wheeler M., Arbogast T., Dawson C., Parashar M., Sepehrnoori K.: A new generation EOS compositional reservoir simulator: Part 1 - formulation and discretization, SPE 37979, 1997

[13] Younis R.M., Aziz K., Parallel Automatically Differentiable Data-Types for Next-Generation Simulator Development, Proc. of Reservoir Simulation Symposium, Houston, TX, February, 2007.

[14] Young, L.C. and Stephenson, R.E., A Generalized Compositional Approach for Reservoir Simulation, SPEJ, October 1983, pp727-742.

[15] Voskov D.V., Younis R.M., Tchelepi H.A., General Nonlinear Solution Strategies for Multiphase Multi-component EoS Based Simulation, SPE 118996, presented at 2009 SPE Reservoir Simulation Symposium, Woodlands, TX, February 2-4.

[16] Chien, M.C.H., Yardumian, H.E., Chung E.Y. and Todd, W.W., The Formulation of a Thermal Simulation Model in a Vectorized, General Purpose Reservoir Simulator. - SPE 18418, 1998.

[17] Michelsen, M.L. and Mollerup, J.M., Thermodynamic Models: Fundamentals and Computational Aspects, Tie-Line Press, Holte, Denmark (2004).

[18] C.M. McDonald, C.A. Floudas, Global Optimization for the Phase Stability Problem, AIChE J., 41, 7 (1995) 1798-1814.

[19] D.V Nichita, S. Gomez, E. Luna, Phase stability analysis with cubic equations of state by using a global optimization method, Fluid Phase Equilib., 194-197 (2002) 411-437.

[20] Voskov D., Tchelepi H., Tie-simplex based mathematical framework for thermodynamical equilibrium computation of mixtures with an arbitrary number of phases, Fluid Phase Equilibria, 283, 1-2 (2009) 1-11.

[21] T. LaForce, K. Jessen, F. M. Orr, Four-component gas/water/oil displacements in one dimension: part II, example solutions, Transport in Porous Media, 72, 1, (2008) 83-96.

[22] Iranshahr A., Voskov D., Tchelepi H., Efficient and Robust Phase Equilibrium Computations for Thermal Compositional EoS Based Simulation, SPE Journal, 15, 2, (2010) 537-548.

[23] Iranshahr A., Voskov D., Tchelepi H., Compositional Space Parameterization for Compositional Reservoir Flow Simulation, AIChE Annual Meeting, Nashville, TN, November 8-13, 2009.

[24] Iranshahr A., Voskov D., Tchelepi H., A Generalized Negative Flash Procedure for Phase Equilibrium Computations Related to Carbon Dioxide Sequestration, AIChE Annual Meeting, Nashville, TN, November 8-13, 2009.

(246c) New Stochastic Approach to Construct Porous Media Geometry Models and Simulation of Non-Darcy Flows

Feng Xiao, Michael Newman and Xiaolong Yin, Petroleum Engineering, Colorado School of Mines, Golden, CO

We present a new method to construct two- and three-dimensional porous media geometry models from Voronoi diagrams. The parameters of the algorithm can be adjusted to produce isotropic and anisotropic fully percolated geometries with both low and high effective porosities (10% to 50%). The permeability calculated using D2Q9 and D3Q19 lattice Boltzmann methods with multiple relaxation time (MRT) collision operators matched well with existing correlations. The tortuosity of the models, calculated from diffusion of tracers, ranged from 1.2 to 1.5. As an application, we studied regime transition from Darcy to non-Darcy flows. In Darcy flows, the superficial velocity through the porous media is a linear function of applied pressure gradient; in non-Darcy flows, due to the inertia of the fluid, the superficial velocity becomes a nonlinear function of the applied pressure gradient. The observed variation in the apparent permeability with the Reynolds number agreed well with existing correlations. The impact of porous media texture on flow regime transition was investigated by randomly removing grains. Porous media geometries with grains removed showed transitions to non-Darcy flows at lower Reynolds numbers and the deviations had higher slopes than those without removed grains at a similar porosity. Flow patterns within the pore space were examined in Darcy and non-Darcy regimes: bypassing streams and steady state vortices were observed when Re was increased to about 10; unsteady flows were observed when Re reached about 20.

See more of this Session: Simulation of Complex Subsurface Processes I
See more of this Group/Topical: Topical D: Chemical Engineering in Oil and Gas Production and Other Complex Subsurface Processes

(246d) Modeling and Analysis of Wormhole Formation in Carbonates with Diverting Acids

Ram R. Ratnakar, Department of Chemical and Biomolecular Engineering, University of Houston, Houston, TX, Nitika Kalia, Halliburton, Duncan, OK and Vemuri Balakotaiah, Department of Chemical and Biomedical Engineering, University of Houston, Houston, TX

We develop a semi-empirical complete rheological model for diverting acids that accounts for the effect of change in temperature, shear rate/pressure drop, and pH/concentration. A previously developed two-scale continuum model is modified to include non-Newtonian and diverting behaviors and used with the rheological model to analyze the stimulation of single/dual core set-ups with both Newtonian and diverting acids at constant injection rate. The acidization curve is found to be shallower for diverting acids as compared to Newtonian acids. We show that an efficient stimulation of dual core set-ups with Newtonian acids can be achieved if it is operated in uniform dissolution region based on total injection rate. Finally, we show that when dual core set-ups are stimulated with diverting acids, flow is diverted from high-perm core to low-perm core and from low-perm core to high-perm core several times until the exit concentration exceeds a certain value that depends on the fluid characteristics.

See more of this Session: Simulation of Complex Subsurface Processes I
See more of this Group/Topical: Topical D: Chemical Engineering in Oil and Gas Production and Other Complex Subsurface Processes

A Coupled Control-Volume Finite Element Geomechanical Model for Discrete-Fracture Reservoir Simulators

Nan Zhao, University of Utah, Salt Lake City, UT
Milind D. Deo, University of Utah, Salt Lake City, UT
John McLennan, University of Utah, Salt Lake City, UT

Abstract

The effects of geomechanical parameters on reservoir properties are usually not considered in conventional reservoir simulation.This may have ramifications in reservoirs where compaction drive is essential or for reservoirs where fracture flow is dominant and is strongly pressure sensitive. Discrete-fracture network representations of a fractured reservoir are examples of the latter; geomechanical effects are expected to have significant impact on production performance. In this paper, we systematically explore a number of different methodologies for incorporating stress dependence into reservoir properties. A finite element, geomechanical model based on force balance principles is interfaced with a control-volume finite element, discrete-fracture reservoir simulator. The geomechanical module that has been developed is able to consider any standard thermo-poroelastic constitutive relationship or stress-strain correlation, and accounts for both fracture and matrix constitutive properties. The pressure and saturation information from the flow model are communicated to the geomechanical model and the matrix/fracture properties are updated. Several different approaches for coupling are demonstrated and compared; from explicit exchange to communication between modules. Examples from fractured tight gas formations are used to demonstrate the model and for assessing the importance and effectiveness of each type of coupling. It is shown that the relevant geomechanical information is incorporated into the reservoir model in a computationally efficient manner by allowing timely communication, not necessarily after every time step. The impact of aperture changes of the fractures is shown to be considerable - resulting in significant local changes in effective permeability and connectivity during production. The influence of fracture aperture change, associated with the geomechanical module on capillarity and other rock-fluid phenomena lead to a number of interesting observations concerning water block, fluid trapping, producible rates and ultimate production. It is significant in dealing with dynamic changes in fracture networks in fractured reservoirs.

Introduction

A finite element geomechanical model coupling with a conventional reservoir simulator, which has the capability to coupling with various reservoir models, is introuced in the Utah Finite Element simulator (UFES). The new simulator is designed to run the black oil reservoir model with geomechanics as well as thermal typed model. And it provides several choices of coupling scheme include soft coupling and hard coupling. All thes features makes UFES a simulator that can handle various of applications.

UFES Frame Work and Governing Equations

The UFES is a highly modularized simulation frame work, and it has several specific mofules with different features. The various combations of different modules enable the UFES to handle multiple applications. The major module of UFES is the reservoir module and the geomechanical module, under each module a variety of submodules is employed to handle specific problem.

Reservoir Model

The UFES maily employed two reservoir models, the black oil model[1] and the compositional k value thermal model. Both models are based on discreted frature network representation of reservoir and work with several discretization methods. The governing equation for a black oil model is:

$$\nabla \cdot (\frac{Kk_{rl}}{B_l \mu_l} \nabla \varphi_l) = \frac{\partial}{\partial t}(\phi \frac{S_l}{B_l}) + q_l \tag{1}$$

φ_l is the potenial of the species l:

$$\varphi_l = p + \gamma Z \tag{2}$$

K is the absolute preambility and k_{rl} is the relative preambility. B is the volumetric factor. ν is the viscousity.q is the source term of each species, in the reservoir simulation this term is generally refer to the injection or production well.

The governing equation for compositional model[2] is:

$$\frac{\partial(U_p + U_{rock})}{\partial t} = \nabla \cdot \bar{\bar{\mathbf{K}}}_{\mathbf{c}} \nabla T - \nabla \cdot (\sum_{p=1}^{Nf} \hat{H}_p \rho_p \vec{v}_p) + Q_e + Q_{loss} \tag{3}$$

where U_p and U_{rock} are

$$U_p = \sum_{p=1}^{Np} \phi S_p \rho_p \hat{U}_p \tag{4}$$

$$U_{rock} = (1-\phi)\bar{U}_{rock} \tag{5}$$

Geomechanical Model

The geomechnical model is based on the assumption that the equilibrium state of mechanical is enforced in the reservoir. And the equation is:

$$\nabla \cdot \sigma + \rho \vec{b} = 0 \tag{6}$$

where σ is a tensor which represents stresses (symmetric):

$$\begin{bmatrix} \sigma_{x,x} & \sigma_{x,y} & \sigma_{x,z} \\ \sigma_{y,x} & \sigma_{y,y} & \sigma_{y,z} \\ \sigma_{z,x} & \sigma_{z,y} & \sigma_{z,z} \end{bmatrix}$$

Effective stress law is applied: $\sigma = \sigma' - \alpha\delta_{i,j}P$ where σ' is the effective stress,and P can be write as the saturation weighted pore pressure[3] ($P = \sum_{i=1}^{n} s_i p_i$).

$$\nabla \cdot (\sigma' - \alpha\delta_{i,j}P) + \rho \vec{b} = 0 \tag{7}$$

The density is defined as: $\rho = (1 - \phi^*)\rho_s + \phi^* \rho_f$ Different models can be applied to the stress like linear elastic model, linear poro-elastic model and so on.In the simulation, a poroelastical model is applied:

$$\sigma_{i,j} = 2G\varepsilon_{i,j} + 2G\frac{\nu}{1-2\nu}\varepsilon_{k,k}\delta_{i,j} - \alpha p \delta_{i,j} \tag{8}$$

ε is the volumetric strain and defines as $\varepsilon_i = \frac{\partial u_i}{\partial x_i}$ and $\varepsilon_{k,k} = \sum_{i=1}^{i=3} \frac{\partial u_i}{\partial x_i}$. G is shear module, α is Biot-Willis coefficient and ν is Poission's ratio.

Coupling schemes

In general, there are two types of coupling schemes, soft coupling and hard coupling. Soft coupling means the geomechanical effect is not directly invoved in the reservoir simulation. Hard coupling method integrate geomechanical model in the reservoir simulator. Both of these two methods need some parameters to connection geomechanical model and reservoir model. One set of the parameter is porosity and permeabilty:

First the ture porosity of the reservoir is introduced[4]:

$$\phi^* = 1 - (1 - \phi_0)e^{-\varepsilon_v}$$

or in linear form:

$$\phi = (\frac{1 - \phi_0}{1 - \phi^*})\phi^*$$

The equation above is come from the continuity of the rock:

$$\frac{\partial \rho_s (1 - \phi)}{\partial t} + \nabla((1 - \phi)v_s) = 0$$

where $\nabla v_s = \nabla \frac{du_s}{dt}$, $\nabla u_s = \varepsilon_v$
The permeability can be ralated to the geomechanic factor[4]:

$$\frac{k}{k_0} = \frac{1 + \frac{\varepsilon_v}{\phi_0}}{1 + \varepsilon_v}$$

where $\varepsilon_v = \frac{\delta V}{V} = \varepsilon_{xx} + \varepsilon_{yy} + \varepsilon_{zz}$ and $\varepsilon_{ij} = \frac{1}{2}(\frac{\partial u_i}{x_j} + \frac{\partial u_j}{x_i})$ Through these information, the impact of geomechanical effect is introduced in the reservoir model.

On the other hand, the geomechanical model is related to the reservoir model by the effective stress formulation. Also the geomechanical model is assumed to be at equilibrium all the time through the reservoir life. Based on the coupling parameter, soft and hard coupling can be applied.

Soft Coupling

To take into account of the geomechanical effect with out much change of the reservoir model. The "artifical" permeability and porosity can be introduced in the soft coupling scheme. In this way, the reservoir permeability and porosity is set to be a function of pressure (may plus saturation) based on the external information of this reservoir. This external information can be some real data and some pure mechanical simulation result of reservoir.

A soft coupling scheme is developed based on Itasca 3DEC simulation result for the multiphase flow simulation In the simulator the permeability and porosity is treaed from a look-up table generated from the geomechanical information of 3DEC. The pore pressure is using as a standard to maching these two simulation.

Hard Coupling

For the hard coupling approach, the geomechanical model is integrated in the reservoir simulator. The geomechanical equation set and flow equation set can be solved either "separately" or "together" depend on the real problem.

For the geomechanical model, a general virtual work formulation can be derived:(neglect the body force for convenience, acttually it just gravity in most case which only in z dirction:

$$\int_\Omega \delta\vec{\varepsilon}^T \Delta\sigma' d\Omega - \alpha \int_\Omega \delta\vec{\varepsilon}^T \Delta p d\Omega - \int_{\Gamma_\sigma} \delta\vec{u}^T \Delta\vec{t} d\Gamma = 0$$

Use the strain stress equation:

$$(\int_\Omega B^T DBd\Omega)(\Delta\vec{u}) - \alpha(\int_\Omega B^T\vec{m}Nd\Omega)(\Delta p) - \int_{\Gamma_\sigma} \delta\vec{u}^T \Delta\vec{t} d\Gamma = 0$$

The coupling parameter is the volumetric strain.

References

1. Fu, Yao (2007), "Multiphase Control Volume Finite Element Simulation of Fracured Reservoirs", PhD thesis, University of Utah

2. Huang,Chung-Kan (2009), "Development of A General Thermal Oil Reservoir Simulator Under A Modularized Framework", PhD thesis, University of Utah

3. Zienkiewicz, O. C. and Shiomi, T(1984), "Dynamic behaviour of saturated porous media; the generalized biot formulation and its numerical solution", Internation Journal for Numerical and Analytical Method in Geomechanics, 87:71-96

4. L.K. Thomas and L.Y. Chin and R.G. Pierson and J.E. Sylte(2002), "Coupled Geomechanics and Reservoir Simulation" in SPE Annual Technical Conference and Exhibition

Multiphase Compositional Space Parameterization for General-Purpose Thermal-Compositional Simulation

Alireza Iranshahr, Denis Voskov and Hamdi Tchelepi, Energy Resources Engineering, Stanford University, Stanford, CA

Introduction

Thermal and gas injection processes are among the most commonly used techniques to improve the recovery of oil. In order to obtain reliable predictions based on the numerical simulation of such processes, a thermal-compositional model is used, where the partial differential equations for the conservation of mass, energy, and momentum are solved together with the equations of thermodynamic equilibrium (Aziz and Wong, 1989).

Coupling thermodynamics with the conservation equations is a major challenge in the development of a general-purpose thermal-compositional simulation framework. The partial differential equations are highly nonlinear and tightly coupled to the phase equilibrium relations. In order to deal with thermal-compositional models of practical utility, accurate treatment of the thermodynamic phase behavior is crucial (Farouq-Ali and Abou-Kassem, 1989).

Simplified compositional (or modified black-oil) models have been developed in order to account for the thermodynamic phase behavior in reservoir flow simulation (Bolling, 1987; Whitson et al., 1988). One of the most common approaches is the so-called 'constant K-values' method, where it is assumed that the component equilibrium factors, K-values, are functions of temperature and pressure only, and therefore, thermodynamic equilibrium can be efficiently modeled using tables of K-values (see Aziz et al. (1987) for a practical example). However, the accuracy of the constant K-values approach is limited to immiscible displacements at relatively low pressure, where the assumption of composition-independent K-values is valid. In this work, we employ an Equation of State (EOS) method to represent the phase behavior associated with thermal-compositional flow simulation.

Phase equilibrium calculations can be the most expensive kernel in a compositional flow simulation. Given an overall composition at some pressure and temperature conditions, the number of phases and their compositions must be determined for every control volume (gridblock) at each time step. The conventional approach is based on: (1) performing a phase-stability test for single-phase gridblocks (Michelsen, 1982a), and (2) solving the fugacity constraints (Michelsen, 1982b) together with the coupled nonlinear flow equations for gridblocks that have more than one phase. Robust algorithms for stand-alone calculations of multiphase stability and flash procedures have been described by several authors (see Michelsen and Mollerup (2004) for instance). However, previous developments for phase behavior computations coupled with flow and transport have basically focused on methodologies that are suitable for only two-phase compositional simulation.

With the objectives of improved performance and stability of EOS-based phase behavior for two-phase compositional simulation, a tie-line based Compositional Space Parameterization (CSP) framework has been developed (Voskov and Tchelepi, 2009b). An effective way of utilizing the concept of CSP for general-purpose simulation is the Compositional Space Adaptive Tabulation (CSAT) framework in which tie-line parameterization is used to accelerate the convergence of EOS-based phase-behavior calculations (Voskov and Tchelepi, 2009a). CSAT is based on theoretical findings from Method of Characteristics (MoC) solutions of gas injection problems under ideal assumptions (Franklin M. Orr, 2007; Entov, 1997). The MoC solutions show that only a few (key) tie-lines determine the structure of the solution route in compositional space. The MoC-based theory has been recently extended to thermal (Zhu, 2003) and three-phase (LaForce et al., 2008b) displacements.

CSAT adaptively parameterizes the compositional space around a given composition (i.e., calculates the tie-lines through the composition at discrete pressure levels). The tie-line tables are then used to identify the phase-state of compositions encountered in the course of the simulation. Linear interpolation in pressure is performed to obtain a tie-line for a composition with the specified pressure value (Voskov and Tchelepi, 2009b).

The initial development of CSAT for immiscible displacements has been generalized for multi-contact miscible problems (Voskov and Tchelepi, 2009a). It is shown that adaptive tabulation of the critical tie-lines significantly improves the efficiency and accuracy of multi-contact miscible simulation. Generalization of CSAT to thermal problems, where tabulation and interpolation as a function of temperature and pressure is performed, has been presented in Iranshahr et al. (2010). Extension to multiphase systems is described in Voskov and Tchelepi (2009c). The wide range of the numerical examples presented in these references shows that CSAT is an effective method for phase behavior modeling in general-purpose compositional simulation.

In this paper, we present an EoS-based negative-flash method for systems that can form an arbitrary number of phases. The proposed algorithm is guaranteed to converge, and it is integrated with our multiphase CSAT framework for phase behavior calculations. We also describe the continuity of compositional space parameterization using tie-simplexes. Based on these ideas, the CSAT framework is able to capture the complex multiphase behavior in the vicinity of a given composition at different temperature and pressure conditions. We also describe a generalization of the CSAT framework for three-phase systems, and we demonstrate accurate phase-state identification in complex multicomponent systems.

EoS Based Multiphase Negative-flash

We start with the formulation of the multiphase negative-flash framework based on an Equation of State (EoS). Let N_p represent the maximum number of the phases that can coexist for a given system at the specified temperature and pressure. Here, we consider the partitioning of mixture made up of N_c components into N_p phases at equilibrium. The equations can be written as

$$\hat{f}_{i,0} - \hat{f}_{i,j} = 0, \quad i = 0, \ldots, N_c - 1, \; j = 1, \ldots, N_p - 1, \tag{1}$$

$$z_i - \sum_{j=0}^{N_p-1} x_{i,j}\nu_j = 0, \quad i = 0, \ldots, N_c - 1, \tag{2}$$

$$\sum_{i=0}^{N_c-1} (x_{i,0} - x_{i,j}) = 0, \quad j = 1, \ldots, N_p - 1, \tag{3}$$

$$1 - \sum_{j=0}^{N_p-1} \nu_j = 0, \tag{4}$$

where z_i denotes the overall composition of component i. The fugacities in Eqs. (1), $\hat{f}_{i,j}$, are computed using an EoS model (Peng and Robinson, 1976). The above $N_p(N_c + 1)$ equations are solved for the fraction (ν_j) and composition ($x_{i,j}$) of each phase. The pure phases define the vertices of a tie-simplex in the compositional space. Noting that the actual number of phases at equilibrium may be smaller than N_p, the phase fractions are not restricted to $[0, 1]$.

Newton's method can be used to solve Eqs. (1-4). Convergence of the Newton iterations depends on the initial guess. A good initial guess is obtained using Successive Substitution Iterations (SSI) (Whitson and Michelsen, 1989). During each iteration of SSI, component fugacities are held fixed, and Eqs. (2-4) are solved simultaneously. Here, we present a generalization of the two-phase negative-flash approach (Whitson and Michelsen, 1989). The overall mole fraction of a component, z_i, is given by:

$$z_i = x_{i,0}\nu_0 + \sum_{j=1}^{N_p-1} x_{i,j}\nu_j, \qquad i = 0, \ldots, N_c - 1, \tag{5}$$

where we have arbitrarily chosen a phase as the 'base' phase, and we denoted its index by 0. Using

$$\nu_0 = 1 - \sum_{j=1}^{N_p-1} \nu_j, \tag{6}$$

$$x_{i,j} = K_{i,j}x_{i,0}, \qquad i = 0, \ldots, N_c - 1; \quad j = 1, \ldots, N_p - 1, \tag{7}$$

one obtains

$$z_i = x_{i,0}\left(1 + \sum_{j=1}^{N_p-1} \nu_j(K_{i,j} - 1)\right) = x_{i,0}m_i(\boldsymbol{\nu}), \qquad i = 0, \ldots, N_c - 1. \tag{8}$$

In order to determine the equilibrium partitioning of all components among all phases, one must solve the following coupled $N_p - 1$ equations

$$\begin{aligned} f_j(\boldsymbol{\nu}) &= \sum_{i=0}^{N_c-1} (x_{i,0} - x_{i,j}) \\ &= \sum_{i=0}^{N_c-1} \frac{z_i(1 - K_{i,j})}{m_i(\boldsymbol{\nu})} = 0, \qquad j = 1, \ldots, N_p - 1, \end{aligned} \tag{9}$$

for $N_p - 1$ unknowns, $\boldsymbol{\nu}(\equiv \{\nu_j\},\ j = 1, \ldots, N_p - 1)$.

Next, we define the domain of the full problem. For this purpose, let us assume that we have an admissible set of K-values, so that the domain of Eqs. (9) is bounded. From Eqs. (8), we conclude that $m_i(\boldsymbol{\nu})$ must be strictly positive for each component (denoted by index i). The set of 'physical' phase fractions for this problem can be represented by

$$\Omega = \{\boldsymbol{\nu} \in \mathbb{R}^{N_p-1} \mid 0 < m_i(\boldsymbol{\nu}),\ i = 0, \ldots, N_c - 1\}. \tag{10}$$

Given an admissible set of K-values, Ω is a full-dimensional bounded convex polytope in $\mathbb{R}^{N_p-1}$ (Brondsted, 1983). Next, let us consider a subproblem of the full negative-flash problem. Suppose that for an N_p-phase negative-flash problem, we want to solve for the first M_p phase fractions given that the last $N_p - M_p - 1$ phase fractions are already known:

$$\min \nu_{j,\Omega} < \nu_j < \max \nu_{j,\Omega}, \qquad j = M_p + 1, \ldots, N_p - 1, \tag{11}$$

where ν_j is a value in the range of the j-th phase fraction for the full-dimensional Ω. The equations to be solved simultaneously for $\boldsymbol{\nu}(\equiv \{\nu_j\},\ j = 1, \ldots, M_p)$ are:

$$f_j(\boldsymbol{\nu}) = \sum_{i=0}^{N_c-1} \frac{z_i(1 - K_{i,j})}{m_i(\boldsymbol{\nu}) + n_i} = 0, \qquad j = 1, \ldots, M_p, \tag{12}$$

where n_i is a fixed number given by

$$n_i = \sum_{j=M_p+1}^{N_p-1} \nu_j(K_{i,j} - 1), \qquad i = 0, \ldots, N_c - 1. \tag{13}$$

Note that $m_i(\boldsymbol{\nu})$ is a function of the M_p phase fractions only. Let

$$\Omega^{M_p} = \{\boldsymbol{\nu} \in \mathbb{R}^{M_p} \mid 0 < m_i(\boldsymbol{\nu}) + n_i,\ i = 0, \ldots, N_c - 1\}, \tag{14}$$

represent the set of physical phase fractions for this subproblem. It can be shown that a solution for the negative-flash subproblem exists, and that it is unique (Iranshahr et al., 2009a). Moreover, the subproblem has all the properties of a full negative-flash problem, so that a recursive procedure can be applied to solve the multiphase negative-flash problem.

Implementation of the Multiphase Negative-Flash

For three-phase systems, Ω is a convex polygon, and we solve for two phase fractions (ν_1, ν_2). Here, we describe a bisection based algorithm. In order to improve the speed of convergence, the bisection strategy can be combined with a gradient based scheme. Consider the following algorithm:
Step 1: Given the set of K-values, find the boundaries of the physical plane, which yield $\nu_{2,min}$ and $\nu_{2,max}$.
Step 2: Set $\nu_{2,mid} = (\nu_{2,min} + \nu_{2,max})/2$.
Step 3: Solve $f_1(\nu_1, \nu_{2,min}) = 0$, and $f_1(\nu_1, \nu_{2,mid}) = 0$ for $\nu_{1,min}$, and $\nu_{1,mid}$, respectively using bisection.
Step 4: If $f_2(\nu_{1,min}, \nu_{2,min}) \times f_2(\nu_{1,mid}, \nu_{2,mid}) > 0$, $\nu_{2,min} = \nu_{2,mid}$; else $\nu_{2,max} = \nu_{2,mid}$.
Step 5: Stop if $\nu_{2,max} - \nu_{2,min} < \varepsilon$; otherwise, go to Step 2.
Figure (1) illustrates the first four iterations of the procedure for a synthetic 15-component system (with arbitrary K-values). The SSI method for solving two-phase negative-flash problems can be generalized to three-phase systems by using this algorithm as a constant K-value flash solver at each iteration. For simplification in the following sections, we will describe the developments for multicomponent systems that can form a maximum of three phases at equilibrium.

Multi-Stage Three-Phase Negative-Flash

The three-phase negative-flash introduced in the previous section is not sufficient for determining the phase-state of a given composition. This is because the two-phase behavior outside the three-phase region may be different from the results obtained from a three-phase negative-flash (see Figure 2). Based on a three-phase negative-flash only, a mixture can be identified as stable in the single-phase region; however, its Gibbs free energy might decrease by splitting into two mixtures. To avoid such problems, a second stage using two-phase negative-flash can be employed.

In order to determine the number and compositions of the existing phases for a given overall composition at specified temperature and pressure conditions, we use a multi-stage negative-flash algorithm. In the first stage, a three-phase negative-flash is performed. If the composition is inside the tie-triangle (three-phase), its state is confirmed. Otherwise, based on the results of the three-phase negative-flash, additional two-phase negative-flash computations are necessary. Here, we assume that the tie-lines are in the plane of the tie-triangle, even though real multicomponent systems show slight deviations from this

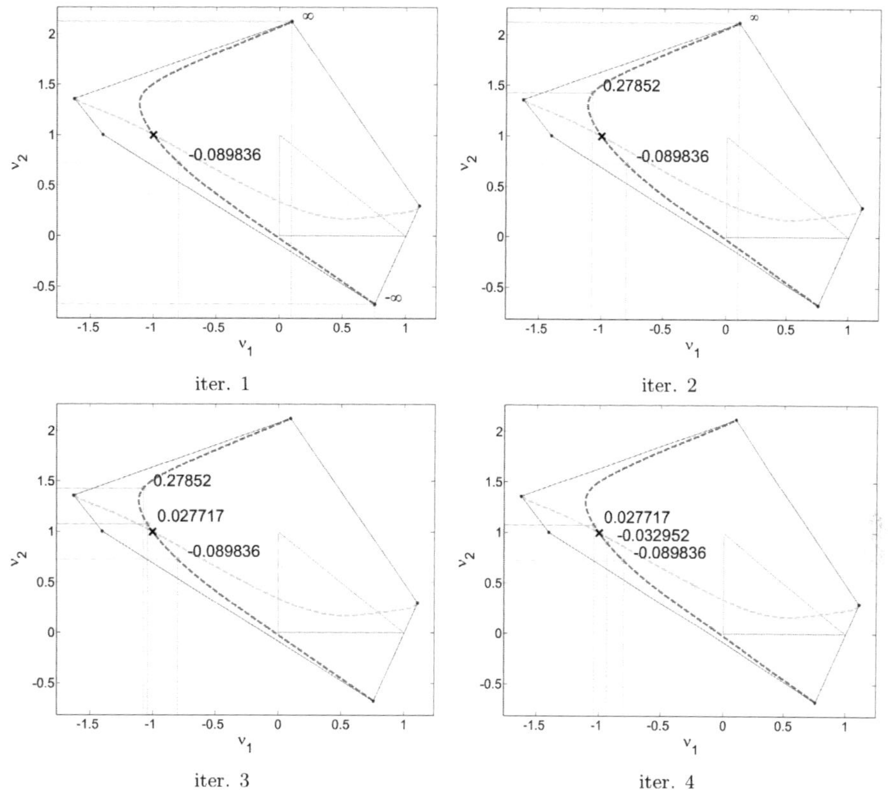

Figure 1 *The first four iterations of the methodology for a 15-component system. Ω is represented by the black polygon, and the blue triangle encloses three-phase mixtures. f_1 and f_2 are zero along the red and green curves, respectively. The numbers represent the values of $f_2(\nu_1, \nu_2)$ at $\nu_{2,min}$, $\nu_{2,mid}$, and $\nu_{2,max}$ where $f_1(\nu_1, \nu_2) = 0$. Note that initially the value of f_2 is indefinite at $\nu_{2,min}$ and $\nu_{2,max}$. These lower and upper bounds become closer as the iterations proceed.*

behavior. The compositions of the tie-triangle edges (two-phase systems) provide good initial guesses for the various two-phase negative-flashes. This is because tie-line compositions change continuously around a tie-triangle edge (Iranshahr et al., 2009b).

The three-phase negative-flash procedure can be used to generate phase diagrams (see Figure 3), and this is demonstrated using two examples: (1) a sour gas example, and (2) a Maljamar reservoir oil (Haugen et al., 2007). The accuracy of our numerical results is verified by a commercial simulator CMG (2009). We consider dilution of both mixtures with pure CO_2. That is, for every mixture which results from the addition of CO_2 to the initial feed, we increase the pressure and record the three-phase behavior. The pressure is increased until the parameterizing tie-triangle degenerates to a tie-line. Figure (3) shows the locus of the three-phase boundary, as well as, the tie-triangle degeneration locus. We note that no phase-stability test is required, since the three-phase negative-flash is capable of determining the bounds of the three-phase region accurately.

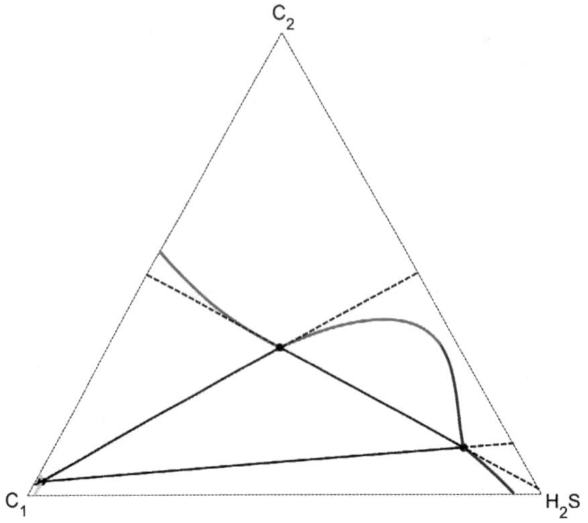

Figure 2 Complete parameterization of $\{C_1, C_2, H_2S\}$ *at* $T = 180\,\mathrm{K}, p = 15$ *bar. The accurate two-phase behavior is shown by the solid curves outside the three-phase region.*

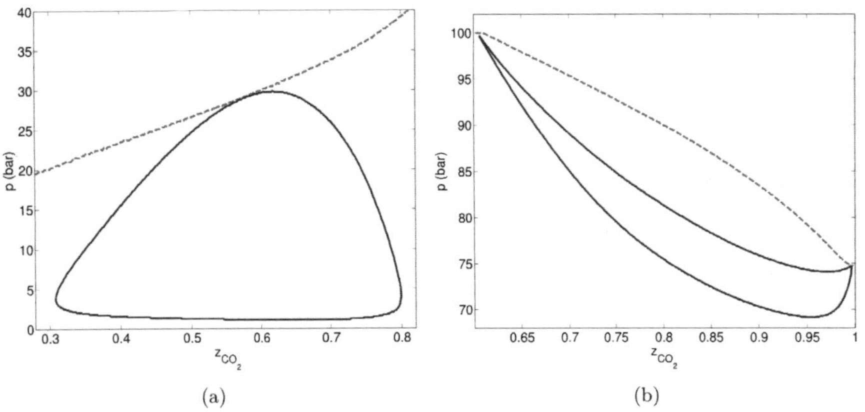

Figure 3 Phase diagram for the three-phase region (solid), and the locus of tie-triangle degeneration (dashed): (a) Sour gas example, $T = 178.8\,\mathrm{K}$*, (b) Maljamar reservoir oil,* $T = 305.4\,\mathrm{K}$.

Continuity of Compositional Space Parameterization

Here, we describe the continuity of parameterization using tie-simplexes. We can show that tie-simplex parameterization in compositional space is a continuous function of pressure, temperature and composition (Iranshahr et al., 2009b).

Continuation of the tie-simplex space justifies a tie-simplex based approach for compositional flow modeling, where interpolation in pressure and temperature is performed during a simulation run using a limited number of tie-simplexes. Figure (4) describes the parameterization continuity for two-phase systems. In the first example (see Figure 4-a), we plot an isothermal table of tie-lines ($T = 375$ K) for a fixed composition. The mixture is initially in the two-phase region. With increase in pressure, the composition becomes single-phase, and the parameterizing tie-line eventually degenerates at $p = 145\,bar$. We also show how parameterization of a four-component mixture changes with temperature at fixed pressure (see Figure 4-b). For this example, the composition is initially single-phase and enters the two-phase region with increase in temperature. The parameterizing tie-line becomes critical at $T = 550$ K.

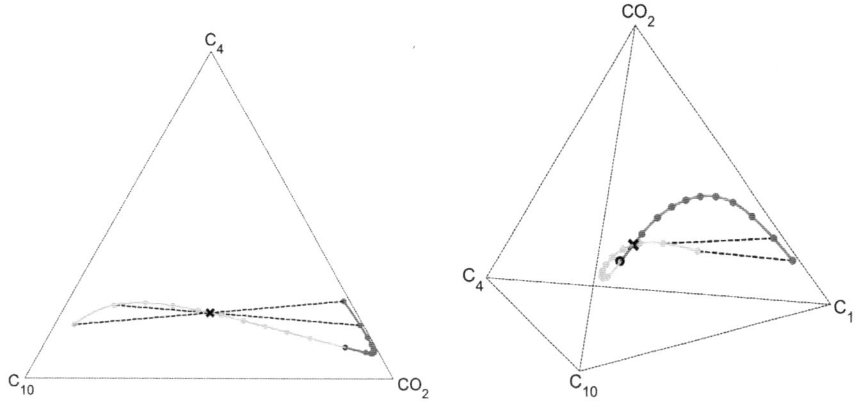

(a) $T = 375$ K, $p = (10, 145)\,bar$ (b) $p = 100\,bar$, $T = (250, 550)$ K

***Figure 4** Continuity of compositional space parameterization for two-phase systems*

Three-Phase Compositional Space Parameterization

We start with an overview of the two-phase CSAT framework (Voskov and Tchelepi, 2009b,a). Given the overall composition of a gridblock, all the tie-line tables are checked (after interpolation in temperature and pressure) to determine if there is any tie-line that parameterizes the composition. The phase-state of the mixture is determined if its composition and the interpolated tie-line are close enough. Otherwise, a new tie-line table is parameterized and added to the list of existing tables. For multi-contact miscible displacements, critical tie-lines (at different pressure values) that parameterize the solution path are adaptively calculated and stored. An initial check in the table of critical tie-lines determines if the composition belongs to the super-critical or sub-critical regions.

For two-phase systems, a single tie-line parameterizes the two-phase region around a given overall composition. However, for three-phase systems, the entire plane defined by a tie-triangle needs to be parameterized. Note that only a few tie-triangle planes parameterize the solution path in a three-phase compositional displacement (LaForce et al., 2008a; Voskov and Tchelepi, 2009c). For a given overall composition, we calculate a tie-triangle (Δ^2), and parameterize three curves: Γ_1^2, Γ_2^2, and Γ_3^2. In general, Γ_f^n parameterizes the n-phase region on the f-th side of a tie-simplex. Each one of the Γ^2 curves parameterizes the two-phase behavior on one of the tie-triangle sides.

Here, we demonstrate the construction of a general three-phase thermal-compositional table for a given composition (for parameterization of two-phase systems, see Iranshahr et al. (2010)). We assume that the starting pressure at each temperature level is below the minimal degeneration pressure for the three-phase region. For the smallest pressure, the tie-triangle that parameterizes the given composition is calculated and stored. Starting from each edge of the tie-triangle, we parameterize the two-phase region by solving for each Γ_f^2.

The three-phase region degenerates with increase in pressure. If the pressure is large enough, we solve for the minimal degeneration pressure of the tie-triangle that parameterizes the given composition. An iterative procedure is used to find a pressure value such that the area of the tie-triangle, A, satisfies:

$$\varepsilon_{min} < A < \varepsilon_{max}, \tag{15}$$

for some small values of ε_{min} and ε_{max}. The plane of the degenerate tie-triangle is then parameterized at higher pressure values. We note that immediately after the degeneration of the tie-triangle, a two-phase region disappears from the plane, or the length of a side of the tie-triangle becomes infinitesimal. However, all the parameterization techniques developed for sub-critical pressures are applicable for higher pressure values.

Tie-lines on a tie-triangle side may also degenerate. For such systems, parameterization of Γ_f^2 is started from the non-degenerate end. Then, we determine how far the parameterization of the tie-lines along Γ_f^2 should proceed until the length of the critical tie-line, ℓ, satisfies:

$$\varepsilon_{min} < \ell < \varepsilon_{max}, \tag{16}$$

for some small values of ε_{min} and ε_{max}. The calculated critical tie-line marks the boundary between extendible and non-extendible regions on the f-th tie-triangle side.

Phase-State Identification in Parameterized Compositional Space

Based on theoretical findings from the Method of Characteristics, only a few tables of tie-simplexes are required for a single simulation run (LaForce et al., 2008a; Voskov and Tchelepi, 2009c). In the second stage of the CSAT framework, we use tabulated tie-simplexes to identify the phase-state of mixtures that require phase stability testing (Voskov and Tchelepi, 2009a).

In the following developments, we assume that a side tie-line lies on the extension of a tie-triangle. The systems that we have considered in our studies exhibit this behavior. We note that CSAT still provides a very good initial guess for conventional phase-behavior calculations in systems that violate this assumption.

The tables of tie-simplexes are used to interpolate in temperature and pressure, then $\mathbf{z}$ is projected onto the closest tie-simplex (Voskov and Tchelepi, 2009c). In the general case, $\mathbf{z}$ is in M_p-phase state, if it is inside a Δ^{M_p-1}. Moreover, $\mathbf{z}$ is single-phase, if it is outside all tie-simplexes. These state identification rules are based on the fact that an M_p-phase negative flash can only determine the state of a mixture in the M_p-phase region. Next, we describe state identification procedures in degenerate steam systems. Consider a three-component steam example at $T = 550\,\mathrm{K}$ (see Figure 5-a). The three-phase region becomes critical at $p = 110\,bar$. With further increase in pressure, the gas-oil region shrinks until it disappears completely from the plane at $p = 148.9\,bar$.

From the continuity of parameterization in degenerate space, the critical gas-oil tie-lines change continuously with pressure. We calculate and store the critical tie-lines at discrete pressure values between the minimal three-phase degeneration pressure and the gas-oil

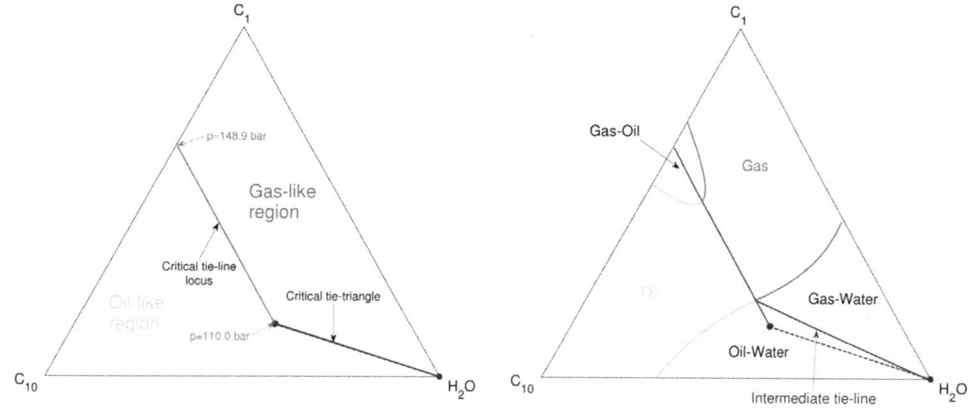

(a) (b)

Figure 5 *(a) Critical locus for* $\{C_1, C_{10}, H_2O\}$ *at* $T = 550\,\mathrm{K}$*; the locus of critical oil-gas tie-lines is extended from* $p = 110\,bar$ *to* $p = 148.9\,bar$*. (b) Phase diagrams at* $p = 137\,bar$

disappearance pressure. Note that the locus of the critical tie-lines is almost linear on the plane of the tie-triangle. The discretized locus of critical tie-lines as well as the critical tie-triangle is then used to identify if a single-phase mixture is gas- or oil-like. This technique is based on the assumption that the phase-state of a composition on the extension of a critical tie-line does not change with an increase in pressure (Voskov and Tchelepi, 2009a). We use this critical boundary to separate water-oil and water-gas tie-lines with an intermediate tie-line, which connects the water phase to a point on the critical tie-line locus (see Figure 5-b). If the water tie-lines shrink in size, the critical tie-triangle serves as the intermediate tie-line.

CSAT Examples

We show that the generalized CSAT framework accurately captures degeneration of the multiphase regions across a wide range of pressure. In an isothermal three-phase table, only two sets of tie-lines parameterize the plane of a degenerate tie-triangle. With further increase in pressure, a two-phase region disappears from the plane. We use an iterative procedure to calculate the pressure at which the two-phase region disappears. That is, we find a pressure value such that the disappearing two-phase region is represented by a trivial tie-line (the length of which satisfies Eq. 16). Figure 6 describes the general structure of an isothermal three-phase table. Note that once the multiphase region is represented by only one set of tie-lines, we use the standard two-phase CSAT approach for phase-state identification (Voskov and Tchelepi, 2009b).Figure (7) demonstrates the accuracy of CSAT close to the pressure at which the three-phase region disappears. Our computations are compared with results from a reliable commercial software CMG (2009).

Parameterization of four-phase systems is significantly more complicated than three-phase problems. We first calculate the tie-tetrahedron that parameterizes a given composition. Four curves, Γ^3, parameterize the three-phase region around the tie-tetrahedron. Moreover, different two-phase regions in the system are calculated using six surfaces, Γ^2. This way, all the multiphase regions on the extension of the tie-tetrahedron are parameterized. Figure (8) shows different sets of tie-triangles around a tie-tetrahedron.

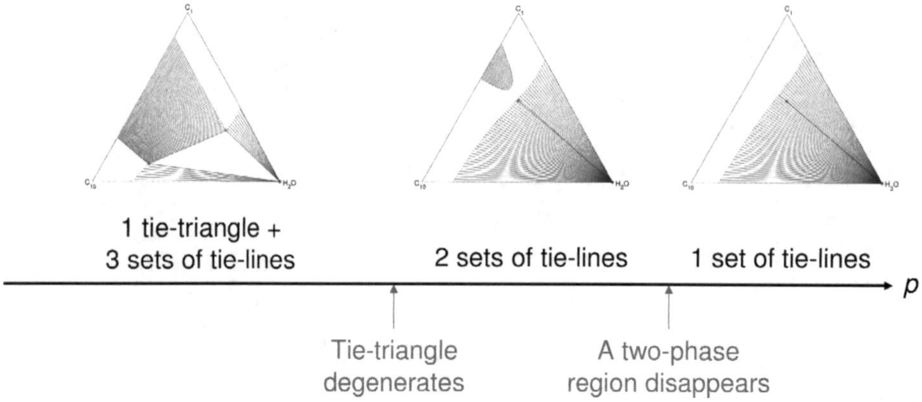

Figure 6 *General structure of an isothermal three-phase table. The three-component systems are calculated at $T = 520\,\mathrm{K}$ and $p = 75, 180, 250$ bar, respectively.*

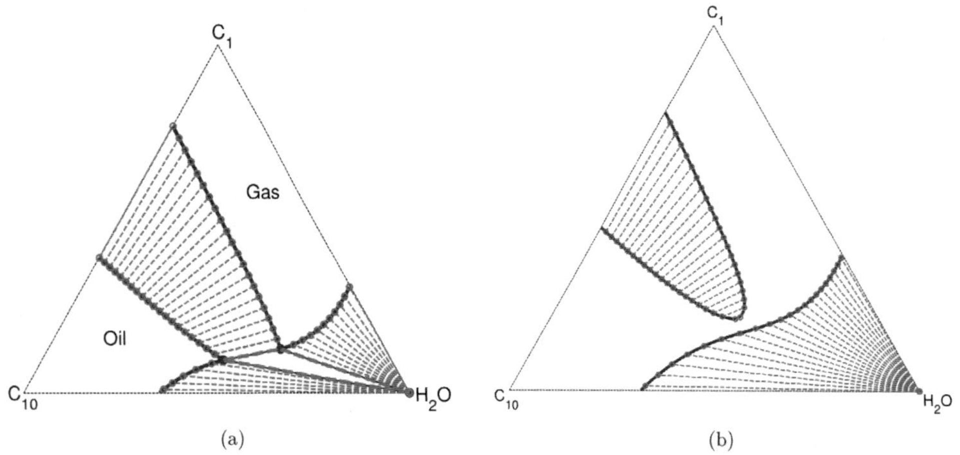

Figure 7 *Full parameterization of $\{C_1, C_{10}, H_2O\}$ at $T = 550\,\mathrm{K}$ and (a) $p = 100$ bar, (b) $p = 113.8$ bar. In (b) only two sets of tie-lines parameterize the space. The calculations are performed using the PR-EoS, tie-triangle degenerates at $p = 110.0$ bar and the oil-gas region disappears at $p = 148.9$ bar. Compare the phase diagrams in blue with the standard calculations (red tie-lines)*

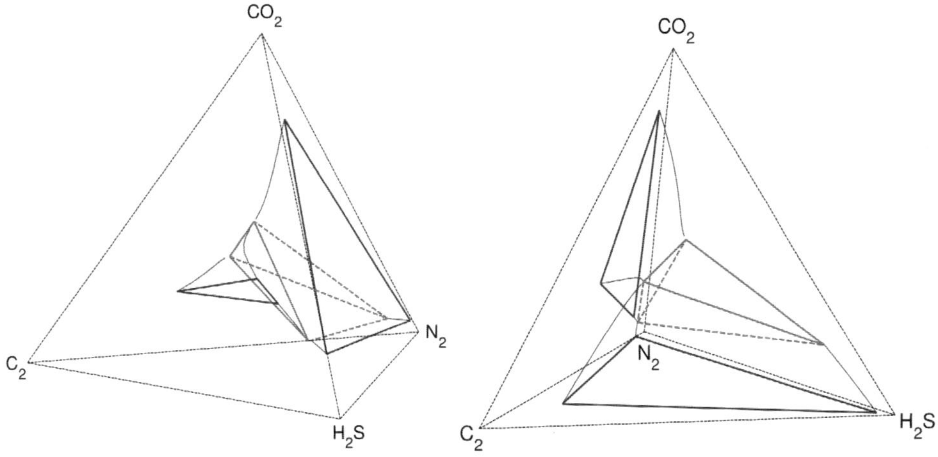

Figure 8 *Four-phase compositional space parameterization at* $T = 140\,\mathrm{K}$ *and* $p = 200\,bar$. *Each figure shows the tie-tetrahedron and two sets of side tie-triangles.*

Next, we demonstrate accurate phase-state identification for several steam systems. Consider the three-component steam example in Figure (7-a). We choose two compositions and run CSAT to identify the state of the compositions along the dilution line (see Figure 9-a). For the first composition, CSAT constructs a three-phase table as described earlier. The phase-state of all the compositions are then determined using this table. Note that a single table parameterizes the entire three-component system. A two- or three-phase composition is parameterized by a tie-line, or a tie-triangle, respectively. A single-phase composition is outside all tie-lines and the tie-triangle. Additional two- or three-phase negative-flash calculations may be required in order to identify the phase-state of mixtures that are too close to the phase boundaries accurately (Voskov and Tchelepi, 2009a).

With increase in pressure, the three-phase region degenerates, and the gas-oil region exhibits critical behavior (see Figure 9-b). Here, we also use CSAT to determine the phase state of a dilution line. Note that the locus of the critical tie-lines, which are at different pressures, separates single-phase gas and oil compositions (see Figure 5). Next, we consider a four-component steam system (see Figure 10-a for the full parameterization). We choose two compositions in this system and run CSAT along the dilution line. As shown in the figure, only a few parameterized planes are required for phase-state identification.

Conclusions

We formulated the multiphase EOS-based negative-flash framework. The Successive Substitution Iterations (SSI) and Newton methods for solving the composition-dependant phase-splitting problem can be generalized for systems with an arbitrary number of phases. Noting that a unique solution of the negative-flash procedure exists for each iteration of SSI, we employ a robust solution framework that is guaranteed to converge. We presented several challenging multiphase multi-component systems to demonstrate the effectiveness of the generalized negative-flash method. Our results indicate that the N_p-phase negative-flash can be used to identify accurately the phase-state of mixtures that partition into N_p equilibrium phases. For mixtures that partition into a lower number of phases at given

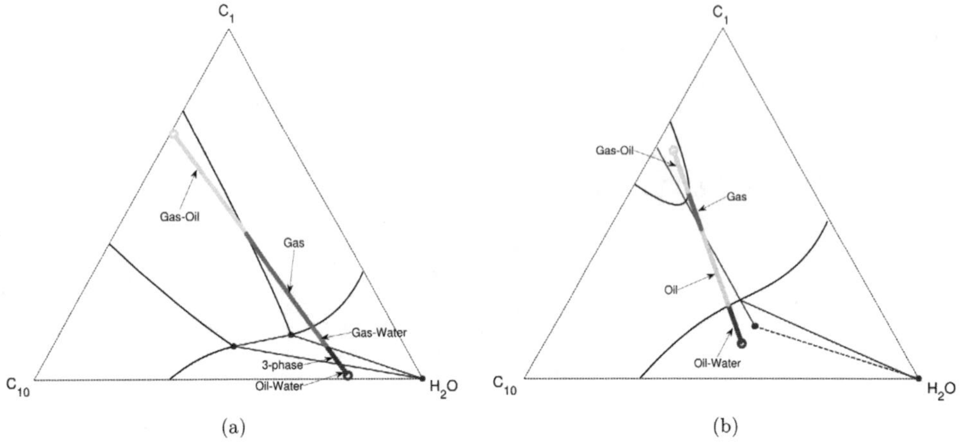

Figure 9 *Phase state identification at* $T = 550$ K *for the dilution line between (a)* $\{C_1(0.7), C_{10}(0.29), H_2O(0.01)\}$ *and* $\{C_1(0.01), C_{10}(0.19), H_2O(0.8)\}$ *at* $p = 100\, bar$*, and (b)* $\{C_1(0.65), C_{10}(0.3), H_2O(0.05)\}$ *and* $\{C_1(0.1), C_{10}(0.4), H_2O(0.5)\}$ *at* $p = 137\, bar$.

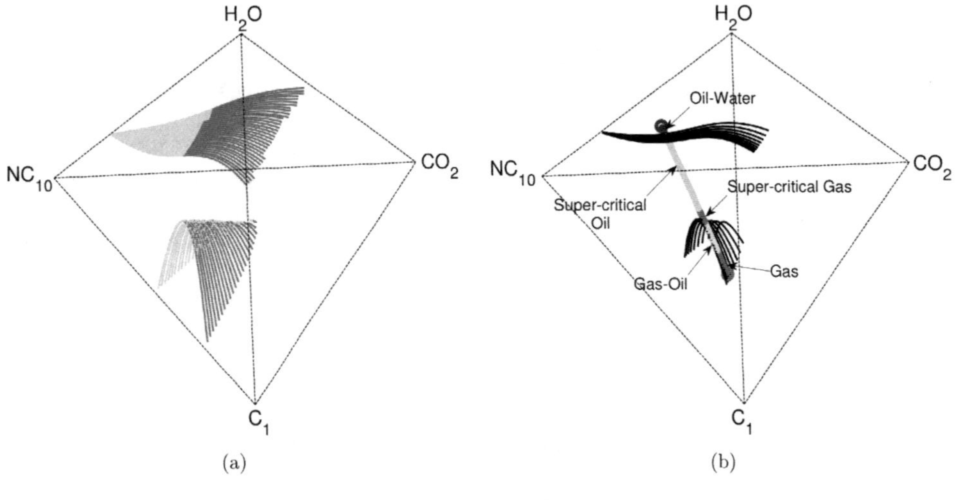

Figure 10 *(a) Full parameterization of* $\{C_1, CO_2, C_{10}, H_2O\}$ *at* $T = 550$ K *and* $p = 137$ *bar, (b) Phase state identification for the dilution line between* $\{C_1(0.05), C_{10}(0.45), CO_2(0.42), H_2O(0.08)\}$ *and* $\{C_1(0.47), C_{10}(0.265), CO_2(0.04), H_2O(0.225)\}$.

pressure and temperature conditions, we proposed a multi-stage negative-flash procedure.

We also described the theory of parameterization continuity for the multiphase Compositional Space Adaptive Tabulation (CSAT) framework. We extended the CSAT framework for thermal-compositional systems that can form an arbitrary number of phases. We presented the details of three-phase compositional space parameterization (e.g., multiphase degeneration, two-phase disappearance) over wide ranges of pressure. We described phase-state identification procedures in degenerate steam systems. Our numerical examples indicate that CSAT is as reliable as conventional procedures for phase-state identification in multiphase systems.

Acknowledgements

Thanks to the SUPRI-B Industrial Affiliates program on reservoir simulation at Stanford University for financial support.

References

Aziz, K., Ramesh, A. and Woo, P. [1987] Fourth spe comparative solution project: Comparison of steam injection simulators. *Journal of Petroleum Technology*, **39**, 1576–1584.

Aziz, K. and Wong, T. [1989] Considerations in the development of multipurpose reservoir simulation models. *proceedings of the 1st and 2nd International Forum on Reservoir Simulation, Alpbach, Austria.*

Bolling, J. [1987] Development and application of a limited-compositional, miscible flood reservoir simulator. *SPE Symposium on Reservoir Simulation, 1-4 February 1987, San Antonio, Texas.*

Brondsted, A. [1983] *An Introduction to Convex Polytopes.* Springer-Verlag New York Inc.

CMG [2009] *WinProp users's guide.*

Entov, V. [1997] Nonlinear waves in physicochemical hydrodynamics of enhanced oil recovery. *Multicomponent flows, Proc. of the International Conference Porous Media: Physics, Models, Simulations, Moscow, Russia.*

Farouq-Ali, S. and Abou-Kassem, J. [1989] Simulation of thermal recovery processes. *proceedings of the 1st and 2nd International Forum on Reservoir Simulation, Alpbach, Austria.*

Franklin M. Orr, J. [2007] *Theory of Gas Injection Processes.* Tie-Line Publications.

Haugen, K., Sun, L. and Firoozabadi, A. [2007] Three-phase equilibrium calculations for compositional simulation, spe 106045. *proceedings SPE Reservoir Simulation Symposium, Woodlands, TX.*

Iranshahr, A., Voskov, D. and Tchelepi, H.A. [2009a] A generalized negative flash procedure for phase equilibrium computations related to carbon dioxide sequestration. *proceedings AIChE Annual Meeting, Nashville, TN.*

Iranshahr, A., Voskov, D. and Tchelepi, H.A. [2009b] Theory and practice of tie-simplex parameterization for compositional reservoir simulation. *proceedings AIChE Annual Meeting, Nashville, TN.*

Iranshahr, A., Voskov, D. and Tchelepi, H.A. [2010] Tie-simplex parameterization for eos-based thermal compositional simulation. *SPE Journal*, **15**, 537–548.

LaForce, T., Jessen, K. and Orr, F. [2008a] Analytical solutions for four-component gas/water/oil displacements: Part ii. example solutions. *Transport in Porous Media*, **72**, 83–96.

LaForce, T., Jessen, K. and Orr, F. [2008b] Four-component gas/water/oil displacements in one dimension: Part i. structure of the conservation law. *Transport in Porous Media*, **71**, 199–216.

Michelsen, M. [1982a] The isothermal flash problem: Part i. stability. **9**, 1–19.

Michelsen, M. [1982b] The isothermal flash problem: Part ii. phase-split calculation. **9**, 21–40.

Michelsen, M. and Mollerup, J. [2004] *Thermodynamic Models: Fundamentals & Computational Aspects.* Tie-Line Publications.

Peng, D. and Robinson, D. [1976] A new two-constant equation of state. *Ind. Eng. Chem. Fundam.*, **15**, 59–64.

Voskov, D.V. and Tchelepi, H.A. [2009a] Compositional space parameterization: Multicontact miscible displacements and extension to multiple phases. *SPE Journal*, **14**, 441–449.

Voskov, D.V. and Tchelepi, H.A. [2009b] Compositional space parameterization: Theory and application for immiscible displacements. *SPE Journal*, **14**, 431–440.

Voskov, D.V. and Tchelepi, H.A. [2009c] Tie-simplex based mathematical framework for thermodynamical equilibrium computation of mixtures with an arbitrary number of phases. *Fluid Phase Equilibria*, **283**(1-2), 1 – 11.

Whitson, C., da Silva, F. and Soreide, I. [1988] Simplified compositional formulation for modified

black-oil simulators. *proceedings of the SPE Annual Technical Conference and Exhibition, SPE 18315.*
Whitson, C. and Michelsen, M. [1989] The negative flash. *Fluid Phase Equilibria*, **53**, 51–71.
Zhu, J. [2003] *Multicomponent Multiphase Flow in Porous Mdedia with Temperature Variation or Adsorption.* PhD Thesis, Stanford University.

Accurate Representation of EoS Related Non-Linearities for Compositional Flow Simulation

Rustem Zaydullin, Denis Voskov and Hamdi Tchelepi, Energy Resources Engineering, Stanford University, Stanford, CA

Introduction

Compositional simulation is necessary for modeling Enhanced Oil Recovery (EOR) processes. To simulate compositional processes we need to solve a coupled system of nonlinear conservation laws for multi-component multiphase flow and transport with a nonlinear system of thermodynamical constraints that represents the phase behavior of the mixture. An Equation of State (EoS) model is usually employed to describe the phase behavior (Aziz and Wong, 1989; Lake, 1989). For a given temperature, pressure, and overall composition, EoS computations define the phase state (stability test) and composition of each phase (flash calculation). Since EoS computations must be performed for each computational cell and for each global Newton iteration, they can consume a significant fraction of the total computational cost.

There are several ways to improve the efficiency of compositional flow simulation. In cases where the phase behavior is a weak function of composition, a model based on constant K-values can be used, in which the K-values depend on pressure only (Bolling, 1987). Another possibility is to limit the number of components and represent the phase behavior using simplified relations (Todd and Longstaff, 1972; Pope and Nelson, 1972).

A general approach for phase-behavior representation of the two-phase compositional systems of practical interest was suggested by Entov et al. (2001). Their method is based on projection of the original compositional problem into tie-line space, where simple polynomial approximations of the tie-line space are used to capture the important features of the two-phase region (Voskov, 2002). Voskov and Tchelepi (2009) used linear interpolation in tie-line space, which improved the accuracy and versatility of the approach. They refer to the approach as CSP (Compositional Space Pramaterization) based flow simulation.

The CSP approach is based on parameterizing the entire compositional space. It is known that gas-injection displacement processes involve a limited number of (key) tie-lines (Franklin M. Orr, 2007). Moreover, it was shown that for purely hyperbolic 1D problems, the projection of the solution into tie-line space is invariant with respect to the hydrodynamic parameters of the problem (Entov, 1997; Voskov and Entov, 2001). Later it was demonstrated that for compressible, multi-dimensional displacements in porous media, projection of the solution in tie-line space is nearly invariant, even though a larger number of tie-lines may be involved due to the numerical dispersion associated with the computed solutions.

In Voskov and Tchelepi (2009b, 2008), the fact that only a few tie-lines are needed was used to improve phase behavior computations for general purpose compositional simulation. The Compositional Space Adaptive Tabulation (CSAT) approach adaptively stores discrete tie-lines during a simulation. The collected tie-lines are used to precondition standard EoS behavior computations. However, since CSAT uses a discrete representation of only part of the compositional space, it cannot replace the EoS computations completely.

Here we present an adaptive version of the full parameterization (CSP) approach. During a simulation, we adaptively compute several supporting tie-lines and use adaptive triangulation of the tie-line space. The governing equations are cast in terms of the tie-line parameters and solved numerically. In order to achieve this, several modifications of the standard compositional framework were made, including the flash calculations. Linear interpolation based on tie-line parameters replaces EoS-based computations and leads to significant gains in computational efficiency. Based on this approach, it is also possible to compute the eigenvectors and eigenvalues that describe the displacement behavior, which can be helpful in constructing improved linear and nonlinear solvers. The applicability of this CSP method is demonstrated for several challenging compositional cases of practical interest.

Conventional Compositional Simulation Approach

The system of equations for a two-phase isothermal compositional problem can be written as:

$$\frac{\partial}{\partial t}\left(\phi\sum_{j=1}^{2} x_{i,j}\rho_j S_j\right) - \text{div}\sum_{j=1}^{2} x_{i,j}\rho_j\left(\mathbf{k}\frac{k_{rj}}{\mu_j}\nabla p\right) + \sum_{j=1}^{2} x_{i,j}\rho_j q_j \quad = \quad 0,\ i = 1,\ldots,n_c. \qquad (1)$$

Here $x_{i,j}$ is the concentration of component i in phase j, S_j is the saturation of phase j, ρ_j – is the phase molar density, $\boldsymbol{k}$ is a permeability tensor, k_{rj} is the relative permeability, μ_j is phase viscosity, and p is pressure.

Instantaneous thermodynamic equilibrium is the usual assumption for compositional flow simulation. Normally, there are two possibilities to deal with the thermodynamic constraints: 1) solve the mass conservation equations and the thermodynamic equilibrium relation simultaneously, 2) couple the thermodynamic relations with the mass conservation equations sequentially. These approaches have their own advantages and drawbacks (Voskov et al. (2009)). Here, we follow the second approach.

The overall composition, $\mathbf{z}$, can be written as

$$z_i \quad = \quad x_{i,o}L + x_{i,g}(1-L), \quad i = 1,\ldots,n_c, \qquad (2)$$

$$L \quad = \quad S_o\rho_o/(S_o\rho_o + S_g\rho_g), \qquad (3)$$

where L is the liquid fraction. The accumulation term of Eq. (1) can be written as

$$\phi\sum_j x_{i,j}\rho_j S_j = \phi\rho z_i, \qquad (4)$$

where $\rho = S_o\rho_o + S_g\rho_g$. We solve system of Eq. (1) using p and z_i, $i = 1, n_c - 1$, as the primary unknowns.

A fully implicit method (FIM) is used to solve the system of equations (1). The Newton-Raphson method is used, in which a Jacobian (linear)system is assembled and solved for each nonlinear iteration.

Since a composition $\mathbf{z}$ at a given p and T in any grid-cell can be in either a single or two-phase state, the phase equilibrium computation involve a two-step procedure: 1) for any single-phase cell, perform a phase stability test (Michelsen, 1982a) to check if the phase state has changed to two phases or to a different single-phase state; 2) for any two-phase cell, perform flash calculations (Michelsen, 1982b) to compute the phase fractions and compositions at equilibrium.

The flash calculations involve solution of the following nonlinear system

$$z_i - Lx_{i,o} - (1-L)x_{i,g} \quad = \quad 0, \quad i = 1,\ldots,n_c, \qquad (5)$$

$$x_{o,i}\zeta_{i,o}(p,T,\mathbf{x}_o) - x_{g,i}\zeta_{i,g}(p,T,\mathbf{x}_g) \quad = \quad 0, \quad i = 1,\ldots,n_c, \qquad (6)$$

$$\sum_{i=1}^{n_c}(x_{i,o} - x_{i,g}) \quad = \quad 0. \qquad (7)$$

Here $\zeta_{i,g}$ and $\zeta_{i,o}$ represent the fugacity coefficients of component i in the vapor and liquid phases, respectively. For a given p, T, and z_i, we need to find $x_{i,o}$, $x_{i,g}$ and L. Successive Substitution Iterations (SSI) are usually combined with Newton's method to solve equations (5)-(7) (Michelsen, 1982b).

Adaptive Compositional Space Parameterizations (ACSP)

In Voskov and Tchelepi (2009b), the solution is constructed based on the standard compositional variable set $\{p, \mathbf{z}\}$. That approach is based on the assumption that equation (6) is linear within any triangle so that interpolation of phase fractions $x_{i,j}$ can be used. This linearity assumption limits triangles to a very small size, and that leads to an excessively large number of triangles to cover the compositional space. In Fig. 1, you can see comparisons between a standard EoS-based solution (saturation) for a 1D displacement of a 4-component system and solutions based on tie-line interpolation using two different triangle sizes. The figure shows that the larger triangles lead to larger errors compared with the reference

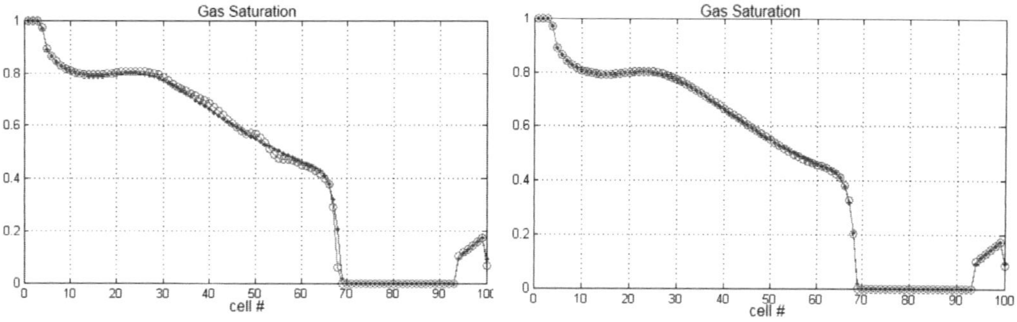

(a) (b)

***Figure 1** Representation of saturation distribution for four component system* $\{C_1, C_{10}, C_4, CO_2\}$ *for large (a) and small (b) size of triangles; here EoS-based solution is shown in solid line, CSP solution in circles.*

EoS-based solution. Moreover, the saturation field obtained with the larger triangles displays small non-monotonic saturation profiles.

Here, we replace the composition vector $\{z_i, i = 1, \ldots, n_c - 1\}$ with a new set, namely, $\{z_j,\ \gamma_i, i = 1, \ldots, n_c - 2\}$. Here z_j is the composition of the primary component and γ is the vector of the tie-line parameter, that is constant for a particular tie-line. In this work γ is the set of coordinates associated with the center point of the tie-line. Given the tie-simplex based γ parameters, pressure, and temperature, our new phase equilibrium procedure returns the liquid and vapor mole fractions, the phase compositions, and all the necessary derivatives.

For each pressure in the interval of interest, we adaptively construct the phase diagram in parameterized space. For each γ value in the specified set, the tie-lines corresponding to each pressure value in the interval of interest are computed and stored in a table. For any given $\{p, \gamma\}$, we compute the phase compositions for the end points of the pressure interval $[p_j, p_{j+1}]$ using interpolation inside the hyper-triangle. Then, using linear interpolation for pressure $p \in [p_j, p_{j+1}]$, we can find $x_{i,o}$ and $x_{i,g}$. We use Eq. (5) for the primary component z_j when computing the liquid fraction, L, and we define a phase state (liquid, vapor, or two- phase) based on it.

Interpolation in parameterized space

Since a tie-line is a continuous function of p, T and composition z (see Iranshahr et al. (2009)), we can use interpolation in order to compute a tie-line inside a triangle for the pressure interval. To represent a tie-line as a function of the parameter γ, we use the natural-neighbor interpolation scheme (Sibson, 1981), where the n-dimensional ($n = n_c - 2$) tie-line space is triangulated using Delaunay tessellation. For the tessellation of scattered data, linear interpolation of any point inside a hyper-triangle is performed, namely,

$$\gamma_i = \sum_{k=1}^{n+1} \lambda_k \gamma^*_{k,i}, \quad i = 1, \ldots, n \tag{8}$$

where λ_k denote the barycenter coordinates of each vertex of a simplex (n + 1 vertices in n-dimensional space), $\gamma^*_{k,i}$ is the coordinates of the kth vertex. We can write the system of linear equations (8) in matrix notation as:

$$\begin{bmatrix} \mathbf{\Gamma} \\ \mathbf{e} \end{bmatrix} \lambda = \begin{bmatrix} \gamma \\ 1 \end{bmatrix}. \tag{9}$$

Here $\mathbf{\Gamma}$ is a matrix of the coordinates of all the vertices $(n+1)$, and $\mathbf{e}$ is a unit vector, which arises from the constraint on the sum of the barycenter coordinates.

By solving the linear system (9), we can determine the barycenter coordinates of the interpolated point. Since the matrix $\mathbf{\Gamma}$ does not depend on the coordinates of the interpolated point γ, the barycenter coordinate is a linear function of the tie-parameter $\boldsymbol{\gamma}$. Therefore, inside each hyper-triangle, derivatives of λ_k, $k = 1, \ldots, n+1$ with respect to γ_i, $i = 1, \ldots, n$ are constants. Knowing the barycenter coordinates of the interpolated point, the mole fractions can be computed as follows:

$$x_{i,p} = \sum_{k=1}^{n+1} \left(x_{i,p}^*\right)_k \lambda_k, \quad i = 1, \ldots, n_c \quad p = \{o, g\}. \tag{10}$$

where $\left(x_{i,p}^*\right)_k$ denote the mole fractions that correspond to γ_k.

Flash calculation in parameterized space

In order to build supporting points in the tie-line space, we need to compute mole fractions (compositions) for a given p, T, and tie-line parameter $\boldsymbol{\gamma}$. For this purpose, we propose a new flash calculation algorithm in the parameterized Γ-space.

The system of nonlinear equations that needs to be solved in Γ-space can be written as

$$\begin{aligned} \gamma_i - g_i(\mathbf{x}_o, \mathbf{x}_g) &= 0, \quad i = 1, \ldots, n_c - 2 && (11) \\ x_{o,i}\zeta_{i,o}(p, T, \mathbf{x}_o) - x_{g,i}\zeta_{i,g}(p, T, \mathbf{x}_g) &= 0, \quad i = 1, \ldots, n_c && (12) \\ \sum_{i=1}^{n_c} x_{i,j} - 1 &= 0, \quad j = o, g. && (13) \end{aligned}$$

The definition of g_i depends on how one defines γ_i. In this paper, we use $g_i = (x_{i,o} + x_{i,g})/2$, $i = 1, \ldots, n_c - 2$. For a given p, T and $\boldsymbol{\gamma}$, we need to compute the corresponding $\mathbf{x}_o$ and $\mathbf{x}_g$.

Similar to the standard flash, we can start from phase relations provided by Wilson's equation

$$K_i = \frac{x_{i,g}}{x_{i,o}} = \frac{p_{c,i}}{p} exp\left(5.37(1 + w_i)\left(1 - \frac{T_c}{T}\right)\right), \quad \forall i = 1, \ldots, n_c. \tag{14}$$

Here p_c and T_c are the critical pressure and temperature, w_i is the acentric factor. With the K-values and $\boldsymbol{\gamma}$ parameters known, we can directly calculate $2n_c - 4$ phase fractions:

$$\begin{aligned} x_{i,o} &= \frac{2\gamma_i}{K_i + 1} \quad \forall i = 1, \ldots, n_c - 2, && (15) \\ x_{i,g} &= K_i x_{i_o} \quad \forall i = 1, \ldots, n_c - 2. && (16) \end{aligned}$$

Next, we need to compute the remaining mole fractions i.e. $x_{i,j}$, $i = \{n_{c-1}, n_c\}$, $j = \{o, g\}$.

From the definition of the K-values we can write:

$$\left(K_{n_c-1} - K_{n_c}\right) x_{n_c,o} = K_{n_c-1} x_{n_c,o} - K_{n_c} x_{n_c,o} = K_{n_c-1} x_{n_c,o} - x_{n_c,g} \tag{17}$$

which can be rearranged as:

$$\begin{aligned} K_{n_c-1} x_{n_c,o} - x_{n_c,g} &= \left(K_{n_c-1} x_{n_c,o} + x_{n_c-1,g}\right) - x_{n_c-1,g} - x_{n_c,g} \\ &= K_{n_c-1}\left(x_{n_c,o} + x_{n_c-1,g}\right) - x_{n_c-1,g} - x_{n_c,g}. && (18) \end{aligned}$$

Taking in account equation (13), the last relation can be written as:

$$K_{n_c-1}x_{n_c,o} - x_{n_c,g} = K_{n_c-1}\left(1 - \sum_{i=1}^{n_c-2} x_{i,o}\right) + \sum_{i=1}^{n_c-2} x_{i,g} - 1 \tag{19}$$

Combining (17) and (19), we can easily obtain

$$x_{n_c,o} = \frac{\sum_{i=1}^{n_c-2} x_{i,g} + K_{n_c-1}\left(1 - \sum_{i=1}^{n_c-2} x_{i,o}\right) - 1}{K_{n_c-1} - K_{n_c}}, \tag{20}$$

(21)

The other compositions can be found from (13) and the K-value definition. Next we substitute the $x_{i,j}$ into the fugacities in Eq.(12) and update the K-values using

$$K_i^{(m)} = \frac{\phi_i^{m-1}(x_{i,o})}{\phi_i^{m-1}(x_{i,g})}, \tag{22}$$

where m is the SSI iteration number and ϕ_i is the fugacity coefficient of component i. The SSI iterations are repeated until convergence. SSI can be accelerated by switching to the Newton method at some threshold value of the residual. Such an approach requires linearization of system (11)-(13) for each Newton iteration and updating of nonlinear unknowns $\mathbf{x}_o$ and $\mathbf{x}_g$ based on the solution of the linear system until convergence is reached.

Simulation results

We compare the behaviors of the ACSP-based formulation described here with the conventional EoS-based approach using four- and five-component systems. A brief description of each system is given below.

- System 1: immiscible gas injection for a four-component system. Initial oil is made up of $\{C_1(20\%),\ CO_2(1\%),\ C_4(29\%),\ C_{10}(50\%)\}$ at an initial pressure of 80 bars (above the bubble point). The injection gas is a mixture of two components $\{CO_2(70\%),\ C_1(30\%)\}$. The injection pressure is 100 bars, which is above the dew point of the injected mixture; the production pressure is 40 bars, which is below the bubble-point pressure of the reservoir fluid. That makes the displacement process highly nonlinear because of the multiple changes in the phase state.
- System 2: immiscible gas injection for a five-component system. The initial oil consists of $\{C_1(1\%),\ CO_2(1\%),\ C_4(20\%),\ C_8(28\%),\ C_{15}(50\%)\}$ at an initial pressure of 80 bars (above the bubble point). The injection gas is a mixture of five components $\{CO_2(60\%),\ C_1(30\%),\ C_4(2\%),\ C_8(3\%),\ C_{15}(5\%)\}$. The injection pressure is 100 bars, and the production pressure is 40 bars.

Two different reservoir models are used. Namely,

- Reservoir 1: 1D homogenous reservoir with 500 cells (gridblock size is 10x100x1 meters); permeability is 1D; porosity is 0.2; relative permeabilities are Corey functions with both exponents equal to two. One injection, one production well at the ends of the reservoir. The reservoir temperature is 373 K.
- Reservoir 2: The distribution of permeabilities and porosity is taken from upper layer of the SPE 10 (Christie and Blunt, 2001). The size of the model is [60x220x1] cells (gridblock size is 20x10x70 meters). Relative permeability is Corey with both exponents equal to two. One injection and one production well are placed at opposite corners of the reservoir. The reservoir temperature is 373 K.

Accuracy check

A limited number of hyper-triangles is used to represent the Γ-space. The spacing between the supported points (tie-lines) can be varied in order to achieve the necessary accuracy. Here we compare the results of EoS-based and ACSP-based simulation with hyper-triangles of different size. To define the size of a hyper-triangle, we use the following criteria

$$\|\gamma_i^* - \gamma_j^*\| < \varepsilon, \tag{23}$$

where γ_i^* is the coordinate of the ith supporting tie-line in Γ-space.

Fig. 2(a) shows the solution of Reservoir-1 with System-1 for a large triangle ($\varepsilon = 0.5$). The ACSP-

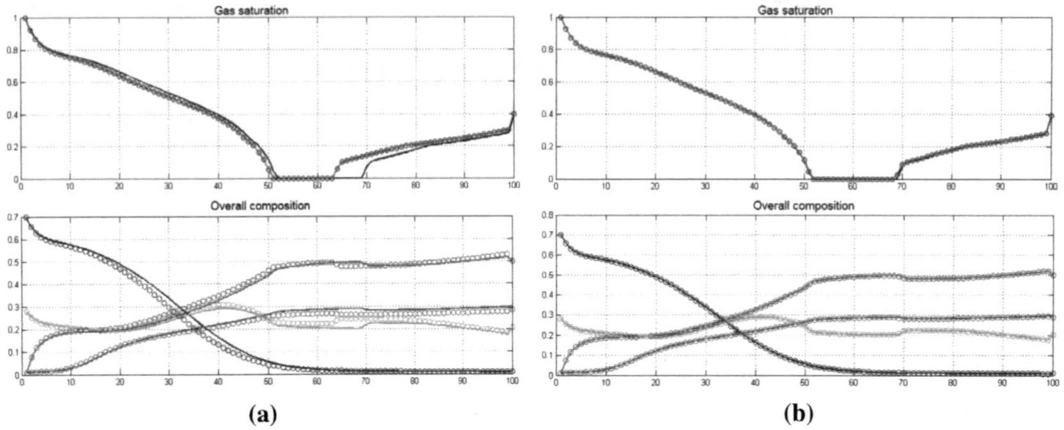

Figure 2 *Representation of saturation and overall composition distributions for four component system* $\{C_1, C_{10}, C_4, CO_2\}$ *with different* ε*. EoS-based simulation is solid lines, ACSP-based simulation - circles.*

based simulation results are shown in circles, and the standard EoS-based results are shown in solid lines. As indicated by the figure, even for very large triangles, the ACSP approach is in good agreement with the EoS-based results excluding a small area close to the displacement front. Simple refinement of the tessellation ($\varepsilon = 0.1$) removes these small errors completely (see Fig. 2(b). The adaptive tessellation results for both of these cases in the Γ-space are shown in Fig. 3(a,b). These figures indicate that we can control the error using tessellation refinement. Note that even when the errors are relatively large, there is no inconsistency in the representation of the physics as opposed to the results in Fig. 2.

2D simulation results of System 1 for the Reservoir 2 model are shown in Fig. 4. The figure shows the saturation distribution, obtained with ACSP-based simulation and difference in saturation after 2000 days. The agreement is excellent except for a few cells around the displacement front. The tesselated Γ-space for this problem is shown in Figure 5. Comparisons between EoS- and ACSP-based simulations for the Reservoir 2 model with System 2 (five component) are shown in Fig. 6. In this case, the dimension of the parameterized space is $n_c - n_p = 3$; hence, the tessellation of Γ-space is represented by tetrahedrons.

Fig. 7 shows the gas saturation (ACSP-based simulation) and difference in saturation after 3000 days. The agreement with EoS-based simulation is excellent.

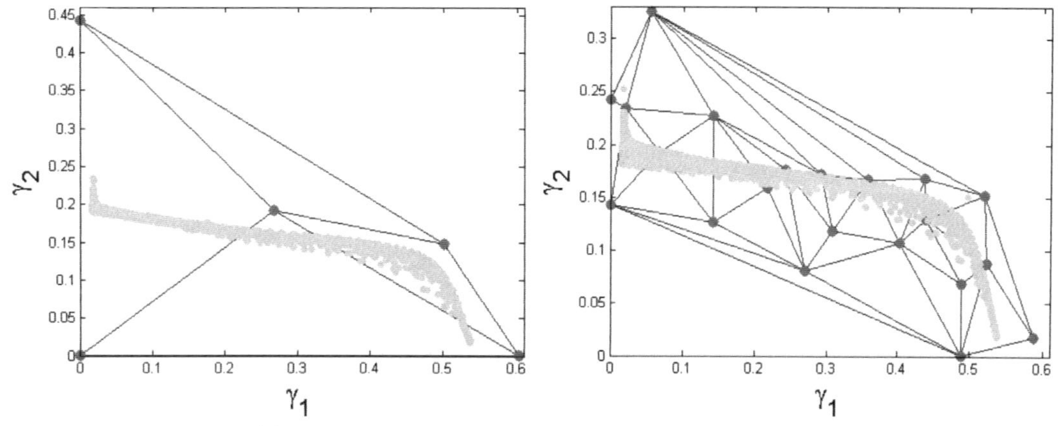

(a) **(b)**

Figure 3 *Triangulated γ space for different tessellations: (a) large triangles and (b) small triangles. Red dots are projections of supported tie-lines. The green dots are projection of the solution.*

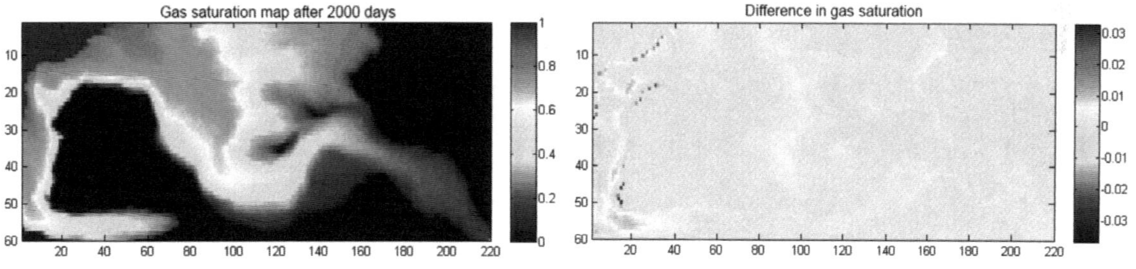

Figure 4 *Saturation and the difference in saturation for EoS-based and ACSP-based simulation after 2000 days for System 1.*

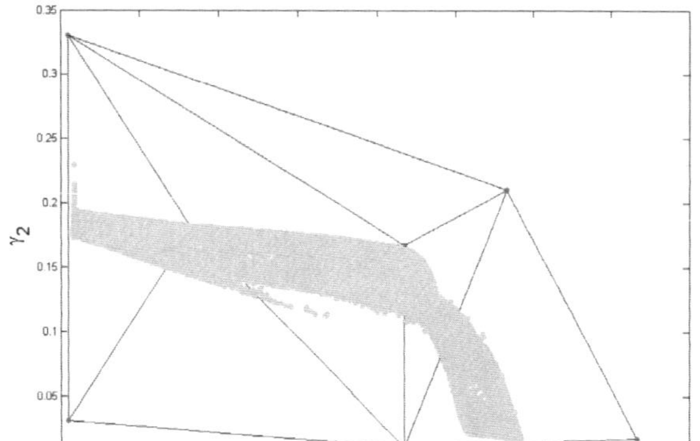

Figure 5 *Triangulated γ space – four-component, two-phase system. Red dots are projections of the supported tie-lines. Green dots are projection of the solution.*

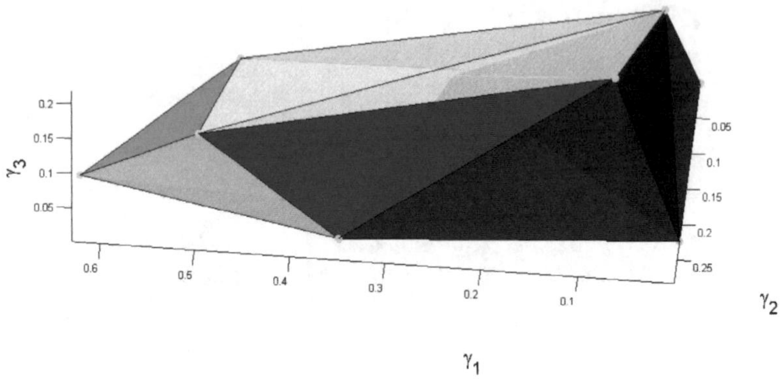

Figure 6 *Triangulated γ space – five-component, two-phase system*

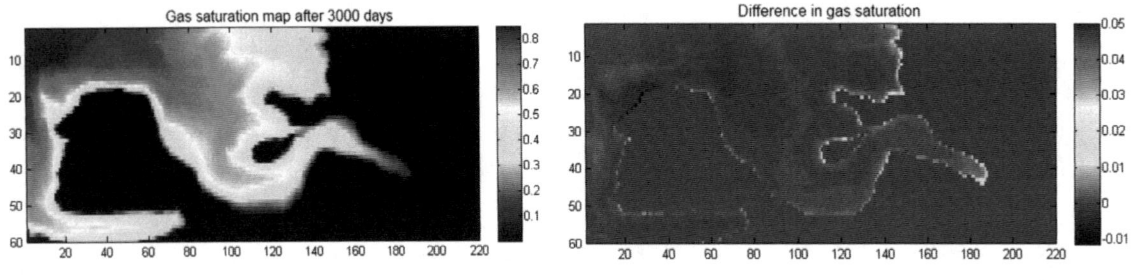

Figure 7 *Saturation and the difference in saturation for EoS-based and ACSP-based simulation after 3000 days for System 2.*

Near-critical representation

The interpolation algorithm described above works well, as long as the solution path is below the hyper-surface of critical tie-lines (Voskov and Tchelepi, 2008). However, in compositional problems of practical interest, part of the solution path may lie in the super-critical region. Here, we discuss how to extend our adaptive parameterization approach to the super-critical region.

The main idea is based on the fact that for a fixed temperature, T, a unique critical tie-line passes through each point in the compositional space (Voskov and Tchelepi, 2009a). Each critical tie-line corresponds to a Minimal Critical Pressure (MCP). As the result, the entire super-critical region is parameterized by critical tie-lines with different MCPs.

Interpolation in the super critical-region can be performed as follows:

- For each supported point, calculate the MCP.
- Find the hyper-triangle that contains the current γ.
- If the current pressure is above the MCP of any vertex, use the critical tie-line defined at these vertexes for the interpolation.
- Otherwise, use the standard tie-line interpolation described above.

This algorithm was tested for several compositional systems of practical interest. In Fig. 8,a you can see an example of interpolation in a large triangle using System 1. We changed the pressure continuously and computed the interpolation errors using a Γ-based flash. Fig. 8,b shows the interpolation error and tie-line length versus pressure. As the pressure increases, the length of the tie-line goes to zero smoothly. The EOS-based tie-line and the interpolated tie-line are in good agreement, even though the size of the triangle is very large. The relative (l_∞) error in the computed phase fraction generally increases with pressure. However, there are no jumps in the pressure values that correspond to the MCP of the supporting point. This fact indicates that the interpolation in the presence of critical tie-lines is reasonably accurate.

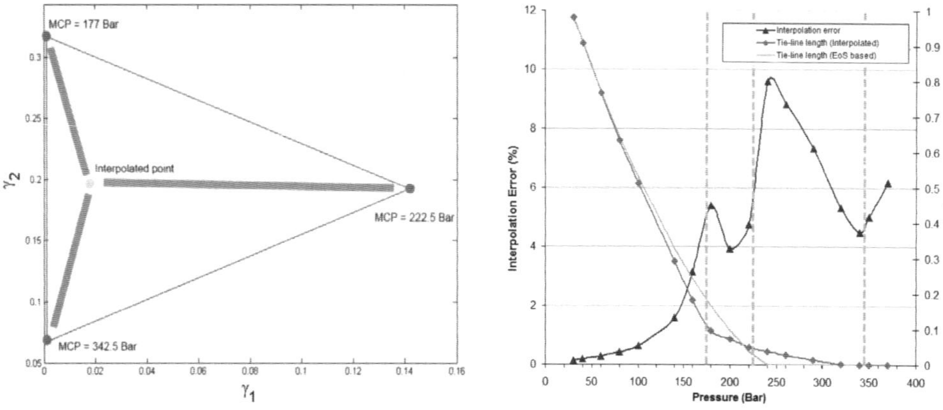

(a) (b)

Figure 8 *Adaptive interpolation for System 1: (a) triangle in the* Γ*-space with corresponding MCP and (b) the interpolation error and tie-line length vs pressure.*

Analysis of the system of conservation law in the hyperbolic limit

It was shown in Entov (1997); Voskov and Entov (2001) that working with the Γ-space can split the full compositional problem into thermodynamic and hydrodynamic sub-problems. The Method of Characteristic theory (Franklin M. Orr, 2007; Entov, 1997) for compositional displacements in one-dimensional systems suggests that solution of the thermodynamic problem is invariant with respect to changes in the properties of the hydrodynamic problem. In practice, it is often the case that even in multi-dimensional problems, projection of the solution in Γ-space follow nearly the same path obtained by the 1D solution (Entov et al., 2001; Voskov and Tchelepi, 2009b).

Based on this fact, we can use characteristics of thermodynamical 1D problems to analyze the behavior of compositional displacements of practical interest. Using Eq. (5) for the primary component j we can find L, and introduce it to the equation for the rest of the components, which leads to

$$z_i = A_i z_j + B_i, \quad i = 1, \ldots, n_c - 2, \tag{24}$$

where

$$A_i = \frac{x_{i,o} - x_{i,g}}{x_{j,o} - x_{j,g}}, \tag{25}$$

$$B_i = x_{i,g} - x_{j,g}\frac{x_{i,o} - x_{i,g}}{x_{j,o} - y_{j,g}}. \tag{26}$$

Note that A_i and B_i depend on the γ-parameter only.

It was shown in Entov (1997); Voskov and Entov (2001) that thermodynamical characteristics of compositional problems are controlled by the matrix $\mathbf{R} = \mathbf{B}'(\mathbf{A}')^{-1}$, where

$$\mathbf{A}' = \left[\frac{\partial A_i}{\partial \gamma_j}\right], \quad \mathbf{B}' = \left[\frac{\partial B_i}{\partial \gamma_j}\right]. \tag{27}$$

Our approach allows for separating the pressure and γ variables during simulation, i.e. any function $F(p, \gamma)$ can be represented as $F(p, \gamma) = \Phi(p)\Psi(\gamma)$. In order to analyze the system of conservation laws in the hyperbolic limit, we need to compute the eigenvalues of $\boldsymbol{R}$ for each entry in the pressure table. If one uses the natural neighbor interpolation (Sibson, 1981) for γ points inside a hyper-triangle, the derivatives of the barycenter coordinates with respect to the tie-line parameter γ_i, $\{i = 1, \ldots, n\}$ are constants and do not depend on γ. Thus, for each hyper-triangle, derivatives of λ_k, $\{k = 1, \ldots, n+1\}$ can be easily computed using Eq. (9) for each pressure value. Next we can substitute these derivatives into Eq. (10) and compute $\mathbf{R}$.

Fig. 9a shows how the eigenvalues for the system shown in Fig. 8 (System 1) change with pressure. In Fig. 9,b the same dependency of eigenvalues on pressure is shown for one of the compositions of System 2. As you can see, all the eigenvalues change continuously with pressure, and this behavior can be a basis for an MOC-based analysis of real systems with large variations in pressure.

Concluding remarks

We presented an adaptive compositional space parameterization (ACSP) method for numerical compositional flow simulation. The methodology builds on the previous work on tie-line (more generally, tie-simplex) based parameterization (Entov et al., 2001; Voskov and Tchelepi, 2009b, 2008). Here, the phase equilibrium relations in the compositional space are recast in terms of tie-line parameters. The approach makes use of the fact that a discrete representation of the tie-simplex space is adequate, and that the solution path associated with compositional displacements are nearly invariant with respect to changes in the hydrodynamic properties.

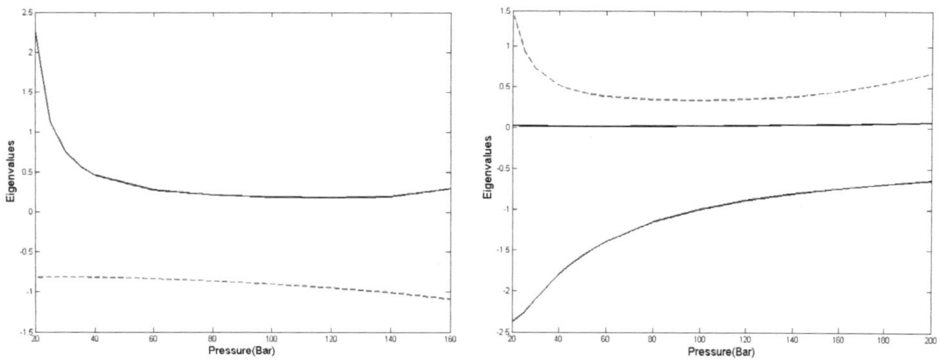

(a) **(b)**

Figure 9 *Eigenvalues vs pressure for: (a) System-1, γ corresponds to $\{CO_2, C_4\}$ and (b) System-2, γ corresponds to $\{CO_2, C_4, C_8,\}$ systems.*

We compared the ACSP- and EOS-based methods. The agreement between the two approaches is quite good, even when a small number of tie-simplexes is used to discretize the compositional space. The ACSP approach is consistent with the conservation laws and appears to be quite stable; as a result, the computations are free of oscillations and the accuracy can be controlled to arbitrary levels by refining the tie-simplex discretization. In terms of computational efficiency, for the problems we tested, the ACSP method reduces the cost of the thermodynamic computations quite substantially compared with the standard EOS based approach. Obviously, further testing for problems with highly detailed reservoir models, larger numbers of components, and wide ranges of operating conditions are needed. Moreover, analysis of solution accuracy versus tie-simplex discretization spacing is necessary and is the subject of ongoing research. While further development and testing is needed, the method can be seen as a tie-simplex Γ-based formulation for a general-purpose compositional simulation.

Finally, we note that the the ACSP based formalism makes it possible to split the compositional problem into thermodynamic and transport parts, and this allows for improved analysis of the nonlinear behaviors associated with multi-component, multiphase displacements.

Acknowledgments

Thanks to the SUPRI-B Industrial Affiliates Reservoir Simulation program at Stanford University for financial support.

References

Aziz, K. and Wong, T. [1989] Considerations in the development of multipurpose reservoir simulation models. *proceedings of the 1st and 2nd International Forum on Reservoir Simulation, Alpbach, Austria.*
Bolling, J. [1987] Development and application of a limited-compositional, miscible flood reservoir simulator. *SPE Symposium on Reservoir Simulation, 1-4 February 1987, San Antonio, Texas.*
Christie, M. and Blunt, M. [2001] Tenth spe comparative solution project: A comparison of upscaling techniques. *SPE Reservoir Evaluation and Engineering SPE 72469.*
Entov, V. [1997] Nonlinear waves in physicochemical hydrodynamics of enhanced oil recovery. *Multicomponent flows, Proc. of the International Conference Porous Media: Physics, Models, Simulations, Moscow, Russia.*
Entov, V., Turetskaya, F. and Voskov, D. [2001] On approximation of phase equilibria of multicomponenthydrocarbon mixtures and prediction of oil displacement by gas injection. *8th European conference on the mathematics of oil recovery, Freiburg, Germany*, 4–27.
Franklin M. Orr, J. [2007] *Theory of Gas Injection Processes.* Tie-Line Publications.
Iranshahr, A., Voskov, D. and Tchelepi, H.A. [2009] Theory and practice of tie-simplex parameterization for compositional reservoir simulation. *proceedings AIChE Annual Meeting, Nashville, TN.*
Lake, L. [1989] *Enhanced Oil Recovery.* Prentice Hall, Englewood Cliffs.
Michelsen, M. [1982a] The isothermal flash problem: Part i. stability. **9**, 1–19.
Michelsen, M. [1982b] The isothermal flash problem: Part ii. phase-split calculation. **9**, 21–40.
Pope, G. and Nelson, R. [1972] A chemical flooding compositional simulator. *SPE Journal*, **18**, 339–354.

Sibson, R. [1981] A brief description of natural neighbour interpolation. *Interpreting Multivariate Data*, 21–36.
Todd, M. and Longstaff, W. [1972] The development, testing, and application of a numerical simulator for predicting miscible flood performance. *Journal of Petroleum Technology*, **24**, 874–882.
Voskov, D.V. and Tchelepi, H.A. [2008] Compositional space parameterization for miscible displacement simulation. *Transp Porous Med*, **75**, 111–128.
Voskov, D.V. and Tchelepi, H.A. [2009a] Compositional space parameterization: Multicontact miscible displacements and extension to multiple phases. *SPE Journal*, **14**, 441–449.
Voskov, D.V. and Tchelepi, H.A. [2009b] Compositional space parameterization: Theory and application for immiscible displacements. *SPE Journal*, **14**, 431–440.
Voskov, D.V., Younis, R. and Tchelepi, H.A. [2009] Comparison of nonlinear formulations for isothermal compositional flow simulation. *proceedings of SPE Reservoir Simulation Symposium, SPE 118996.*
Voskov, D. [2002] *Two-Phase Multicomponent Flows through Porous Media*. PhD Thesis, Gubkin Russian State Oil and Gas University.
Voskov, D. and Entov, V. [2001] Problem of oil displacement by gas mixtures. *Fluid Dynamics*, **36**, 269–278.

(302b) Rate Dependent Oil Recovery in Fractured Basement Reservoirs

Justin Wriedt[1], Milind D. Deo[1] and Thomas Doe[2], (1)Chemical Engineering, University of Utah, Salt Lake City, UT, (2)Golder and Associates, Redmond, WA

In fractured basement reservoirs, where oil is primarily stored in fractures, only large aperture fractures are visible through seismic reflection surveys. Considerable reserves may exist in sub-seismic fractures. A discrete fracture network (DFN) simulation study is undertaken to examine the effect of rate on oil and water displacements. The displacement of oil from fractures in waterflooding is controlled by gravity, viscous, and inertial forces. Multiphase displacement in fractures is controlled by fracture dip and continuity of flow paths in the direction of buoyant flow. Higher capacity for flow in seismic fractures can lead to high water production before sub-seismic oil is recovered. High water cut in oil-water basement reservoirs may make oil production economically unrealistic. The effects of threshold capillary pressure, wettability, relative permeabilities, and storage ratios on rate dependent recovery were investigated in this paper. Results indicate that a high rate of production is ideal for optimal oil recovery when time is considered. Comparing pore volumes injected, equivalent recovery is achieved with fewer pore volumes for lower water injection rates. Delayed water production occurs at low rate; however, overall recovery does not increase. Rate dependent relative permeabilities would make low recovery rates more beneficial, however the physical basis for rate-dependent relative permeabilities in oil-water systems is not known at this time. The viscous forces are dominant in high rate simulations. A Darcy flow assumption in the DFN simulations neglects inertial forces which may become important if a Reynolds number > 1 is achieved at higher displacement rates. Results from this study may be used to develop operational guidelines for waterflooding in fractured basement reservoirs. History matching using a DFN approach will improve the understanding of flow interaction in seismic and sub-seismic fractures.

See more of this Session: Simulation of Complex Subsurface Processes II
See more of this Group/Topical: Topical D: Chemical Engineering in Oil and Gas Production and Other Complex Subsurface Processes

Methodology for the selection of small scale LNG technologies for remote gas production

L. Castillo, C.A. Dorao*
Department of Energy and Process Engineering, Norwegian University of Science and Technology, N-7491 Trondheim, Norway
**email: luis.castillo@ntnu.no*

Abstract

Natural gas fields accessible via proven technology are limited demanding that smaller and less accessible gas field should be considered for commercial exploitation. The Liquefied Natural Gas (LNG) could be a solution for this kind of gas reserves, but in the case of remote gas fields with a low productivity, the large scale LNG technologies are neither practical nor economical. The small and middle offshore scale technologies demand different requirements than the ones related to conventional on-land LNG plants. While a thermodynamic efficient process is considered a most important parameter for the selection of onshore LNG plants, other factors need to be considered in the process selection for offshore LNG plants such as safety, compactness, low equipment count, modularization, easy maintenance, reliability, simple operation, low influence of the motion of the vessel, etc. Offshore LNG plants require overcoming many challenges due to the marine environment (winds, currents and waves). The motion of the vessel can affect the process efficiency. In addition, the limited space demands highly compact and light processes. In general, the main characteristics of the Mixed Refrigerant and Expander technologies made them suitable for developing an offshore LNG project, but which is the most convenient choice might depend on the particular scenario.
The most important aspect in this work is that the costs related to the offshore installation are included as a component of the capital cost of LNG project. The plot area required for each technology plays a significant role in the final comparison and this parameter should not be neglected in the economical analysis of ~~future~~ offshore LNG plants.

Introduction

The Liquefied Natural Gas (LNG) is natural gas in liquid state at atmospheric pressure and temperatures around -161 ° C. The volume is reduced a factor 600 times compared to the standard conditions, which allows large volumes of LNG to be transported by sea in refrigerated ships. In the last decades, the world-wide LNG market has growth significantly allowing business opportunities associated with short-term contracts, i.e. spot market, constituting LNG in an important commodity [1].

LNG plants have been designed for high capacities, base-load plant, exceeding 150 million cubic feet per day (Mcfd) of natural gas. The designs of large capacity LNG plants are focused mainly on the exploitation of vast natural gas fields and towards the construction of major facilities taking advantage of economy of scale. However natural gas fields, accessible via proven technology, are limited demanding that smaller and less profitable gas fields should be considered for commercial exploitation. In particular, half gas reserves in the world (6000 trillion cubic feet) are considered stranded, i.e. very little volume of gas, or far away from consumption centers to be economically viable an exploitation project [2].

In the case of remote gas fields with a low productivity, large scale technologies are neither practical nor economical. Different technologies for liquefying of natural gas have been developed for remote gas fields, being the most used the technologies of two and three cycles of cooling, with cascade or propane pre-cooling plus mixed refrigerants schemes. The small to medium scale liquefaction technologies can be divided in two main groups: mixed refrigerants (MR) and expanders based (E) technologies. The first group allows a heat exchanger between the gas and cooler fluid defined as refrigerant or coolant. The second group forces the gas to develop a reverse thermodynamic cycle in one or several steps. These technologies can be differentiated by the refrigerants, heat exchangers or the number of cycles of cooling involved.

The small and middle offshore scale technologies demand different requirements than the ones related to conventional on-land LNG plants. While thermodynamic efficiency of the process is considered the most important parameter for the selection of onshore LNG plants, others factors need to be considered in the process selection for offshore LNG plants. First of all, offshore LNG plants require overcoming many challenges due to the marine environment (winds, currents and waves). The motion of the vessel can affect the process efficiency. In addition, the limited space demands highly compact and light processes. Therefore, selecting an offshore technology should consider several parameters such as safety, compactness, low equipment count, modularization, easy maintenance, reliability, simple operation, low influence of the motion of the vessel, etc. In general, the main characteristics of the Mixed Refrigerant and Expander technologies made them suitable for developing an offshore LNG project, but which is the most convenient choice will depend on the particular scenario [3]. The goal of this work is to discuss a methodology for guiding the selection of small and middle LNG technology based on an economic model where the cost of the plot area is included in the analysis. The proposed methodology can be extended to include other parameters and thus to have the possibility of investigating the effects of several variables in the selection of a particular LNG technology.

Methodology and results

The present work discusses criterion for selecting a small scale LNG technology based on a ranking matrix and an economical evaluation. The first step in this work was to identify a possible scenario for LNG production. The scenario determines the range of the applicability of the technology. The second step was to identify and classify the available technologies. The last step was to perform an economical analysis in order to identify the main parameters that should be considerate for selecting a particular technology. The Figure 1 shows the steps for developing an approach to decision making in LNG projects.

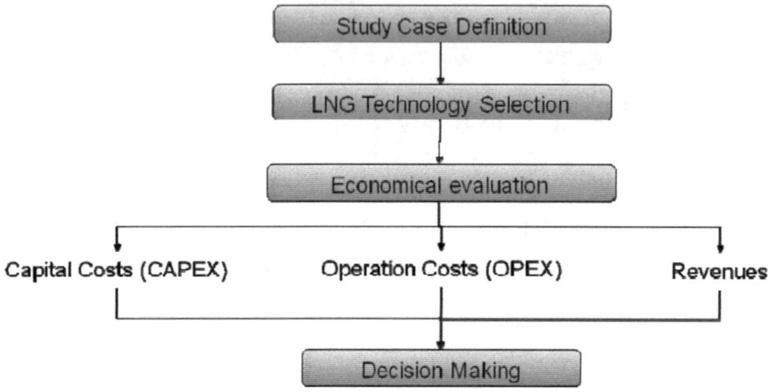

Figure 1: Schematic approach for decision making in LNG projects

Study Case Definition

A generic case is considered where LNG will be located in Venezuela and the main target will be to supply the energy demand of the Caribbean Market, once domestic demand is satisfied. The production capacity might be around 0,5 Mtpa based on typical production rate of the peak shave plants [4].

LNG Technology Selection

Several technologies are available for small scale LNG plants; more details about these technologies could be found in [1,2,5,6]. A first pre-selection of LNG technology could be based on technical analysis as suggested by Castillo et al. [7].

A decision matrix could be constructed in order to evaluate the technologies. The selection of the main parameters could be based on its direct impact in the minimization of investment costs and maximizing efficiency of LNG production. The quality of the information is an important factor that must be took into account in order to obtain reliable results. The weighting stage involves two steps: the construction of the matrix and the weighting to each parameters and sub-parameters. The weights consider the priorities established at the beginning of each particular LNG project. After assigning weights to each parameter and sub-parameters, and defining the appropriate rating, the technologies can be ranked according to the resulting score from the weighted sum of the different parameters measured at the decision matrix. This technique is useful for quick viewing of the strengths and weaknesses of each technology, while allowing comparisons between the options assessed [7].

For simplicity, two groups of technologies are considered: (1) Mixed Refrigerants (MR) with an energy efficiency of 350kWh/tonne, and (2) Expander based with an energy efficiency of 700kWh/tonne. The energy efficiency of the technology can be used for determining the CO_2 production and energy cost.

Economical Analysis

The cost estimates for a LNG project should include many factors, where the capital cost is not the only driver in project economics and must be balanced with the expenditure costs like fuel costs, maintenance costs, and plant reliability. Safety and environmental issues are also important parameters.

The economical analysis for selecting small LNG plant can be related to $\text{CashFlow} = \text{Revenues} + \text{CAPEX} + \text{OPEX}$; where the Cash Flow is the cash into or out of a project during a finite period of time; CAPEX is the Capital Expenditures; OPEX is the Annual Operating Costs and Revenues represents the income due to sale of the products.

Capital cost (CAPEX)

The estimate in the capital cost has been studied during the last decades. Some generic cost estimates are based on historical cost data or simple published estimating formulas [8 , 9]. According to Jamieson et al. [8] investments in the whole LNG value chain, from gas supply through liquefaction, shipping and importation can easily be 4000 - 5000 MUSD for a LNG production of 6 Mtpa. But according to Al-Saadoonat al. [10] , the CAPEX of LNG projects is estimated to be between 200 and 400 USD/tone/year, therefore, LNG plant of 6 Mtpa can cost between 1200 and 2400 MUSD, depending on several factors, including type of technology used, greenfield or expansion project, shipping distance, regasification facility, economies scale, learning curve improvement, and local infrastructure availability.

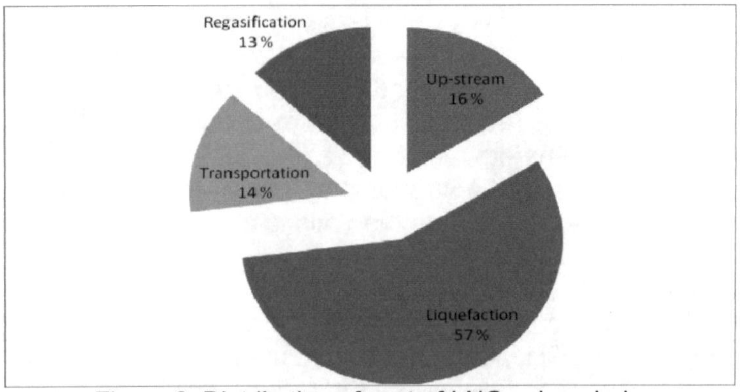

Figure 2. Distribution of cost of LNG value chain

The Figure 2, shows the distribution of cost of LNG value chain. The major cost is related to liquiefaction plant above 50% [9,11]. The liquefaction CAPEX depends on some factors such as technology used, facilities, i.e. gas treatment and storages. In this work, the main focus is on the comparison of the liquefaction technologies and thus only this cost is considered.

According to Barclay et al. [2], the offshore LNG plants has different requirements to the on-load plants, while thermodynamic efficiency is attributed to be the most important criteria for selecting technologies onshore, others parameters can be more important for offshore projects. In particular, the area required impacts significantly in the cost of the offshore installations, and this area is related to the technology.

Figure 3 shows the relation between plot area, power consumption, train capacity and complexity for LNG plant. The Expander technologies require higher power consumption than MR technologies; but they demanded lower plot area and complexity than MR technologies. Moreover, the expander technologies require higher power consumption than MR technologies, but according to Jamieson et al. [8], the liquefaction OPEX can only represent about 5% of the CAPEX. The application of MR technologies for offshore developments requires a reduction of the plot area while maintaining the efficiency of the process, while expander technologies need to improve the thermodynamic efficiency while maintaining a reduced plot area.

Due to the fact the complexity and plot area increase the CAPEX; and CAPEX is an intensive investment, both parameters should be considered into an economical evaluation. The thermodynamic efficiency and plot area play major roles into the economical model because there is a commitment between them on offshore applications.

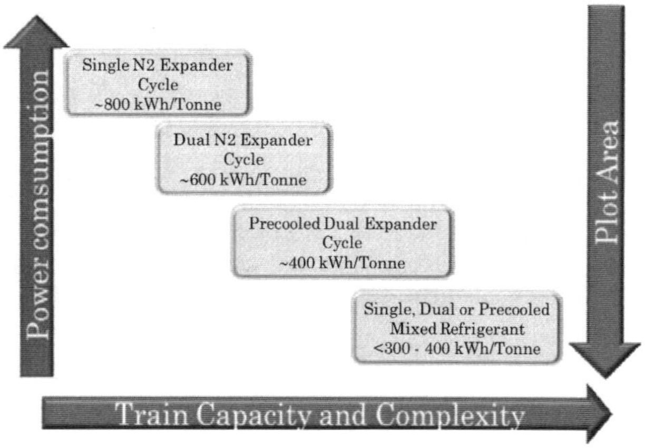

Figure 3. Relation between plot area, power consumption, train capacity and complexity

The CAPEX is assumed to have two main contributions, i.e. $CAPEX = CAPEX_B + CAPEX_A$. $CAPEX_B$ is the CAPEX related to process and can be determined following the work of Garcia et al. [12]. $CAPEX_A$ is the cost related to the cost for installing to process offshore. The estimation of this cost is quite difficult but it can be related to the area required for installing the process at offshore, i.e. $CAPEX_A = Area(P) \times Cost(Area)$, where A(P) is a function of the plot area and depends on the production, and Cost(A) is cost per area unit.

Hence, a simple approach consists in considering that $Area(P) = m \times P + b$; where, m and b depends on the particular technology. Based on this model, a base case was defined, where the plot area as a function of the production is the same for each group of technology; and others three options were defined where the plot area, for MR technology, as a function of the production are higher than expander technology. Finally, the plot area cost is assumed as a fix value and relative to the 25% of $CAPEX_B$, i.e. Cost(A) = 1500 MUSD$/m^2.

Operation Cost (OPEX)

The OPEX of an industrial plant can be resumed in: personnel, energy, maintenance and other costs. In this work, the OPEX can be defined based on make-up of refrigerants; energy costs, CO_2 production, and others fixed cost.

The cost related to CO_2 production is considered as an operational cost. According to world regulations about CO_2 emission, the company should consider the production of CO_2 in their operations. The reference cost under the current European Union carbon dioxide emission quota regimen is about 15 USD/tonne CO_2 [13,14]. Even thought, in the LNG value chain, CO_2 is produced in several stages, in this work only CO_2 related to energy consumption for using small turbine (offshore application) is considered, and it is about 0,7 $KgCO_2$/kWh [15,16].

Hence, the main objective is to compare two groups of the LNG technologies, the others fixed costs, which are the same for both technologies groups, will be assumed as the costs for an LNG plant of 1 Mtpa (about 15 MMUSD/year [12]).

Revenues

In this work, neither NGL or condensates sales are considered. The LNG prices are referenced at Henry Hub Index because the LNG plant is assumed to be located in South America, and thus an average price of 5 USD/MMBtu [3, 17] is considered.

Economical analysis

The economical analysis is based on the following scenario (input data)

Table 1: Input data for the studies cases

Project	
Localization Plant	Offshore
Remote reserve (TCF)	0,1 to 0,5
Production (MTPA)	0,5
Timelife project (year)	25
Discount rate (%)	15
Technology	
MR energy consumption (kWh/ton)	350
Expander energy consumption (kWh/ton)	700
CO2 production ($KgCO_2$/kWh)	0,7
Market	
$CAPEX_B$ (USD/TPA)	600
Electricity cost (USD/kWh)	0,04
LNG price (USD/MMBTU)	5
CO2 price (USD/tonne of CO_2)	15
Ethane price (USD/tonne)	1400
Propane price (USD/tonne)	620

Internal Return Rate (IRR)

The IRR for each technology corresponding to the base case is presented in Figure 4. The MR technology has a higher IRR than the Expander technology at 15% of the discount rate, when the function of plot area is the same. The difference between the IRR for MR and Expander technologies is about 7%, and thus it is not possible to determine a priori which technology is, economically speaking, more convenient for developing the project.

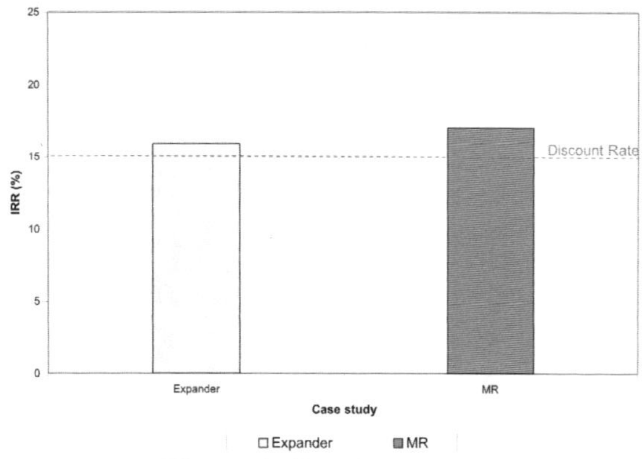

Figure 4. IRR Base Case

Figure 5 displays the economical evaluation for the three options where the plot area for the MR technology changes according to the models proposed. For the three cases evaluated the IRRs are under the discount rate (Figure 5). Considering the Expander technology as a reference, the effect of varying the plot areas for the MR case shows an important impact.

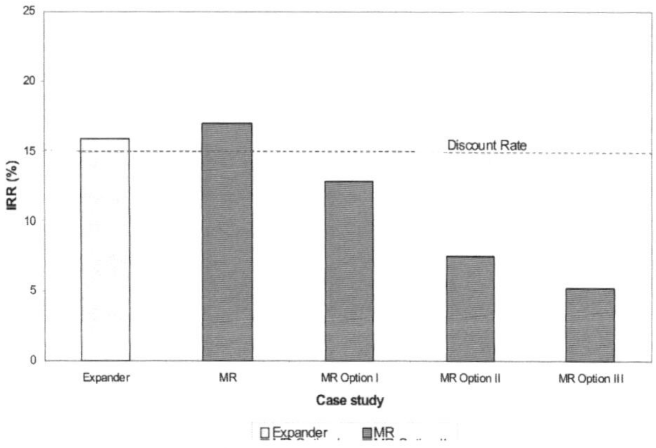

Figure 5. IRRs to the different options

The selection of a particular LNG technology for offshore developments presents the major challenge that thermodynamic efficiency is not necessarily controlling the decision. Parameters like safety, compactness, low equipment count, modularization, easy maintenance, reliability, simple operation, low influence of the motion of the vessel, etc. can impact the overall process and total investment. In the studied case, it was shown that in the economical model the area cost can play a major role in the CAPEX.

Conclusion

In this work a procedure for defining a selection criterion for remote small LNG plants was presented. The approach consists in identifying a possible scenario for remote gas production. The effect of the plot area for offshore LNG plants was considered and in the analysis it was observed that the area plays a major role in economical evaluation of the project. However limited information is available regarding the dependency of the plot area for different technologies and production rates.

References

1 L. Castillo, R. Nadales, C. Gonzálezand A. Viloria, Liquefied Natural Gas (LNG): Markets and Technologies of liquefaction, Visión Tecnológica journal, 2010 (accepted).

2 M. Barclay, N. Denton, Selecting offshore LNG processes, LNG journal Oct. (2005) 34–36.

3 L. Castillo and C.A. Dorao, Influence of the plot area in an economical analysis for selecting small scale LNG technologies for remote gas production, Journal of Natural Gas Science and Engineering, 2010 (accepted).

4 S. Pérez and R. Díez, Opportunities of monetising natural gas reserves using small to medium scale LNG technologies, in: IGU 24th World Gas Conference, REPSOL, Argentina, 2009.

5 C. Begazo, E. Carvalho and J. Sinoes-Moreira, Small-scale LNG plant technologies, Hydrocarbon World. 2007.

6 T. Shukri, LNG technology selection, Hydrocarbon Engineering, 2004.

7 L. Castillo, R. Nadales, C. Gonzalez, C.A Dorado and A. Viloria. Technology Selection for Liquiefied Natural Gas (LNG) on Baseload Plant, in: International Gas Conventions & Industrial Gas Exhibitions, Caracas-Venezuela, 2010.

8 Jamieson, D.; Johnson, P. & Redding, P. Targeting and achieving lower cost liquefaction plants, 12th International Conference & Exhibition on Liquefied Natural Gas, 1998.

9 R. DiNapoli. Evolution in LNG Project Costs and Estimation Techniques for New Projects, in: 8th LNG Conference, Los Angeles, 1986.

10 Al-Saadoon, F. & Nsa, A. U. Economic of LNG project, Hydrocarbon Engineering, 2009.

11 A. Fredheim and R. Heiersted. Possibilities for Cost Reduction in Baseload LNG Plant, in: European Applied Research Conference on Natural Gas. 101-114. 1996.

12 E. D. Garcia and F.Sanz, A new business approach to conventional small scale LNG, in: IGU 24th World Gas Conference, REPSOL, Argentina, 2009.

13 J. Jakobsen, G. Tangen, O. Nordbo and M. Molnvik, Methodology for CO2 chain analysis, International journal of greenhouse gas control (2008) 439-447.

14 H.Devold, T. Nestli and J. Hurter, All electric LNG plants, better, safer, more reliable-and profitable, Tech. rep. (2006).

15 H.Herzog, An introduction to CO_2 separation and capture technologies, Tech. rep. (1999).
16 DOE. Carbon dioxide emissions from the generation of electric power in the United States, Tech. rep. (2000).
17 DOE-EIA, Annual energy outlook 2010, Tech. rep. (2010).

(302d) Modeling and Simulation in Tight Gas Sands with Different Fracture Orientation and Studying Impacts of Geo-Mechanical Properties in Tight Gas Sands

Swapnil V. Thakur, John McLennan, Milind D. Deo, Justin Wriedt and Nan Zhao, Chemical Engineering, University of Utah, Salt Lake City, UT

Producing gas from unconventional sources is critical for developing secure sources of energy. In unconventional gas reservoirs like in tight gas sands the permeability values are extremely low. In such cases the natural fractures and fracture networks play an important role in enhancing the production rate by increasing permeability. This paper describes an approach of representing the fracture network and then analyzing the effect of the fracture network on production predictions. The first stage of this process is representing the fracture network as a discrete fracture network. The created discrete fracture network is then introduced into a geo-mechanical simulator. The geo-mechanical simulator is used to create a complex hydraulic fracture in the presence of the existing natural fractures. This system of complex fractures is incorporated into a flow simulator for predicting the production rates and to generate pressure and saturation signatures. This study includes studying the effect of different fracture orientation, fracture spacing on the production values. The simulations are carried to find out how the fractures are created or grown in controlled environments. There is possibility that fracture maybe of irregular shape and may not intersect with the well or with other conductive fractures, but may show conductive features due to the geo-mechanical stress and strains applied on them at reservoir condition. The simulation results for tight gas sands obtained from geo-mechanical simulation coupled with fluid flow are highlighted and discussed in the paper.

See more of this Session: Simulation of Complex Subsurface Processes II
See more of this Group/Topical: Topical D: Chemical Engineering in Oil and Gas Production and Other Complex Subsurface Processes

(302e) An Eulerian-Lagrangian Model to Predict Filter Cake Formation and Expression in Oil Drilling

Jeremiah N. Mpagazehe and C. Fred Higgs III, Mechanical Engineering, Carnegie Mellon University, Pittsburgh, PA

In the oil drilling process, a specially designed slurry forms a filter cake which prevents the premature escape of oil from the well. An understanding of this filter cake's ability to perform in extreme drilling environments is desired. In this work, an Eulerian-Lagrangian model is used to predict filter cake formation and expression. It was developed by combining an Eulerian computational fluid dynamics (CFD) code for the fluid phase with a discrete element method (DEM) model for the particle phase; two-way coupling is achieved between the phases. Individual particle trajectories are determined by fluid drag, electrostatic and Saffman forces. The main results from this work relate to the modeling of the filter cake formation process while varying fluid properties, particle properties, and inlet conditions. Results from the model are validated with experiments.

See more of this Session: Simulation of Complex Subsurface Processes II
See more of this Group/Topical: Topical D: Chemical Engineering in Oil and Gas Production and Other Complex Subsurface Processes

(361a) Massive Energy Storage Design for Renewable Electricity Generation Systems

Benjamin P. Omell and Donald J. Chmielewski, Chemical and Biological Engineering, Illinois Institute of Technology, Chicago, IL

It is widely recognized that a major concern with the installation of renewable energy systems is the fact that wind and solar sources are non-dispatchable. That is, the power produced from renewable systems is dependent on environmental conditions and is likely uncorrelated with the power demand from load centers. In contrast, fossil based sources typically have the ability to track demand, but depending on the technology possess limits with respect to response time (coal plants are slow, combined cycle turbines are faster and simple cycle turbines are fastest). Thus, many have advocated the use of massive energy storage systems to capture energy produced from renewable sources and then dispatch as load centers demand. An additional concern is the fact that renewable sources tend to be located far from the load centers (again in contrast to most fossil sources which are typically placed near demand). This issue is compounded by the fact that due to efficiency concerns the only viable energy storage systems is the pumped hydro-electric system, which in most cases must be located far from load centers. This combination of spatial and temporal mismatch with load center demand has caused many to worry about the stability of the electric power grid. Clearly, wide swings in power throughput over great distances will cause the weakest parts of the grid to fail.

To address both the spatial and temporal issues surrounding renewable power generation we have developed a scheme aimed at analyzing the capabilities of a regional power generation and distribution system. In particular, the scheme proposes a supervisory control law aimed at utilizing a mix of renewable and fossil sources along with energy storage systems to meet power load demands. This proposed control law takes into consideration not only the power output capacity of each source but also the rate at which power output can change (coal plants being slow and turbines being fast) as well as the storage capacity and efficiency losses associated with the pumped hydro system. In addition, the proposed scheme incorporates the maximum and minimum throughput limits on the distribution lines. If time allows, we will present an extension of the scheme aimed at sizing and placing energy storage systems as well as locating new power distribution lines.

See more of this Session: Sustainable Electricity: Generation and Storage
See more of this Group/Topical: Sustainable Engineering Forum

(361b) Sizing of Electrical Energy Storage for Large Scale Renewable Energy Penetration in the Grid and Implementation of a Lab Scale Grid for Experimental Validation

Alexandre F.T. Yokochi[1], Alex Bistrika[1], Hai-Yue Han[2], Ted Brekken[2] and Annette von Jouanne[2], (1)School of Chemical, Biological and Environmental Engineering, Oregon State University, Corvallis, OR, (2)School of Electrical Engineering and Computer Science, Oregon State University, Corvallis, OR

The inclusion of renewable energy into utilities' generation portfolios has become a highly desirable evolution. For example, in the State of Oregon, the renewable portfolio target is to achieve 25% renewables generation by the year 2025 [[1]]. As a consequence of renewable energy targets across the world, wind power is growing at a significant rate. For example, the BPA load management region extending mainly through Washington and Oregon has average load of 10GW with approximately 2.5 GW wind installed, and an approximate additional 5 GW likely within the next 5 years [[2]]. The addition of such a large variable generation will strain the stability of the grid and require increased spinning reserve and cycling of thermal plants [[3], [4], [5]]. The variable nature of the resource is illustrated by Fig. 1, which shows the variation of the power output of a large wind farm in Northern Oregon over one day. The wind farm power data sample points (in blue) are 10 minutes apart and are calculated as the average over those 10 minutes. The baseline forecasting (red line) algorithm is a simple one hour persistence model.

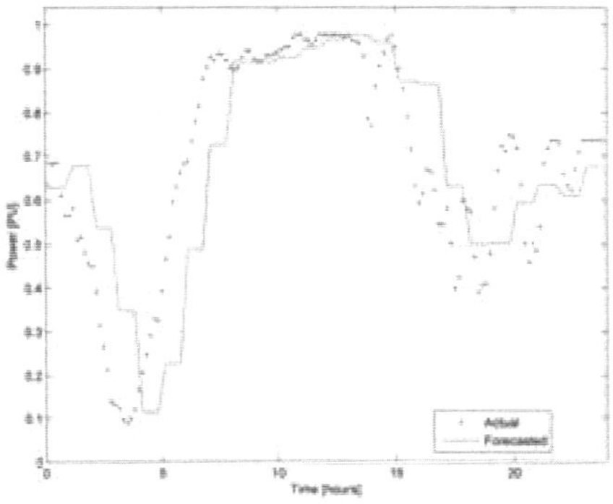

Fig. 1. Example daily energy from wind farm. Blue dots: actual wind farm output; red line: forecast energy output.

To determine the energy storage requirements for such a system, analysis of year long data sets were carry out to determine the economic sizing of a battery system capable of mitigating differences in forecast power greater than 0.04 per unit for 90%of the events occurring in any given year. This analysis suggests that incorporation of about 12% of rated power capacity and about 0.5 hours at full rated power output is necessary to achieve adequate system performance. To test this model, an in-lab research grid was designed and constructed, supported by a 480V, 750 kVA dedicated utility supply. The in-lab grid features emulation of several high-power grid sources and loads, including a wind farm, energy storage system, hydro resources, and local loads. The energy storage system is a 25 kW, 50 kWh Zinc Bromine flow battery. The wind farm is emulated using an Arbitrary Waveform Generator (AWG), which functions as a 120 kVA externally controlled source. Experimental validation of the wind farm sizing is currently under way as is verification of battery system lifetime performance assumptions.

This presentation focuses on the problems posed by present energy forecasting assumptions, the sizing of an energy storage system the capable of mitigating the resource fluctuation using technical and economic constraints, and the development of a lab scale grid for experimental evaluation of the energy storage model developed.

[[1]] Summary of Oregon's Renewable Portfolio Standard, Oregon Department of Energy 2007. http://www.oregon.gov/PUC/Oregon_RPS_Summary_Oct2007.pdf accessed 3/09.

[[2]] "Northwest's wind power could quadruple," http://www.oregonlive.com/news/oregonian/index.ssf?/base/news/1216529717321800.xml&coll=7

[[3]] "Northwest Wind Integration Action Plan," Northwest Power and Conservation Council, http://www.nwcouncil.org/energy/Wind/library/2007-1.htm

[[4]] "Utility Wind Integration State of the Art," Utility Wind Integration Group, http://www.uwig.org/UWIGWindIntegration052006.pdf

[[5]] M. Milligan, "Wind Integration Cost and Ancillary Service Impacts," NREL Technical Report, http://apps1.eere.energy.gov/tribalenergy/pdfs/course wind milligan1.pdf

See more of this Session: Sustainable Electricity: Generation and Storage
See more of this Group/Topical: Sustainable Engineering Forum

(361c) A Steady-State Model of a Sub-Scale Flow Battery System

Arun Pandy, Systems Department, United Technologies Research Center, East Hartford, CT, Rachid Zaffou, Physical Sciences, United Technologies Research Center, East Hartford, CT and Michael Perry, Energy Systems, United Technologies Research Center, East Hartford, CT

As we move towards the next generation of the electricity grid, energy storage devices will play a significant role in how electricity is produced, transmitted and consumed. For example, the penetration of renewable energy sources such as solar and wind energy into the electric grid is likely to increase in the future. However, the intermittent nature of these renewable sources makes them non-dispatchable and as a result these renewable energy sources have to be operated in conjunction with energy storage systems to realize their full potential. At the other end of the value chain the end-users who are subjected to time-of-use pricing of electricity can benefit from energy arbitrage made possible by energy storage technologies. Flow battery technology is one of the energy storage technologies that can make an impact along the entire value chain of electricity production, transmission and consumption. In this work we will focus on the flow battery technology. The flow battery consists of an ion-exchange membrane sandwiched between two porous electrodes. Liquid electrolytes forming the redox couples are passed through these electrodes. The redox reactions in the electrodes occur in opposite directions during the charging and discharging operation as shown in Figure 1. In this work, we present a 2-dimensional steady-state model of a sub-scale flow battery cell. This model solves for the reaction kinetics in the electrodes, transport (convection, diffusion) in electrodes, and ionic and electronic conduction in the electrode and the membrane. Using this mathematical model we investigate the performance of the flow battery as a function of various cell configurations. The variations in the electrode properties and in flow configurations were investigated. The insights obtained from the simulation of these models will be presented and discussed.

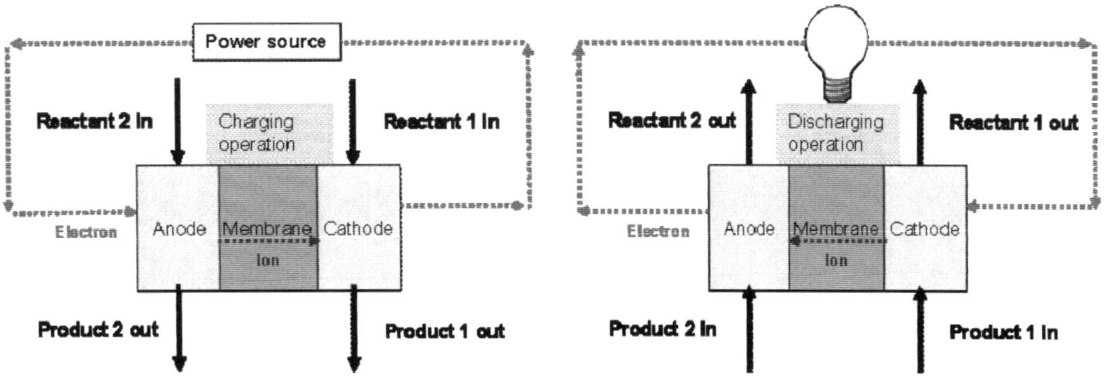

Figure 1: Schematic of charging and discharging operation of the flow battery

See more of this Session: Sustainable Electricity: Generation and Storage
See more of this Group/Topical: Sustainable Engineering Forum

(361d) Renewable Energy Storage Using Hydrated Lime

Frank Zeman, New York Institute of Technology, New York, NY

Providing sufficient energy storage capacity is a key to matching the energy generation profile of renewable energy systems with human use patterns. It is important to recognize that end use energy consumption takes the form of electricity and heat. As such, an energy storage technology that can deliver both forms of energy would be advantageous. Providing a low cost method for energy storage would help displace that oil as wind energy, a local resource, eventually gets built.

We propose a feasibility study on using lime (Calcium Oxide) as an energy storage medium by cycling between the hydrated and anhydrous states. The fundamental chemical equation is listed below. CaO(s) + H2O(g) „\ Ca(OH)2(s) The reversible reaction has an enthalpy of formation of 100 kJ/mol and a free energy of formation of 0 kJ/mol at 520oC. Varying the partial pressure of steam can control the reaction temperature.

The primary energy released can be converted to 520 Watt-hour (Wh) per kilogram (kg) of lime, which can be compared to values of 130 Wh per kg for lithium ion batteries. The energy stored is in the form of thermal energy, as opposed to electrical, but it can be converted, using a Rankine cycle or Stirling engine, back to electricity. While the additional conversion steps reduce overall efficiency, the simplicity and multiple outputs of the proposed system more than compensate. Even with the efficiency losses, the valuable product in a renewable electricity grid is service and not electricity.

See more of this Session: Sustainable Electricity: Generation and Storage
See more of this Group/Topical: Sustainable Engineering Forum

(452a) Fast Pyrolysis Bio-Oils From High Protein Containing Feedstocks

Charles A. Mullen and Akwasi A. Boateng, Eastern Regional Research Center, Agricultural Research Service, USDA, Wyndmoor, PA

Typically biomass considered for conversion to biofuels is comprised mostly of carbohydrates and lignin, with small amounts of other materials such as protein, lipids, and ash. Fast pyrolysis of such lignocellulosic materials generally produces bio-oils that are acidic, thermally unstable and have approximately half the energy content of petroleum fuels. However, the properties of bio-oils produced from biomass that contains a higher percentage of protein are different. Fast pyrolysis was performed on several high protein feedstocks, including distiller's grains with solubles (DDGS) from grain fermentation to ethanol, extracted oil seeds (pennycress, camelina), and animal rendering by-product. We found that during the pyrolytic breakdown of protein, basic organo-nitrogen compounds are formed and therefore the bio-oils have a higher pH and lower acid number than typical bio-oils. They also tend to have higher energy contents than wood pyrolysis oils. In this presentation, the results of these experiments, including mass and energy balances, carbon conversion efficiencies, and detailed product characterizations will be presented. Bio-oil chemical characterizations include elemental analysis, GC/MS, GPC, NMR and stability assessments. The fuel properties of the bio-oils including heating values and viscosities will also be presented. Finally, the bio-char co-products of high protein feedstocks will be compared with bio-chars from wood-like feedstocks.

See more of this Session: Biorefinery - Thermochemical Conversion of Biomass I
See more of this Group/Topical: Forest and Plant Bioproducts Division

Fast pyrolysis of lignocellulosic biomass to produce biooil and biochar : Parameter optimization and physicochemical analysis of them

Kwang Ho Kim, Seoul National University, Seoul, South Korea
In Yong Eom, Seoul National University, Seoul, South Korea
Sun Joo Moon, Seoul National University, Seoul, South Korea
Tai Seung Kim, Seoul National University, Seoul, South Korea
Soomin Lee, Korea Forest Research Institute, Seoul, South Korea
Hwangmyeong Yeo, Seoul National University, Seoul, South Korea
In Gyu Choi, Seoul National University, Seoul, South Korea
Joon Weon Choi, Seoul National University, Seoul, South Korea

Introduction

Fast pyrolysis, which is characterized by its simple process and short reaction time, has the capability of transforming inhomogeneous and recalcitrant lignocellulosics into a liquid form of fuel, called biooil [1]. This process is one of the most promising thermal technologies in the world. Recently, liquid biooil has been highly regarded both as an alternative energy [2] and green chemical resources. The fast pyrolysis process also has another advantage, as compared to other biofuel production processes, all different types of biomass can be considered as potential raw materials for fast pyrolysis [3]. A wide variety of biomass has been studied for pyrolysis, from agricultural residues to lignocellulosic biomass [4-11].

Fast pyrolysis is operated in the absence of oxygen producing gas, liquid and char in different proportions, according to experimental conditions. During fast pyrolysis, temperature, residence time of volatiles, heating rates, and particle size are well known as crucial parameters affecting the yield and fuel properties of biooil, as well as its chemical composition. Uzun et al. [12] studied the effect of temperature and nitrogen gas flow rates on the yield of biooil using soybean residues after the extraction of biodiesel. Zheng [13] produced biooil from rice husks in different temperature ranges and analyzed the fuel properties of the biooil.

In this study, yellow poplar wood, a representative fast-growing species of woody biomass, was pyrolyzed at various temperatures and residence times, using a fluidized bed-type reactor. The goal of this study was to elucidate the effect of temperature and residence time during fast pyrolysis on the conversion rate of woody biomass to biooil and to analyze the physicochemical properties of the biooil, as well as the biochar, which is a co-product of fast pyrolysis of woody biomass

Materials and methods

Feedstock

Yellow poplar wood *(Liriodendron tulipifera)* was used in this experiment as the raw lignocellulosic biomass. Bark-free stems were chipped and filtered through a 0.5mm aperture screen. The wood meal was air dried to about 8% moisture content.

Pyrolysis

Fast pryolysis of yellow poplar wood meal was performed using a fluidized bed reactor. A schematic diagram of the fluidized bed reactor is shown in figure 1. Before the experiment, the reactor was purged for 30 min with nitrogen to create an oxygen-free atmosphere. A sample weighing 150g was fed into the reactor for 1hr. After the experiment, the produced biooils were kept in the refrigerator at 4°C for further analysis.

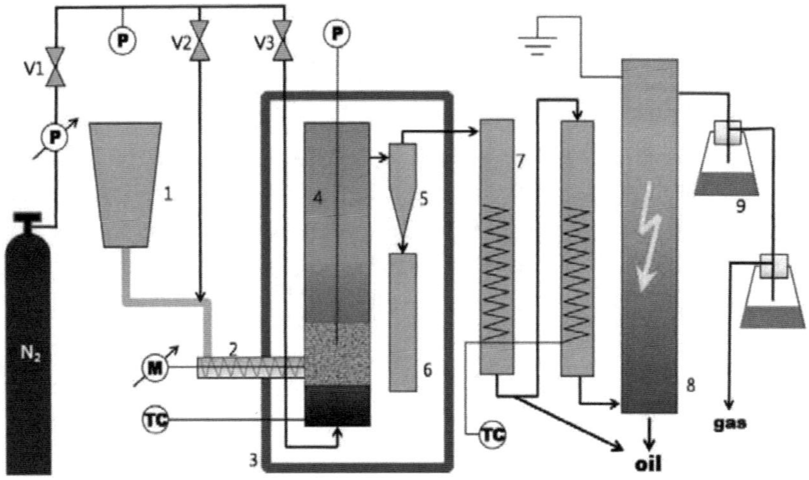

Figure 1. Schematic diagram of fluidized bed reactor (1: feeder, 2: screw feeder, 3: heater, 4: reactor, 5: cyclone, 6: char collector, 7: cooler, 8: electrostatic precipitator, 9: filter)

Analytical methods

Several physical and chemical analyses were performed to characterize biooil and biochar. The list of analytical methods for biooil is shown in Table 1. Elemental analysis and calorific value of biochars were measured with same equipment applied to biooil analyses. The surface morphology of the biochars was investigated by a field-emission scanning electron microscope (FE-SEM, SUPRA 55VP of Carl Zeiss).

Table 1. Analytical methods

Property	Method	Unit
Water content	Karl-Fischer titration	%(w/w, wet basis)
pH	pH meter	pH unit
Viscosity (@ 40°C)	Capillary type viscometer	cSt
Calorific value	Bomb type calorimeter (Parr 6400, Parr instrument company)	MJ/kg
Elemental analysis	Elemental analyzer (CHNS-932 Leco Corp.)	%C, %H, %N (w/w, wet basis)
Gas chromatography	Column type : DB 5 Dimensions : 30 m × 0.25 mm × 0.25 μm Injector : 250 ℃, split ratio 1:20, FID detector : 280 ℃ Oven : 45°C(4min. hold), 3 °C/min. to 280 °C (20min. hold)	

Results and discussion

Influence of pyrolysis temperature

Fast pyrolysis of yellow poplar wood meal was carried out in the temperature range of 400°C to 550°C at intervals of 50°C, and each yield of pyrolytic products (gas, biochar and biooil) is shown in Figure 2. This experiment revealed that the quantitative composition of the pyrolysis products was clearly affected by temperature. The yield of biooil, which was the target product of the fast pyrolysis of woody biomass, gradually increased up to 500°C, then slightly decreased at 550°C. The yield of biooil was maximized at approximately 68.5% at 500°C, based on the weight of yellow poplar. On the other hand, when the temperature of fast pyrolysis increased, the yield of non-condensable gases also continuously increased from 20.9% at 400°C to 28.2% at 550°C, while the formation of char decreased from 28.9% at 400°C to 7.9% at 550°C.

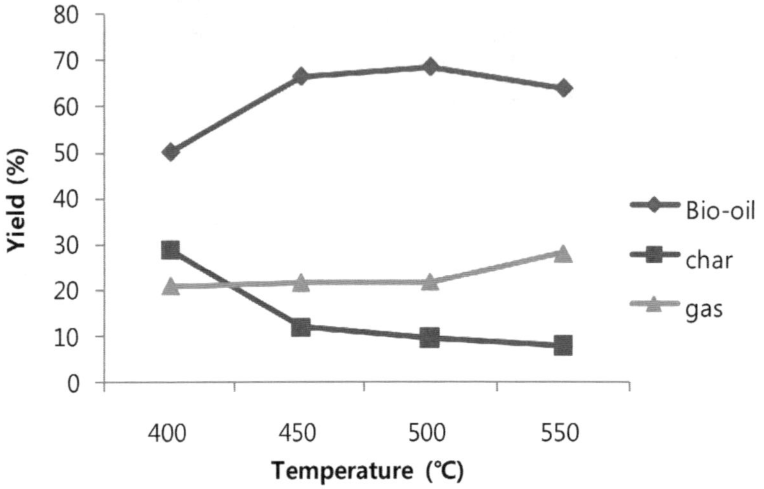

Figure 2. The distribution of product yields as a function of pyrolysis temperature (wet basis, residence time: 1.2sec)

According to previous studies [14-16], the highest yields of biooil from woody biomass feedstock were obtained at temperatures between 500°C and 520°C, which means that the energy required for biomass degradation to low molecular weight compounds was not sufficient below 500°C, but char formation was possible. In other words, a lower pyrolysis temperature prevents full decomposition of the biomass [7]. However, the components of biomass could decompose too much at temperature ranges over 500°C to be condensed into biooil. Higher pyrolysis temperatures cause secondary cracking of primary pyrolytic products, which leads to a decrease in biooil yield and an increase in gas yields [14, 15].

Influence of residence time

Residence time of pyrolytic products in the reactor is another critical parameter in the fast pyrolysis process. Pyrolysis was performed at different residence times, between 1.2s and 7.7s, which was adjusted by the nitrogen gas flow rate. The residence time is defined as the void volume of the reactor divided by the inlet gas flow rate at reactor conditions [8]. Four different nitrogen flow rates,

2.5ℓ/min, 5.0ℓ/min, 10.0ℓ/min, and 15.0ℓ/min, were chosen to examine the effect on pyrolytic product distribution. Each rate indicates 7.7s, 3.8s, 1.9s and 1.2s of residence time, respectively. In this case, the temperature was fixed at 500°C, at which the yield of biooil was the highest. The distribution of product yields, as a function of residence time at 500°C, is shown in Figure 3.

The yield of biooil was influenced by the residence time of the pyrolytic products. There was little difference in the yield of biooil at the residence times of 1.2s and 1.9s. However, when the residence time was prolonged to 3.8s and 7.7s, biooil production was clearly reduced to 61.8% and 51.6%, respectively. In the case of char and gas formation, however, their yields increased from 8.9% to 17.9%, and from 21.8% to 30.5%, respectively. This result could be explained by the residence time prolonging the primary degradation products to further exposure of secondary cracking reactions, thus converting the biomass to non-condensable gases and not biooil [7].

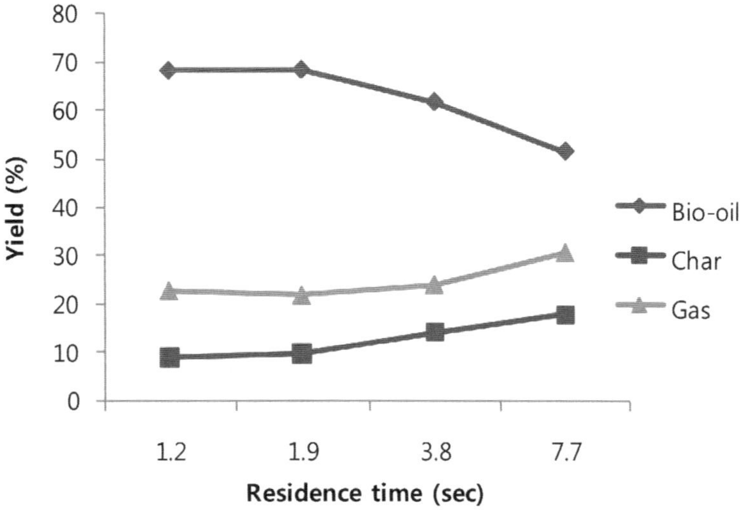

Figure 3. The distribution of product yields as a function of residence time (wet basis, pyrolysis temperature: 500°C)

Physicochemical properties of biooil

The water content in the biooil varied from 21.6 wt% to 42.6 wt%, depending on pyrolysis conditions. Considering the water content (approximately 8%) of the yellow poplar wood meal, the amount could be explained by the considerable amount of water formed during the fast pyrolysis of biomass, especially by a dehydration reaction [17]. Several kinds of acids, such as acetic acid and benzoic acid were formed from the wood components during fast pyrolysis, and pH values of the biooil were between 1.9 and 2.9 [17]. The heating values of the biooils, produced under different temperatures and residence times, were found to be between 10.4 and 17.2 MJ/kg. The relatively low heating values of the biooils were attributed to the high water content and oxygen-containing degradation products.

Characterization of biochar

Elemental composition and calorific value of the pyrolysis biochar was analyzed, and is presented in Table 2. As can be seen in table, the biochar was composed of more than 80% carbon in all experimental runs having a calorific value of 27.1~29.9MJ/kg, which is comparable to coal (approximately 30MJ/kg). The amount of carbon in biochar increased with decreasing pyrolysis temperature and increasing residence time. Like routine carbonization processes, the formation of biochar with a high carbon content and calorific value was favored at lower temperatures and longer residence times.

Table 2. Elemental analysis of biochar coproduced from fast pyrolysis of yellow poplar

Elemental analysis (wt%)	Residence time fixed at 1.9 sec				Temperature fixed at 500°C			
	400°C	450°C	500°C	550°C	7.7s	3.8s	1.9s	1.2s
Carbon	83.4	83.1	82.7	82.0	84.3	83.3	82.7	80.0
Hydrogen	2.0	1.4	1.1	0.8	1.6	1.3	1.1	1.0
Nitrogen	trace	trace	trace	trace	trace	trace	trace	trace
Higher heating value (MJ/kg)	29.9	29.0	28.4	27.8	29.8	28.9	28.4	27.1

A field-emission scanning electron microscope (FE-SEM) was applied to investigate the effect of thermal treatment on the biomass morphology. Figure 4 shows the FE-SEM images of feedstock and biochar obtained from pyrolysis. Significant structural differences can be seen when comparing both images. The biochar looked distorted, and its surface was rough, due to the thermal treatment.

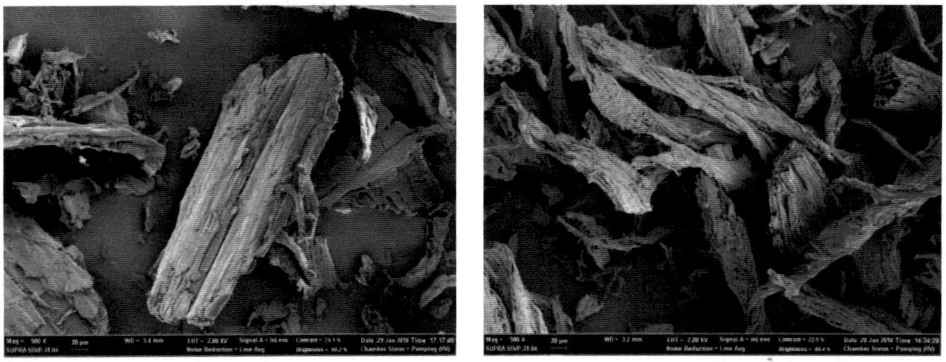

Figure 4. SEM pictures (×500) of yellow poplar wood meal (left) and biochar (right)

Conclusion

The yellow poplar wood meal was pyrolyzed to convert it into biooil and biochar under different experimental conditions using a fluidized bed reactor, and their physicochemical properties were characterized. The mass balance of pyrolytic products, biooil, biochar and gases, was significantly influenced by pyrolysis temperature and residence time. The maximum yield of biooil was 68.5%, based on a biomass weight at 500°C and with a residence time of 1.9s. Pyrolysis of yellow poplar wood for a prolonged residence time, as well as below 500°C, resulted in lower yields of biooil but activated the formation of co-products.

The yield of biochar was 8.0~28.9% on a mass basis. Biochar distribution was highly dependent on the pyrolysis conditions. The analytical results showed that the elemental composition of biochar was affected by experimental conditions. The carbon content of biochar increased with decreasing pyrolysis temperature and increasing residence time, which is similar to the carbonization mechanism.

References

1. Westerhof R.J.M., Kuipers N.J.M., Kersten S.R.A., Swaaij W.P.M. (2007) "Controlling the water content of biomass fast pyrolysis oil." *Industrial & Engineering Chemistry Research*, 46(26): p. 9238-9247.
2. Scott D.S., Majerski P., Piskorz J., Radlein D. (1999) "A second look at fast pyrolysis of biomass - the RTI process." *Journal of Analytical and Applied Pyrolysis*, 51(1-2): p. 23-37.
3. Bridgwater, A.V. (2003) "Renewable fuels and chemicals by thermal processing of biomass." *Chemical Engineering Journal*, 91(2-3): p. 87-102.
4. Putun, A.E. (2002) "Biomass to bio-oil via fast pyrolysis of cotton straw and stalk." *Energy Sources*, 24(3): p. 275-285.
5. Mullen C.A., Boateng A.A., Goldberg N.M., Lima I.M., Laird D.A., Hicks K.B. (2009) "Bio-oil and bio-char production from corn cobs and stover by fast pyrolysis." *Biomass and Bioenergy*, 34(1): p. 67-74.
6. Tsai W.T., Lee M.K., Chang Y.M. (2007) "Fast pyrolysis of rice husk: Product yields and compositions." *Bioresource and Technology*, 98(1): p. 22-8.
7. Xu R., Ferrante L., Briens C., Berruti F. (2009) "Flash pyrolysis of grape residues into biofuel in a bubbling fluid bed." *Journal of Analytical and Applied Pyrolysis*, 86(1): p. 58-65.
8. Piskorz J., Majerski P., Radlein D., Scott D.S., Bridgwater A.V. (1998) "Fast pyrolysis of sweet sorghum and sweet sorghum bagasse." *Journal of Analytical and Applied Pyrolysis*, 46(1): p. 15-29.
9. Park H.J., Park Y.K., Dong J.I., Kim J.S., Jeon J.K., Kim S.S. (2009) "Pyrolysis characteristics of Oriental white oak: Kinetic study and fast pyrolysis in a fluidized bed with an improved reaction system." *Fuel Processing Technology*, 90(2): p. 186-195.
10. Garcia-Perez M., Wang X.S., Shen J., Rhodes M.J., Tian F.J., Lee W.J. (2008) "Fast pyrolysis of oil mallee woody biomass: Effect of temperature on the yield and quality of pyrolysis products." *Industrial & Engineering Chemistry Research*, 47(6): p. 1846-1854.
11. Schnitzer M.I., Monreal C.M., Facey G.A., Fransham P.B. (2007) "The conversion of chicken manure to biooil by fast pyrolysis I. Analyses of chicken manure, biooils and char by 13C and 1H NMR and FTIR spectrophotometry." *Journal of Environmental Science and Health B*, 42(1): p. 71-7.
12. Uzun B.B., Putun A.E., Putun E. (2006) "Fast pyrolysis of soybean cake: product yields and compositions." *Bioresource and Technology*, 97(4): p. 569-76.
13. Zheng, J.L. (2007) "Bio-oil from fast pyrolysis of rice husk: Yields and related properties and improvement of the pyrolysis system." *Journal of Analytical and Applied Pyrolysis*, 80(1): p. 30-35.
14. Bridgwater, A.V. (1999) "An overview of fast pyrolysis of biomass." *Organic Geochemistry*, 30(12): p. 1479-1493.
15. Venderbosch, R.H, Prins, W. (2010) "Fast pyrolysis technology development." *Biofuels Bioproducts & Biorefining-Biofpr*, 4(2): p. 178-208.
16. Mohan D., Pittman C.U., Steele P.H. (2006) "Pyrolysis of wood/biomass for bio-oil: A critical review." *Energy & Fuels*, 20(3): p. 848-889.
17. Lu Q., Li W.Z., Zhu X.F. (2009) "Overview of fuel properties of biomass fast pyrolysis oils." *Energy Conversion and Management*, 50(5): p. 1376-1383.

(452c) Types of Carbon in Biochars From Slow Pyrolysis, Fast Pyrolysis and Gasification

Catherine E. Brewer, Center for Sustainable Environmental Technologies, Iowa State University, Ames, IA, Robert C. Brown, Center for Sustainable and Environmental Technologies, Iowa State University, Ames, IA and Klaus Schmidt-Rohr, Iowa State University, Ames, IA

Biochars, sustainably produced charcoals used for soil amendment and carbon sequestration applications, contain both organic (carbon) and inorganic (ash) constituents. The composition of the inorganic fraction is mostly dependent on the mineral matter present in the feedstock. The types and quantities of carbon in the organic fraction of the biochars, however, are highly dependent on processing conditions such as temperature, heating rate, reaction atmosphere (inert or oxic), and residence time. This study compares the types of carbons in biochars from the slow pyrolysis, fast pyrolysis, and gasification of switchgrass, corn stover and various hardwoods. Among the analytical techniques used to characterize the chars are CHNOS elemental analysis, proximate analysis, Fourier-transform infrared spectroscopy (FTIR), and quantitative ^{13}C solid-state nuclear magnetic resonance spectroscopy (NMR) using direct polarization magic angle spinning (DP/MAS) techniques. Three biochars from the fast pyrolysis of corn stover were also used to conduct a soil incubation study to measure carbon bioavailability based on CO_2 evolution rates. Overall findings suggest that biochars can be tailored to specific applications and that not all chars created from the thermochemical conversion of biomass will be suitable for use as biochars.

See more of this Session: Biorefinery - Thermochemical Conversion of Biomass I
See more of this Group/Topical: Forest and Plant Bioproducts Division

Product and Economic Analysis of Direct Liquefaction of Swine Manure

Mitchell James Minarick, University of Illinois, Urbana, IL, USA. Additional authors: Yuanhui Zhang, Zhichao Wang, Guo Yu, and Ted Funk.

Abstract

Hydrothermal processing of biomass is a technology that has shown promising results in treating waste and producing oil. A batch hydrothermal processing system that was previously developed at the University of Illinois at Urbana-Champaign was used to treat swine manure and it successfully converted up to 70% of swine manure volatile solids into oil and reduced manure chemical oxygen demand by up to 75%. A continuous reactor was developed and it resulted in similar conversion rates to the batch process and it indicated the potential for the process to be scaled-up. These results were also documented previously.

Following these results, a slurry pump was designed and built to enable longer test durations and lower repair costs, and it was used to conduct 24-hour continuous tests and produce quantities of oil suitable for more detailed analysis.

The hydrothermal process utilizes manure containing around 25% solids content which is not available in typical swine confinement facilities so a preliminary investigation of dewatering methods was conducted to determine the best dewatering method in terms of the resultant oil yields and the effects on fertilizer values. The best dewatering method appears to be a coarse filtering process or the use of a low-cost additive. A coarse filtering process followed by the hydrothermal process results in an oil product as well as a fertilizer product that retains the majority of its nitrogen value with a reduced level of phosphorus as compared to the unprocessed swine manure.

Energy and economic analyses of the dewatering and hydrothermal processes were conducted for swine manure as well as general analyses for other domestic biomass feedstocks. The feasibility of the conversion of swine manure was judged to be high, dependent mainly on the validity of the assumptions for the value of the oil product.

Introduction

Using waste materials to produce biofuel sidesteps the controversial food versus fuel debate and gives the production process a competitive advantage over methods using traditional feedstocks since they can be acquired much more cheaply, if not for free. It makes sense to research methods of biofuel production from waste streams alongside research of biofuels produced from more valuable feedstocks.

Waste Conversion technologies

The conversion of biomass to fuel generally follows one of two routes: biochemical or thermochemical. A summary of the primary biomass conversion technologies is provided by He (2000) in Figure 1 below.

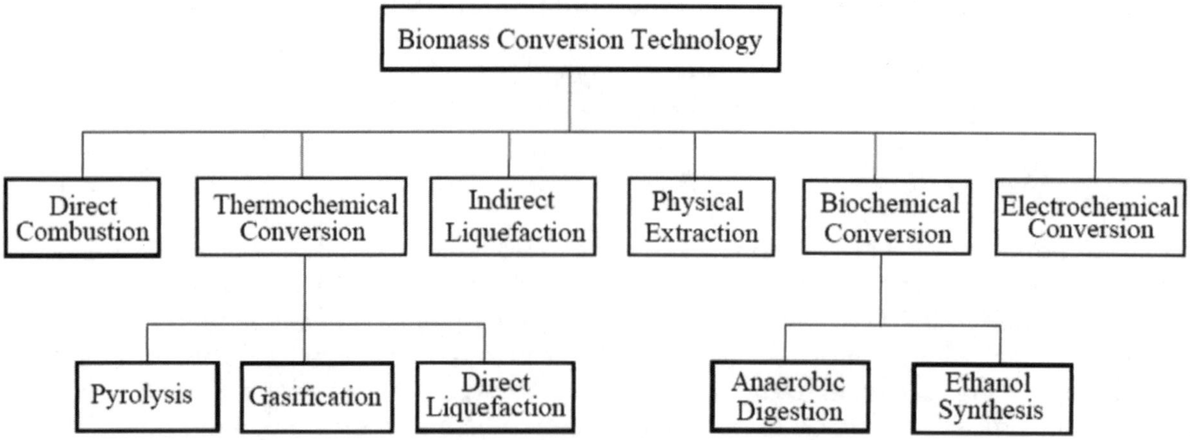

Figure 1. Biomass conversion technologies.

Direct liquefaction is a type of thermochemical conversion that has been studied extensively during the past 10 years at the University of Illinois. Similar to pyrolysis, direct liquefaction is well-suited for the conversion of livestock waste because it does not require a dry feedstock. Previous research teams headed by Dr. Yuanhui Zhang have concluded that direct liquefaction could be a viable solution for the disposal of swine manure, and they have attempted to determine the parameters that optimize the oil yield.

Ocfemia (2005) applied these parameters to a continuous reaction process to determine the feasibility of scaling the system up beyond the batch scale. His system had a capacity to process up to 48 kg of manure slurry per day, and successfully produced oil at equal or greater efficiencies than the batch process used by Brian He.

Water-separation considerations

The desired solids content of the feedstock for the hydrothermal process is around 25% according to He (2000). This is approximately the solids content contained in fresh (that is, taken directly from the confinement floor) swine manure. However, according to Fulhage (2001) most swine manure is handled as a liquid, utilizing shallow storage pits beneath slotted floors that allow the producer to apply manure to fields when it is most beneficial. This slurry has a higher water content than what is optimal for the hydrothermal conversion process (He, 2000) and since additional water translates into additional heating and processing costs, it is necessary to remove as much water as possible without affecting the oil yield.

For this technology to be useful on a farm, it is necessary to develop Ocfemia's continuous process (2005) into a working prototype that utilizes swine manure taken directly from the production facility. In a typical production facility, the manure as well as urine and spilt food and water falls through the slotted floor to the shallow-pit, where it is stored until it is pumped into an external tank for disposal every few weeks. Initial tests (Wang, 2007) showed that conversion into oil of the manure slurry as taken from the pit was impaired when compared to the conversion of manure taken from behind the pig, so it is necessary to determine the effect of different water separation methods on oil yields in order to dovetail thermochemical conversion technology with manure collection systems that are currently in place.

Methods

Water Separation Methods

As previously discussed, swine manure is commonly stored in a shallow pit, where it becomes diluted to less than 10% solids. The thermochemical process requires a feedstock with a solids content of around 20%, so a practical means for attaining higher solids content needed to be determined.

Evaporation is the simplest method of water separation, but also the most time consuming. The swine manure was simply stored in open containers with a fan circulating air over the tops of the containers. The solids content was measured periodically until it reached the desired level.

Coarse filtration was accomplished using a standard aluminum window screen as the filter medium. The screen had square openings measuring 1.13 x 1.30 mm, and it was used as a liner inside a round laundry basket that served as the filter holder. The manure from the pit was placed in the filter assembly and the laundry basket was rolled by hand until the flow of aqueous product out of the filter ceased.

The difficulties associated with this water separation method are discussed in section 3.3.4, but in summary fine filtration was defined as any filtration with a smaller pore size than the coarse filter (1.13x1.3 mm). The portion of the swine manure that was subjected to the fine filtration was the aqueous portion of the swine manure that was not already removed by the coarse filter.

For the low-speed centrifugation, a vertical-shaft solid-basket centrifuge manufactured by Lavin Centrifuge (model 12-413V) was used to subject the aqueous product from the coarse filtration to 2,250 Relative Centrifugal Force (2,250 RCF=2,250 times the force of gravity) for 5 minutes. The solids were collected and the aqueous portion was saved for further separation.

The aqueous solution remaining after the low-speed centrifugation was placed in capped 500mL Nalgene polycarbonate centrifuge bottles (Part #3140-0500, 69.5 mm diameter, 158.8 mm length, 13,700 maximum allowed RCF) and carefully weighed to make sure they would be balanced in the centrifuge. The centrifuge was a Dupont Sorvall centrifuge equipped with a GS-3 rotor and was used to subject the samples to 10,816 RCF for 15 minutes. The solids formed a dense layer on the bottom of the centrifuge bottles, so the aqueous solution was easily separated and both portions were saved.

Thermochemical Conversion Process

The thermochemical conversion was conducted using similar conditions as those used in the experiments by He (2000) and Ocfemia (2005). The temperature was the main control variable and was held at 305°C for 30 minutes of residence time in a Parr 2-L reactor. The oxygen was evacuated by pressurizing the reactor with 100 psi of nitrogen gas and then releasing the pressure. This was repeated three times before filling the reactor with 100 psi of nitrogen gas which was used as a process gas throughout the experiment.

These operating conditions resulted in a peak reactor pressure of approximately 1650 psi when the reactor was at its peak operating temperature. No catalyst was used during these tests.

Oil Analysis

The thermochemical process results in a raw oil product that is very crude and unrefined. The water content of the bio-oil was determined using ASTM standard D95-99 and subtracted from the oil yields. The toluene solubility was determined using ASTM standard D4072-98 and used to define a 'refined oil yield' which makes up roughly 35-45% of the raw oil. The refined oil yield gives an idea of both the quantity and quality of the oil and is a more robust measurement of the oil production.

Assumptions

Energy

The following assumptions were used when calculating energy balances:

The bio-oil has a heating value of 30,000 kJ/kg (Ocfemia, 2005).

The energy contained in the feedstocks before undergoing TCC was negligible or unavailable. In other words, it is assumed that the only energy input is heat via electricity, and the chemical energy in the feedstock is not considered. In most cases the feedstock had too high of a water content to be burned so this was an accurate assumption for the feedstocks considered.

The reactor is perfectly insulated. Although perfect insulation is not truly possible, the amount of heat loss through the reactor walls could be negligible with good insulation.

The internal energy values of the feedstocks are equal to those of water (i.e. the energy needed to heat/pressurize the feedstock is equal to that needed for water). This is a reasonable and conservative assumption because the feedstocks contained 75-95% water, and the specific heat of water is higher than most materials. For instance, the specific heat of water is over 4 times that of sawdust. Thus, more energy is needed to heat pure water to a given temperature than would be needed to heat the same amount of water mixed with sawdust. These figures for the change in internal energy are therefore higher than the actual values, and the actual energy needed to heat the different solutions should be less than (but fairly close to) what would be calculated using the following values for water:

Reaction conditions are approximately 100 bar (1450 psi) and 303°C,
Internal energy: Uf (303°C) = 1350.5 kJ/kg
Uf (22°C) = 92.32 kJ/kg
ΔU for reaction = 1258 kJ/kg
Energy used to pre-process the feedstock is negligible.

No heat is recaptured with a heat exchanger unless otherwise stated. When a heat exchanger is used in calculations, the effectiveness is assumed to be 75%.

Economic

For the following calculations (unless otherwise specified), the cost of electricity was estimated to be $0.05/kW-hr. If the reactor was only operated during certain times of the day, this is a reasonable estimate for industrial power usage, although in practice this value will vary.

Additionally, the value of crude oil varies greatly. It is reasonable to assume that the value of the bio-oil will be related to the value of crude oil since they are competitive products. Consequently, the value of the bio-oil can be expected to vary significantly based on the fluctuations in the price of crude oil, which can be volatile. For example, in just a 12-month span from winter of 2008 to winter of 2009 the price of oil varied from $40-$140. This variability certainly has an effect on the value of the bio-oil and on any alternative fuel for that matter.

For most of the calculations that follow, the price of oil is assumed to $50/barrel unless otherwise specified. If this price were assumed to be higher, it would result in more favorable results for the thermochemical process.

Estimating the actual value of the bio-oil is difficult. It is hard to find the true market value of an oil product which does not have an established market yet. For the purpose of calculation, the value of the refined bio-oil (i.e. the toluene soluble portion) was estimated to be 60% of the value of petroleum light crude oil. This value was chosen in an attempt to be conservative so as not to overestimate the value of the bio-oil. This value might well be higher since the crude oil is unrefined and the toluene-soluble portion of the bio-oil has already been subjected to one refinement step (hence the choice of the term 'refined oil yield' to describe it). This assumption is admittedly difficult to defend. Without an existing market, the value of the bio-oil will not be precisely known. However, taking 40% of the value off of the unrefined petroleum crude hopefully allows adequate room for this uncertainty.

Most petroleum chemists divide crude oil into four fractions: saturates, aromatics, resins and asphaltenes. Asphaltenes are, broadly speaking, the fraction of oil that are insoluble in heptane or pentane and soluble in benzene/toluene. Appleford (2004) analyzed the bio-oil from the thermochemical conversion processing of swine manure and found that the oil was primarily resins and asphaltenes as seen in Table 1 below where the first two columns of results represent crude oil from swine manure:

Table 1. Properties of bio-crude oil from swine manure vs. natural crude oil.

Fraction (%)	**C-305-1500-60**	**N-305-1500-60**	**Pooled Sample**	**Natural Crude Oil***	**Natural Bitumen***
Saturates	2.93	2.48	1.32	>30	15
Aromatics	1.84	1.67	1.60	>30	34
Resins	44.65	45.87	35.87	<20	34
Asphaltenes	44.73	43.39	54.81	<10	17
Water + Loss	5.85	6.59	6.40	varies	varies

As seen in Table 1, the bio-crude from the swine manure contains around 50% asphaltenes and around 35% resins. Therefore an alternative valuation for the bio-oil can be obtained by assuming the products will be used as asphaltanes rather than as a refined oil product. Asphalt prices, like oil prices, fluctuate dramatically. Table 2 shows the price of asphalt used for paving for 2007 and 2008 (California Department of Transportation, 2009):

Table 2. Monthly price of asphalt for 2007-2008.

Month-Year	$/ton	Month-Year	$/ton
Jan-07	$287.10	Jan-08	$497.80
Feb-07	$267.70	Feb-08	$434.40
Mar-07	$295.30	Mar-08	$510.00
Apr-07	$317.40	Apr-08	$504.10
May-07	$314.90	May-08	$567.70
Jun-07	$321.70	Jun-08	$645.30
Jul-07	$355.70	Jul-08	$716.90
Aug-07	$384.30	Aug-08	$630.40
Sep-07	$370.70	Sep-08	$547.40
Oct-07	$393.40	Oct-08	$486.10
Nov-07	$464.50	Nov-08	$288.10
Dec-07	$439.30	Dec-08	$199.40

Importantly, the use of the bio-oil as an asphalt would eliminate many of the expenses, technological hurdles, and uncertainties associated with refinement of the oil and finding a market for the finished product.

The net present value (NPV) of an investment is a standard method for using the time value of money to appraise long-term projects. Essentially it involves calculating the sum of the present values of future cash flows based on an assumed discount rate (or the expected interest rate of investment alternatives), and if the NPV is a positive number the investment is considered favorable to the investment alternatives. The NPV for a given set of cash flows can be calculated with a few assumptions. Labor costs could be limited to one employee at $35,000 per year. Maintenance costs can be estimated at $20,000 per year based on an assumed capital cost of $200,000. The annual discount rate (i.e. the rate of return of investment alternatives) is assumed to be 15% and the life expectancy of the equipment is 10 years.

Results

Figure 2 below shows the summarized results of the TCC tests using swine manure and a variety of different dewatering methods in terms of their refined oil yields. However, refined oil yield is not the only variable determining a successful test. Certainly energy input costs have a significant effect. For a feedstock with low solids content a significant amount of energy is spent heating and pressurizing water in addition to the solids. Manure slurry taken straight from the pit and sent to the reactor without any dewatering suffers from its high water content. Figure 3 below summarizes the relative processing costs associated with oil production based on the different dewatering methods.

Figure 2. Refined oil yields.

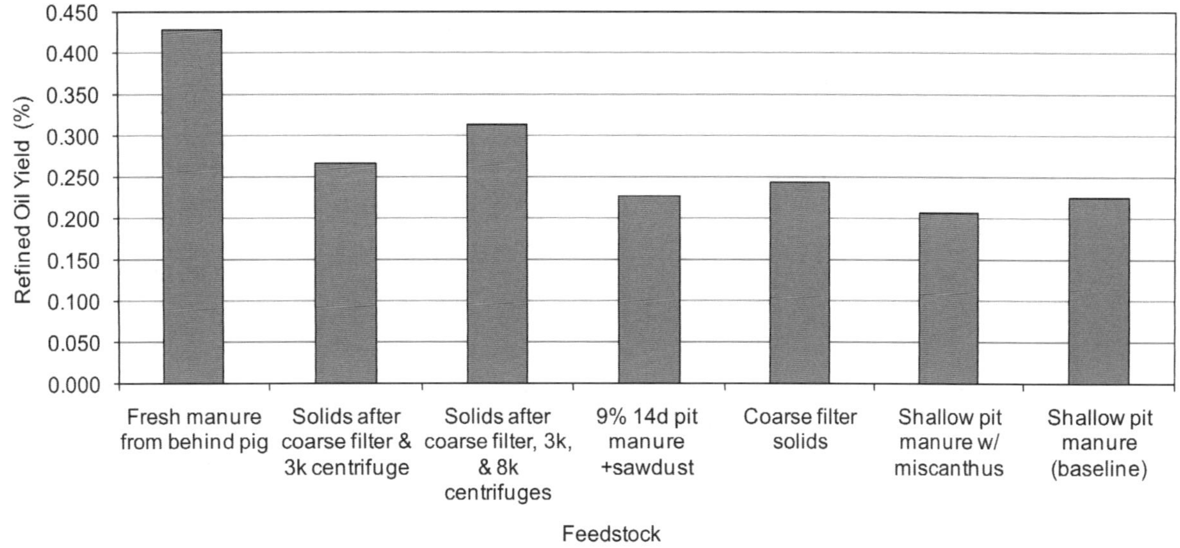

Figure 3. Relative processing costs per oil production.

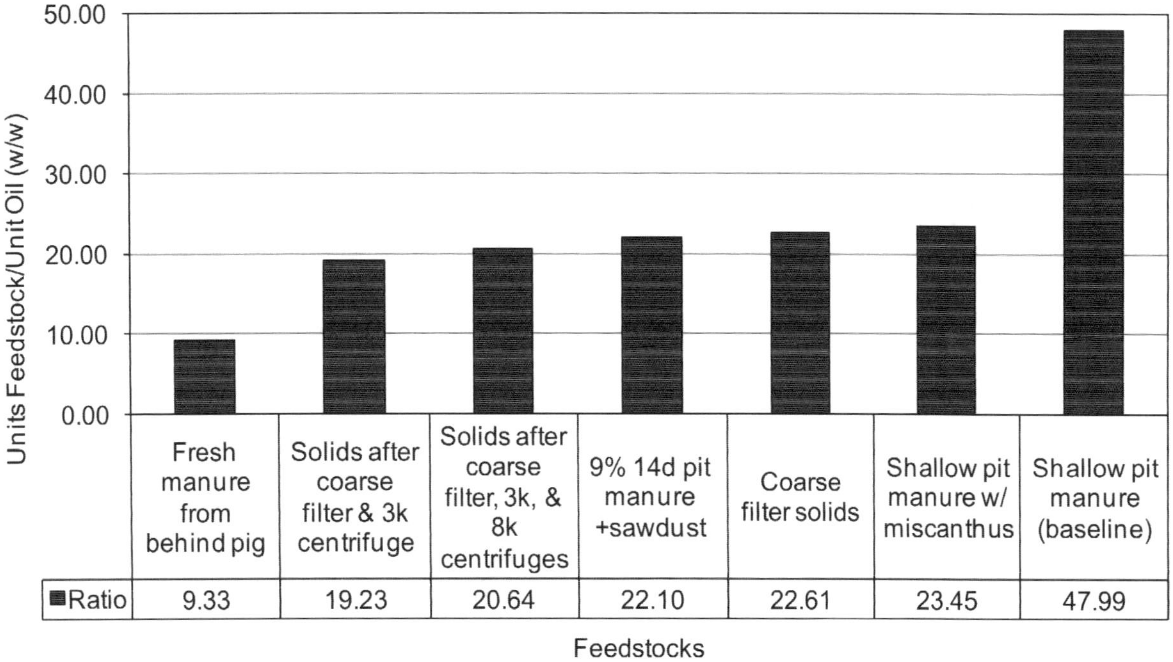

In summary, the best dewatering method appears to be mixing the slurry with additives such as sawdust or miscanthus if the additive is cheap or free. If the additive is not cheap, rough filtration is probably better.

Figure 4 below demonstrates the energy balances of the oil production process in terms of processing energy in versus oil-based energy out, both with and without a heat exchanger. The balances are favorable for all cases except the raw manure straight from the pit. It is important to remember that although most of the ratios seem to show a negligible energy gain, the main value added by the process is not due to an addition of net energy but rather a conversion of the form of the energy, from heat via electricity or combustion to a liquid fuel product.

Figure 4. Energy out versus energy in.

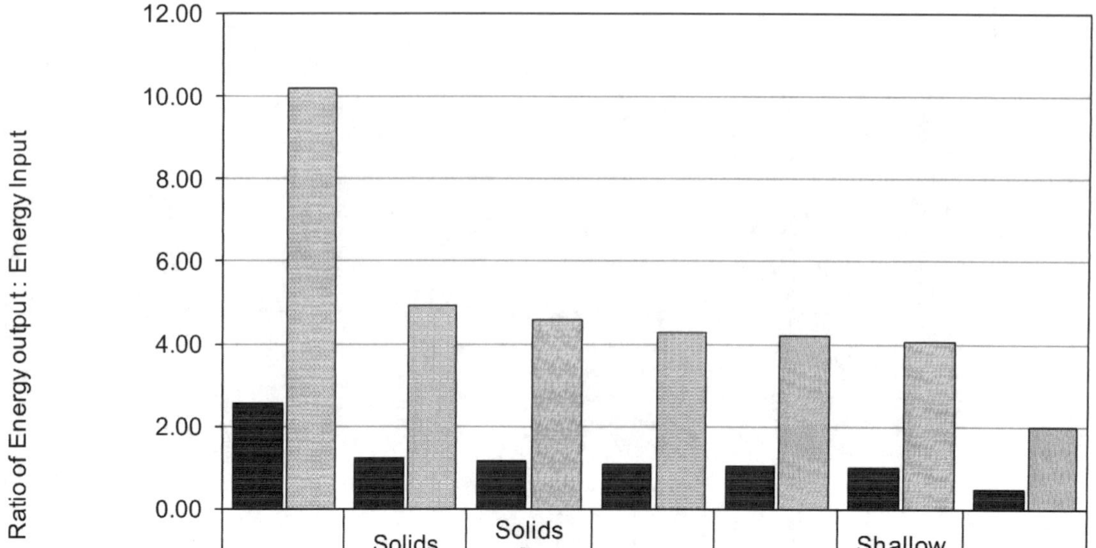

	Fresh manure from behind pig	Solids after coarse filter & 3k centrifuge	Solids after coarse filter, 3k, & 8k centrifuges	9% 14d pit manure +sawdust	Coarse filter solids	Shallow pit manure w/ miscanthus	Shallow pit manure (baseline)
■Energy output : energy input	2.55	1.24	1.16	1.08	1.05	1.02	0.50
■Energy out:in w/ heat exch.	10.22	4.96	4.62	4.32	4.22	4.07	1.99

Feedstocks

Figure 5 below shows that the fertilizer values of the aqueous products after the TCC test are not generally destroyed by the conversion process, with the exception of phosphorus content. The bars on the left side of each set of data represent the fertilizer values contained in the manure before processing. Nitrogen is fairly well preserved in most forms, except in Case 4 with the miscanthus as an additive. Potassium was retained well in all cases. This is an important piece of information because it shows that a farmer can process manure thermochemically to obtain bio-oil, and still retain over half of the original fertilizer value. Additionally, less phosphorus is applied to the field after the TCC process, lowering the risk of phosphorus runoff.

Figure 5. Fertilizer values remaining in aqueous solution.

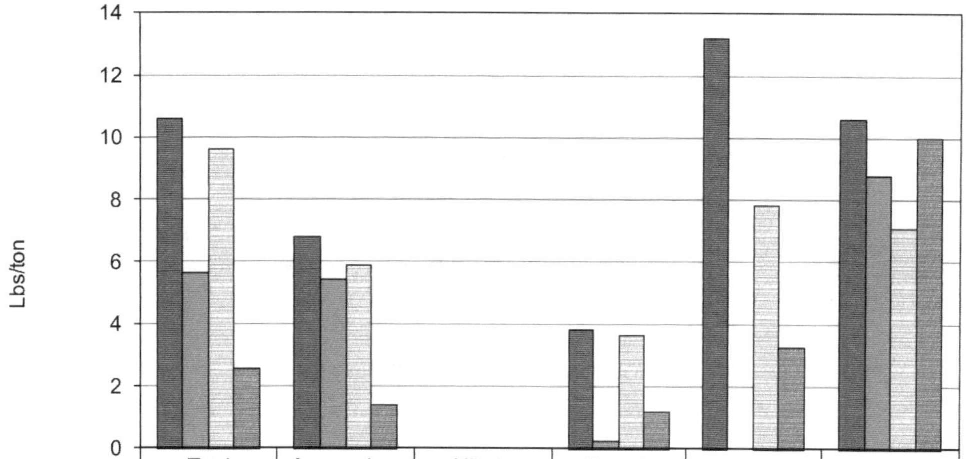

	Total Nitrogen	Ammonium Nitrogen	Nitrate Nitrogen	Organic Nitrogen	Phosphorus	Potassium
Raw Manure from pit, Case 1	10.60	6.80	0.00	3.80	13.20	10.60
Raw Manure after TCC, Case 2	5.64	5.41	0.00	0.23	0.00	8.80
Raw Manure after coarse filter & TCC, Case 3	9.61	5.90	0.00	3.66	7.84	7.09
Raw Manure w/ Miscanthus after TCC, Case 4	2.56	1.40	0.00	1.17	3.26	10.02

Components

Economics of Swine Manure Conversion

For an operation with 10,000 hogs, a feedstock supply that does not cost anything to acquire, and an empirically-determined (Wang, 2007) refined oil yield of 42.9% for manure from the floor behind the pig, 6,552 barrels of refined oil per year could be produced.

Assuming the value of refined oil is equal to 60% of the value of crude oil and a barrel of crude oil is worth $50, the annual gross income for the operation is $196,560. The cost to process this amount of biomass assuming a fresh manure solids content of 25%, an electrical energy cost of $0.05/kW-hr, and no heat recapture is $178,101 annually. Thus an annual income of $196,560-$178,101=$18,459 would be expected in this case study. This is probably not even enough income to pay for the associated labor costs, and a system like this is not economically feasible.

Factors that would decrease this amount include: oil prices lower than $50/bbl, refined oil worth less than 60% of crude oil, oil conversion rate of less than 42.9%, energy input costs more than $0.05/kW-hr, labor and capital costs, heat losses to the atmosphere, etc. Factors that would increase this amount include: oil prices higher than $50/bbl, refined oil worth more than 60% of crude oil, energy input costs less than $0.05/kW-hr, and heat recapture via a heat exchanger. Of these factors, the most significant changes would probably come in the estimates for energy input costs. If the operator was careful to run the reactor at hours of low electrical demand, it is conceivable that the cost of electricity would average less than $0.05/kW-hr. There are other energy input options as well. Natural gas was estimated to cost Midwest customers $1.11/therm or $0.0378/kW-hr during 2006 (Energy Information

Administration, 2007), although the efficiency of the boiler would not be 100% so that value would be increased slightly in a practical application.

The simplest way to reduce energy input costs, however, would be to install a heat exchanger to take some of the energy from the hot exit stream and use it to preheat the cool incoming feedstock.

When heat recovery assumptions are applied to the economic calculations, the numbers improve dramatically. The oil yield and gross income of $196,560 would remain the same, but the energy input costs are quartered due to the assumed energy recovery of 75%. The energy input costs become $44,525 annually. This drives the annual profit up from $18,459 to $152,035 before labor, capital, and maintenance costs and the NPV is then $486,996 which means the project is profitable (i.e. returns more than 15%) if the capital costs are less than the NPV. These results are very promising and would be even more compelling for larger hog farms because with larger volumes they would generate greater annual profits yet the labor and capital costs would remain virtually the same. This result is dependent on multiple assumptions but they were kept conservative and overall the feasibility is judged to be very high.

If the refined oil product was able to be used as an asphalt substitute, it could be conservatively priced at $200/ton. The gross income of the operation then becomes $241,098. This drives the annual profit up to $196,573 and the NPV becomes $710,522. To recap then, the NPVs for the manure from the floor are as follows:

- Manure from floor used to make oil: NPV=negative
- Manure from floor used to make oil, heat exchanger used: NPV=$486,996
- Manure from floor used to make asphalt, heat exch. used: NPV=$710,522

The benefits of using a heat exchanger and exploring the opportunities for using the bio-oil as an asphalt are apparent, and these modifications will be used in the remaining economic analyses.

NPVs using manure from shallow pit

The gross income using manure from a shallow pit and adding sawdust would be $323,479, however this is assuming that the additive does not cost anything. In reality, sawdust can cost up to $30 per ton in some areas, reducing the gross income by $137,394 to $186,085 per year. The operating costs would then be $141,815 and the gross profit is only $44,270, which is not even enough to pay for labor and maintenance expenses.

If the excess water was removed with rough filtration instead of using sawdust as an additive, then the gross income would be $42,201 with processing costs of $17,179 annually. The ratio of the income to the expenses is 2.46. This is a very acceptable ratio, but the gross profit of the TCC process ($42,201-$17,179=$25,022) is not even large enough to support a full-time employee. The high profitability ratio, however, indicates good potential for scaling up the system. If the thermochemical conversion system was mounted on a portable trailer and used to process the manure from 5 farms, the net profit would jump to $125,110 and the NPV would be $351,866.

To recap then, the NPVs for the manure from the shallow pit are as follows:

- Manure from pit, heat exchanger, asphalt product and $30/ton sawdust: NPV=negative

- Manure from pit, heat exchanger, asphalt product, rough filtration: NPV=negative after labor/maintenance costs are added, but the profitability ratio is high, meaning more volume is needed for profitability.
- Manure from pit, heat exchanger, asphalt product, rough filtration throughout 5 farms: NPV=$351,866

These results are very promising because they represent conservative calculations for the net present values of the types of system that would be used with existing production facilities without modification. The numbers would be even more compelling for larger hog farms, again because larger volumes would generate greater annual profits yet the labor and capital costs would remain virtually the same.

The primary downside risk to these NPV figures is the difficulty in determining an accurate value of the bio-oil. As stated previously, the refined bio-oil was sometimes valued at 60% of petroleum crude oil in these calculations, but that number is simply an estimate and not backed by firm data (since there really isn't any firm data extant). Other times the refined bio-oil was considered as a substitute for petroleum asphalt selling for $200/ton. This is probably more accurate than the other approximation since the oil product more closely resembles asphalt, but firmer data regarding the suitability of the bio-oil as a substitute for asphalt would solidify these estimates considerably.

References

Appleford, J. M. 2004. Analysis and optimization of the thermochemical conversion process to produce oil from biomass. Masters thesis. Urbana, IL: University of Illinois at Urbana-Champaign, Department of Agricultural Engineering.

ASAE Standards. 2005. D384.2: Manure production and characteristics. St. Joseph, Mich.: ASAE.

ASTM. 2004a. ASTM D95-99. Standard test method for water in petroleum products and bituminous materials by distillation. In: Annual Book of ASTM Standards. West Conshohocken, PA: Am. Soc. for Testing Materials.

ASTM. 2004b. ASTM D473-02. Standard test method for sediment in crude oils and fuel oils by the extraction. In: Annual Book of ASTM Standards. West Conshohocken, PA: Am. Soc. for Testing Materials.

ASTM. 2004c. ASTM D4072-98. Standard test method for toluene-insoluble (TI) content of tar and pitch. In: Annual Book of ASTM Standards. West Conshohocken, PA: Am. Soc. for Testing Materials.

California Department of Transportation. 2009. California statewide paving asphalt price index. Sacramento,CA. Available at: http://www.dot.ca.gov/hq/esc/oe/asphalt_index/PREVIOUS_YEARS/07-08_AC_INDEX.pdf.

Dong, R. 2007. Personal communication. Department of Agricultural and biological engineering. Urbana, IL.: University of Illinois at Urbana-Champaign.

Elliott, D. C., et al. 1994. Chemical processing in high-pressure aqueous environments: continuous-flow reactor process development experiments for organics destruction. *Industrial and Engineering Chemistry Research*. 33: 566-574.

Energy Initiative Administration. 2007. Natural gas annual 2006. Washington, D.C.: United States Department of Energy, Energy Initiative Administration.

Fulhage, C., and D. Pfost. 2001. Swine manure management systems in Missouri. Columbia, MO.: University of Missouri-Columbia Extension.

He, B. 2000. Thermochemical conversion of swine manure to produce oil and reduce waste. PhD dissertation. Urbana, IL: University of Illinois at Urbana-Champaign, Department of Agricultural Engineering.

Hrubant, G. R., R. A. Rhodes, and G. H. Sloneker. 1978. Specific composition of representative feedlot wastes: a chemical and microbial profile. Washington, D.C.: Science and Education Administration, U.S. Department of Agriculture.

Kreith, F., and D. Y. Goswami. 2004. The CRC handbook of mechanical engineering. 2nd edition. CRC Press. Boca Raton, FL.

Moran, M. J., and H. N. Shapiro. 2008. *Fundamentals of engineering thermodynamics*. 6th edition. John Wiley and Sons. Hoboken, NJ.

Ocfemia, K. 2005. Hydrothermal process of swine manure to oil using a continuous reactor system. PhD dissertation. Urbana, IL: University of Illinois at Urbana-Champaign, Department of Agricultural Engineering.

Ott, L. S., B. L. Smith, and T. J. Bruno. 2008. Advanced distillation curve measurement: application to a bio-derived crude oil prepared from swine manure. *Fuel* 87: 3379-3387.

Reece, M. 2004. Special issue paper 9: Densities of oil products. Energy Statistics Working Group Meeting, International Energy Agency. Paris, France.

Scurlock, J. 2009. Bioenergy feedstock characteristics. Oak Ridge, TN. Oak Ridge National Laboratory, Bioenergy Feedstock Development Program.

Speight, J. G. 2001. *Handbook of Petroleum Analysis*. New York: Wiley Interscience.

Sweeten, J. M. 1992. Livestock and poultry waste management: A national overview. In J. Blake et al., eds., National Livestock, Poultry and Aquaculture Waste Management, pp. 4–15. ASAE Publication 03–92. American Society of Agricultural Engineers, St. Joseph, MI.

USDA-ERS. 2009. Livestock, dairy, & poultry outlook. Washington, D.C.: USDA Economic Research Service.

USDA-NASS. 2008b. Quarterly hog and pigs report. Washington, D.C.: USDA National Agricultural Statistics Service.

USDA-NASS. 2009. Crop production 2008 summary, p. 13. Washington, D.C.: USDA National Agricultural Statistics Service.

USDOE and USDA, 2005. Biomass as feedstock for a bioenergy and bioproducts industry: the technical feasibility of a billion-ton annual supply, p. 7. Washington, D.C.: USDOE and USDA.

U.S. Environmental Protection Agency Office of Solid Waste. 1999. Biosolids Generation, Use, and Disposal in the United States. EPA530-R-99-009, p. 2.

U.S. Environmental Protection Agency Office of Solid Waste. 2008. Municipal Solid Waste in the United States: 2007 Facts and Figures. EPA530-R-08-010, p. 1.

Vogel, K. P., and R. A. Masters. 1998. Developing switchgrass into a biomass fuel crop for the midwestern USA. Lincoln, NE. University of Nebraska. Agricultural Research Service, United States Department of Agriculture.

Wang, Z. 2007. Unpublished data. Department of Agricultural and Biological Engineering. Urbana, IL.: University of Illinois at Urbana-Champaign.

Wang, Z. 2007. Personal communication. Department of Agricultural and Biological Engineering. Urbana, IL.: University of Illinois at Urbana-Champaign.

Yu, G. 2008. Unpublished data. Department of Agricultural and Biological Engineering. Urbana, IL.: University of Illinois at Urbana-Champaign.

(452e) Effect of Torrefaction on the Fast Pyrolysis Using Forest Resources

Jiajia Meng, Sunkyu Park, David Tilotta and Junyeong Park, Forest Biomaterials, North Carolina State University, Raleigh, NC

Bio-oil has a great potential for multiple uses including heat/electricity generation and chemical production. This thermochemical technology appears to be a practical, effective, and environmentally sustainable means of producing renewable fuels while simultaneously reducing emissions of greenhouse gases. In addition, pyrolysis has a key advantage over biochemical methods in that it is less sensitive to variation in feedstocks, which makes supply-chain development less challenging. However, the quality of pyrolysis bio-oil is not sufficient to directly produce biofuels and chemicals due to high oxygen content, high acidity, and poor long-term stability. In this study, we investigated the potential of integrating fast pyrolysis with torrefaction to produce high quality bio-oil from forest resources. The benefits of this integration along with the preliminary experimental results will be discussed.

See more of this Session: Biorefinery - Thermochemical Conversion of Biomass I
See more of this Group/Topical: Forest and Plant Bioproducts Division

LIQUID PRODUCTS OF LIQUID PHASE-PYROLYSIS

Nikolaus Schwaiger[1], Verena Mertlitz[1]; Kerstin Zahel[2] , Peter Pucher[2], Edgar Ahn[2], Matthäus Siebenhofer[1],

[1]Graz University of Technology; Institut für Thermische Verfahrenstechnik und Umwelttechnik; Inffeldgasse 25/C/II 8010 Graz/Austria;

[2]BDI-BioEnergy International AG; Grambach/Austria

Ligno-cellulosic feed is expected to contribute significantly to production of liquified combustibles in future, because of the quantity and the variety of feed material. Therefore the pyrolytic degradation properties of spruce wood and its building blocks glucose and cellulose were investigated. The interaction between heat carrier, biomass and biomass products in the condensed liquid and vapour phase was investigated.

The process was carried out in a semi-batch reaction vessel under isothermal conditions at a temperature of T = 623 K and at ambient pressure. For optimum heat transfer the pyrolysis process was carried out in liquid heat carrier phase. Gaseous products as well as condensibles were removed on top of the reaction vessel. Condensation temperature was T = 20°C. After recording flux and composition the gaseous phase was incinerated in a flare.

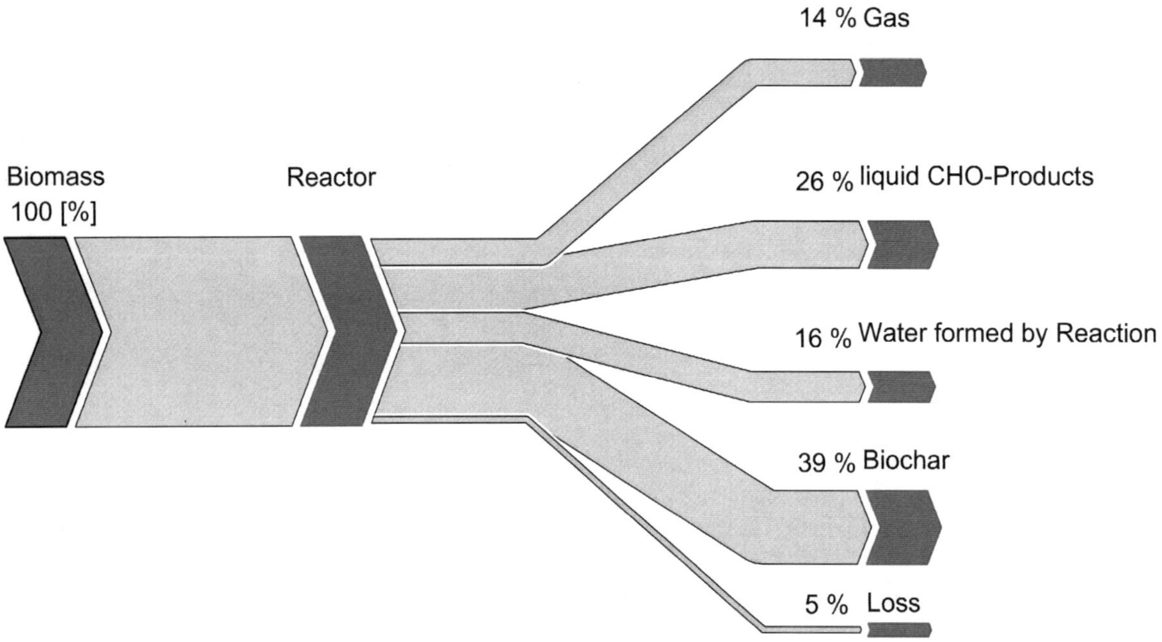

Figure 1: Biomass mass balance

The condensed products were analyzed by GC-MS and NMR spectroscopy as well as elemental analysis (CHNO).

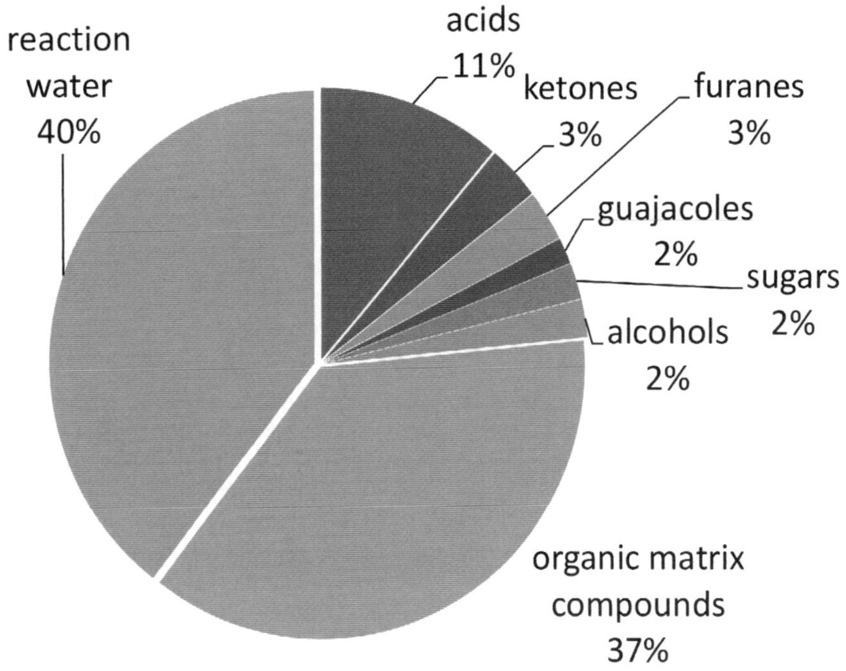

Figure 2: Composition of liquid products measured by GC-MS and Karl Fischer Titration

The solid product residues were pretreated by solid-liquid extraction and afterwards analyzed by elementary analysis, FTIR-ATR, Raman spectroscopy, scanning electron microscope and energy dispersive X-ray spectroscopy.

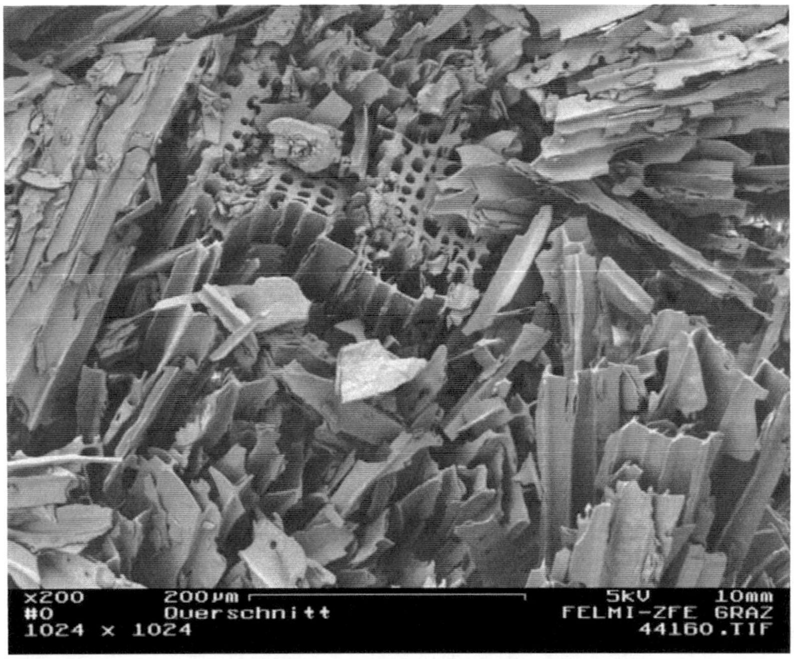

Figure 3: Scanning electron microscope photo of a Biochar cross section

Several operation parameters were investigated. Representatively the effect of particle size is briefly mentioned.

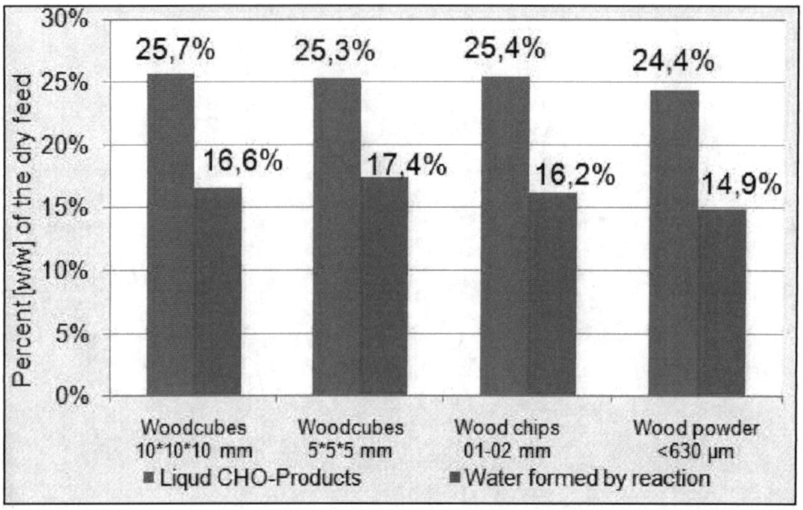

Diagram 1: Effect of feed particle size on the amount of CHO-products and the amount of water formed by thermal dehydration of softwood

Following the results shown in Diagram 1 variation of particle size does not have an effect on the amount of liquid CHO-products, when pyrolysing softwood. Water formation decreases for feed particles of smaller particles.

As shown in Diagram 2 the total amount of liquid CHO-products seems to be constant and independent of the feed particle size. The chemical composition of liquid phase pyrolysis products differs.

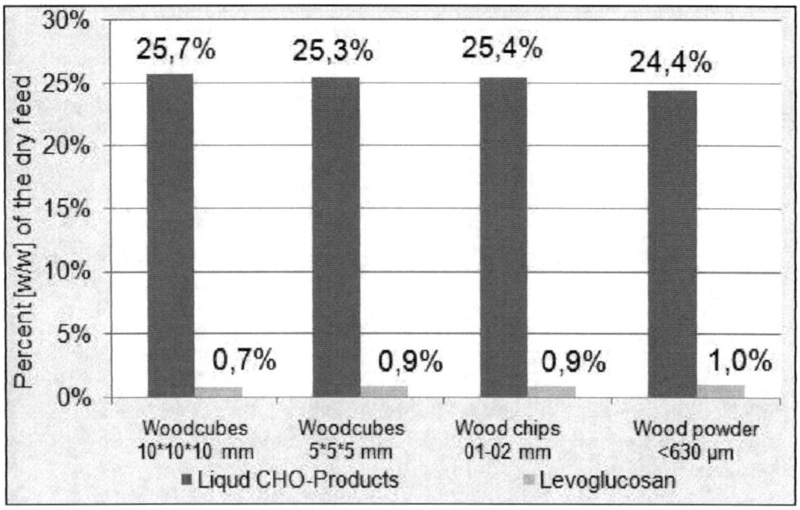

Diagram 2: Effect of feed particle size on non aqueous condesibles and products

Based on the detailed mass balance the effect of operation parameters on pyrolytic degradation of softwood, product formation and energy balance was determined.

FRAMEWORK FOR SELECTING THIN-CAKE CANDLE FILTER TECHNOLOGY FOR REMOVING SOLID CONTAMINANT FINES FROM RECIRCULATING ACID GAS SCRUBBING FLUID STREAMS

Barry A. Perlmutter, President & Managing Director
BHS-Filtration Inc.
9123-115 Monroe Road
Charlotte, North Carolina 28270
Phone: 704.845.1190
Fax: 704.845.1902
E-mail: barry.perlmutter@bhs-filtration.com

ABSTRACT

The need for filtration systems in scrubbing units has been well documented. Filtration provides the benefits of reduced hot spots in the regenerator reboiler, reduced heat exchanger fouling and reduction of foaming in the absorbers. This paper discusses thin-cake candle filter technology that is used to remove trace amounts of solid contaminant fines from the recirculating fluid stream. These contaminants originate from various sources and are generally less than 3 - 5 microns in size, which makes their removal very difficult. Candle filter technology and the process of thin-cake building, is a new approach that is employed for high-efficiency and cost-effective fines removal.

Introduction

Sulfur dioxide (SO_2) removal systems are in many types of applications such as fluid catalytic cracking (FCC) and Coker units at refineries, acid plant tail gas, spent acid recovery plants, smelters, pulp mills, natural gas, power generation, cogeneration, chemical process plants as well as process vent streams in sulfur plants. While the plant location and type of removal process may vary, there is a need to remove fine particulate matter from the recirculating liquid streams for efficient scrubbing operation and to protect absorbers, condensers and reboilers, heat exchangers, pumps, downstream purification systems, etc. The particulate matter can be catalyst fines and other contaminants in the feedstock. It is difficult or impossible to remove particles of this size in settling tanks, hydrocyclones or centrifuges, so the particles must be removed by filtration. The use of thin-cake candle filter technology has been proven to be a cost-effective and reliable approach to removing the contaminants, recovering the scrubbing liquids and drying the cake for easy landfill disposal.

Problem Definition

Various catalyst and carbonized particles are carried into the gas and are captured by the scrubbing fluid. The flow rate of the scrubbing can be as high as 150 m^3/hr so the fluid must be regenerated and reused to make the process economical. The fine particles are less than 3 to 5 microns and they will accumulate in the scrubbing system unless they are removed from the scrubbing fluid. These particles cause fouling in heat exchangers and foaming in the trays of the scrubbers, so the particles must be removed from the scrubbing fluid before it can be reused. The scrubbing fluid can fluid can be water, an amine salt fluid (e.g. CanSolv®) or some other proprietary fluid (e.g. the Labsorb® phosphate salt system). This paper discusses specifically the removal of fines from an amine fluid, but the theoretical basis and the technology can be used for other scrubbing fluids.

Theoretical Basis

The following equation is the basic equation that relates the flow rate through a filter for constant pressure filtration. Various forms of this equation are used to evaluate the lab data gathered and then to scale up the lab data to the required production system.

$$\frac{dV}{A\,dt} = \frac{\Delta P}{\mu(\alpha'(\Delta P)^s \{V[\rho c/(1-mc)]/A\} + r)} = \frac{\text{Driving Force}}{\text{Total Resistance}} \qquad \text{Eqn. 1} \qquad \text{Ref. 1}$$

where: V = slurry volume
- A = filtration area
- t = time
- ΔP = pressure drop across the filter cake and filter media
- μ = filtrate viscosity
- r = resistance of the filter media
- α' = resistance of the filter cake
- s = a compressibility constant and it is 0 for an incompressible cake and it is 1.0 for very compressible cake.
- ρ = filtrate density
- c = the mass fraction of the solids in the slurry
- m = mass ratio of the wet cake to the dry cake

Equation 1 is integrated and rearranged to determine the amount of time required to filter a known quantity of material at a given pressure. The results of the integration are shown in Equation 2.

$$t = \frac{\mu\alpha'(\Delta P)^s [\rho c/(1-mc)]}{2\Delta P}\frac{V^2}{A^2} + \frac{r\mu}{\Delta P}\frac{V}{A} \qquad \text{Eqn. 2}$$

Equation 2 can be further simplified when the cake is incompressible and when the resistance of the filter media is negligible. Equation 2 then reduces to the following:

$$t = (b' / \Delta P) * (V/A)^2, \qquad \text{Eqn. 3}$$

where $b' = \mu\alpha'[\rho c/(1-mc)] / 2$

Equation 3 is the final equation that was used to evaluate the lab data.

Laboratory Tests

Filtration Tests

All filtration and washing tests were conducted in a 400 ml Pocket Leaf Filter that had a filtration area of 20 cm^2 and a fill volume of 400 ml. The filter media for all of the tests was a woven, specially designed media for the scrubbing application; the cloth was precoated with a known amount of Hyflo Super Cell. Filtration tests were conducted at ambient temperature and at 1.0 bar, 2.0 bar, 3.0 bar and 4.0 bar. The filtration data is shown in Figure 1.

Figure 1 – Filtration Time vs. (Filtrate Volume / Filter Area)^2

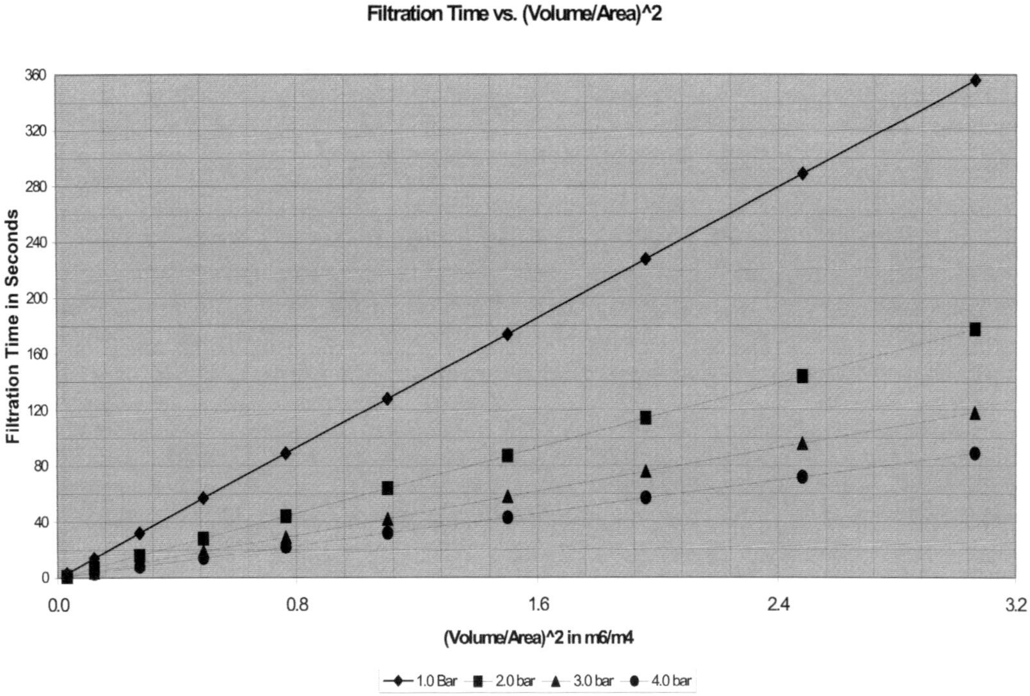

A Least Squares Regression was performed on the lab data to determine b' for Equation 3. The regression had a correlation of fit R2 = 0.99 and determined that b' = 32.38 X 10^{-3} bar * hr * m^6/m^4. The excellent fit of the data to Equation 3 confirms that the assumptions of a negligible media and precoat resistance, and an incompressible cake are correct.

Washing Tests

Cake samples were washed with tap water in a single pass in the Pocket Leaf Filter to determine the amount of washing required to remove the amine salts from the cake. It was determined that a washing was efficient enough to wash virtually all of the amine salts from the cake. Counter-current washing was not tested during these trials.

Drying Tests and Cake Density Measurement

Five of the washed cake samples were pulse dried in the Pocket Leaf Filter with ambient temperature air. The drying was accomplished by pressurizing the Pocket Leaf Filter and then releasing the pressure through the filtrate valve. Each sample was dried by pulse drying. The moisture of the cake was low enough after drying for easy landfill. The cake density was measured after drying in the Pocket Leaf Filter and found to be 1800 kg / m^3.

Scale-up for a production Filter

An existing Scrubbing System had an amine flow rate of 50 m3/hr with an average concentration of fines of 600 ppm. It was desirable to remove the fines by filtration at a constant flow rate. The filter cake was to be washed before the drying to recover the amine salts from the cake. A candle filter was selected for this application and was operated using the following processing steps:

- Step 1 = Filling the Filter With Slurry
- Step 2 = Precoating the Filter
- Step 3 = Filtration
- Step 4 = Cake Washing
- Step 5 = Draining the Filter
- Step 6 = Drying the Cake
- Step 7 = Discharging the Cake

Each Step requires a certain amount of time and the sum of these times is the total cycle time. The minimum size for the filter occurs when the time for the Filtration step is equal to the sum of the other steps, but it is also desirable for most operating plants to load the precoat into the system no more than one time per day. Therefore, the starting point for sizing the system was to determine what size filter would be required to process the required flow rate using a 24 hour Filtration Step. The time required for the other steps was then calculated based on the selected filter.

Required Filtration Area Calculations

Equation 3 was rearranged to solve for the required filtration area assuming a filtration time of 24 hours and an average filtration pressure of 1 bar.

$A = V * [b' /(t * \Delta P)]^{0.5}$ Eqn. 4

Where, V = Flowrate X Filtration Time = 50 m^3/hr X 24 hours = 1200 m^3

b' = 32.38 X 10^{-3} bar * hr * m^6/m^4

t = 24 hours

ΔP = 1.0 bar

$A = 1200\ m3 * [32.38 \times 10^{-3}\ bar * hr * m^6/m^4 /(24\ hours * 1.0\ bar)]^{0.5} = 44.1\ m^2$

A BHS-Filtration Candle Filter Type KT-91/46 has 91 candles and a filtration area of 46.3 m^2. Two KT-91/46 filters were used in series for this application so that one filter would be in the Filtration Step while the other filter would be in the Washing, Draining, Drying, Discharge, Filling, and Precoating Steps.

Cake Mass, Volume and Height Predictions

The cake consists of the precoat, the fines that are collected and the residual moisture in the cake. The KT-91/46 has an area of 46.3 m^2 and the lab tests determined that the precoat loading was adequate for this application. The fines collected in the Filtration Step can be calculated by the fines concentration in the feed, the feed rate, and the filtration time. The fines collected were 600 kg Fines / 10^6 kg Feed X 50 m^3/hr Feed X 1,089 kg/m^3 Feed X 24 hours = 784 kg Fines. The total mass of cake (precoat, fines and moisture) per cycle results in a cake thickness of 14 mm.

Filling Step

The Type KT-91/46 candle filter has a fill volume of 11.0 m^3 and was filled at 50 m^3/hr, so the time for the filling step was predicted to be 11.0 m^3 / 50 m^3/hr = 0.22 hours.

Precoating Step and Required Amount of Precoat

The precoat was pumped through the filter and allowed to recirculate to the precoat tank for 0.75 hours.

Cake Washing

The lab tests determined that the residual amine salts can be removed from the cake. The washing time was = 0.44 hours.

Draining and Drying

The heel liquid in the filter was drained by pressurizing the filter with 1.0 bar air while draining the heel through a heel drain valve near the bottom of the filter. The heel drain required 0.45 hours. The cake was pulse dried in the filter by pressurizing the filter with plant air and then releasing the pressure through the filtrate valves. The drying time required is 0.25 hours.

Cake Discharge

The cake is discharged from the candles by pulsing air back through the candles while the bottom valve is open. The total discharge time is 0.08 hours

Actual Production Filter Results vs. Predicted Results

The actual cycle time for the production filters was recorded and found to be within 2% of the predicted cycle time. The residual salts in the filter cake were undetectable and the residual moisture in the filter cake was suitable for sending to a landfill.

Summary

Various fine particles must be removed from scrubbing fluids to ensure proper operation of the scrubber, heat exchangers, etc. Laboratory tests were conducted on an amine and the results of the laboratory tests were used to scale-up to a production candle filter that would remove 600 ppm of solids from a 50 m^3/hr amine stream. The results in the production filter matched the predictions and demonstrated that candle filter technology is well suited to remove fines from scrubbing fluids.

Reference 1 - Transport Processes and Unit Operations, Second Edition by Christie J. Geankopolis, 1983, pp 750 - 758

(495b) Modeling and Simulation of Fluid Flow During Deep Drilling Process

Isaac K. Gamwo[1], Mohd A. Kabir[2], Jamie L. Brown[3] and Jimmy D. Thornton[3], (1)National Energy Technology Laboratory, U.S. Department of Energy, Pittsburgh, PA, (2)ORISE, U.S Department of Energy, National Energy Technology Laboratory, Pittsburgh, PA, (3)U.S Department of Energy, National Energy Technology Laboratory, Morgantown, WV

Due to ever increasing demands of gas in the United States and worldwide, shallow gas reservoirs are disappearing. The giant gas reserves lurk several miles beneath the Earth's surface where high temperatures and high pressures conditions prevail and render the environment hostile to down-hole tools. It is estimated that 169-187 trillion cubic feet of technically recoverable natural gas are trapped in reservoirs below 15,000 feet. The United States Department of Energy-National Energy Technology Laboratory is attempting to understand the deep drilling process at extreme conditions of temperature and pressure. During drilling process, particulate multiphase drilling fluids are continuously cycled through the drill assembly. The drilling fluid performs many functions simultaneously. Some of the functions are transport of drilled solids from the wellbore and release them at the surface, lubricate the drilling assembly, and seal the well wall in permeable formation by forming a filter cake at the wall. The present research uses computational fluid dynamics technique to simulate the fluid flow during deep drilling process at high temperature (170oC) and high pressure (25,500 psi) as well as the filter cake formation at the well wall. The effects of drilled particles volume fractions (from 0.4 to 1.0) on the fluid flow pattern and filter cake formation have also been investigated.

See more of this Session: Solid/Liquid Separation Processes for Energy and Environmental Systems
See more of this Group/Topical: Separations Division

(495c) Performance of Solid-Liquid Separation Processes On Lignocellulosic Biomass Slurries

David A. Sievers and Daniel J. Schell, National Bioenergy Center, National Renewable Energy Laboratory, Golden, CO

Separation of lignocellulosic slurries into liquor and solids fractions is a critical but often overlooked step in ethanol production from biomass. Many dilute-acid pretreatment-based biorefinery configurations require solid-liquid separation at various points throughout the process. For example, higher fermentation yields are achieved when the liquor fraction is separately conditioned at an elevated pH before recombination with the solids. Separate fermentation of pentose and hexose sugars is another process configuration that requires extraction of xylose-rich liquor from the cellulose solids. Because the performance of these solid-liquid separation steps directly affects product recovery, the efficacy of this unit operation also influences process economics.

Experiments were performed on bench and pilot scale equipment to determine solid-liquid separation efficiency during co-current washing of dilute-acid pretreated corn stover slurries. The three separation techniques assessed were pressure filtration, vacuum filtration and centrifugation. Computer simulations were then developed to simulate multiple stage counter-current washing to determine the impact of staged separation and wash water usage rate on recovery of soluble sugars from pretreated slurries. Equipment configuration, slurry physical properties, and solids washing techniques were identified as the major factors influencing performance, and these issues will be discussed and recommendations will be provided.

See more of this Session: Solid/Liquid Separation Processes for Energy and Environmental Systems
See more of this Group/Topical: Separations Division

(495d) Recovery of Water From Water Alumina Slurries by Cross Flow Filtration

Seyi A. Odueyungbo, Applied Research and Catalysis, Process R&D, Chevron Energy Technology Company, Richmond, CA

The recovery of water from water-alumina mixture was conducted using cross flow filtration. The alumina particles ranged from few to 60 micron. The filtrate flux, the turbidity of the filtrate and the impact of filter cake on the quality of the filtrate is discussed

See more of this Session: Solid/Liquid Separation Processes for Energy and Environmental Systems
See more of this Group/Topical: Separations Division

(529a) Structural & Energy Considerations in the Design of Carbon Capture Units

Angelo Lucia, University of Rhode Island, Kingston, RI, Anirban Roy, Chemical Engineering, University of Rhode Island, Kingston, RI and Mikhail Sorin, CANMET Energy Technology Centre - Varennes, Natural Resources Canada, Varennes, QC, Canada

It is well known that the CO2 stripping column is energy intensive and, in power plant applications, can consume ~30% of the energy produced by the power plant. Amine-based absorption-stripping processes represent a well understood carbon capture technology that can be readily deployed and/or retrofitted to meet the need of greenhouse gas reductions from power plants and other industrial manufacturing operations. Conventional absorber-stripper processing units consist of interconnected absorption and CO2 stripping columns as well as various heat exchangers for re-boiling, condensing and exchanging heat between process streams. Standard process designs consist of a single interconnection that feeds CO2-rich solvent to the stripping column and recycles CO2-lean solvent to the absorber.

In this work, we study monoethanolamine (MEA)-based absorption-stripping processes under various lean loadings. The novel aspects of this work include 1) Data-driven vapor-liquid equilibrium. 2) Structural optimizations with multiple recycle streams. 3) The shortest stripping line approach to determine minimum energy designs.

We use experimental data directly to capture and describe the vapor-liquid phase and chemical equilibrium in CO2-MEA-water mixtures. This avoids the need to choose models and fit model parameters to data and provides the flexibility of using laboratory, pilot plant or operation data in modeling and simulation. We also investigate the structural optimization of the absorber coupled to a three-product stripping column with multiple recycle streams. Here both lean (bottoms) and intermediate (side-stream) CO2-loaded solvent streams are recycled back to the absorption column with the purpose of understanding trade-offs between capital investment and energy costs. The shortest stripping line distance approach of Lucia et al. (2008) and iterative refinement (Lucia & Hassan, 2010) are used to find minimum energy requirements and determine if the resulting designs are pinched or non-pinched. Many numerical examples and geometric illustrations are presented and are validated and compared to results using the Aspen Plus simulator with the KEMEA model of Rochelle (2003).

See more of this Session: CO2 Capture, Control and Sequestration
See more of this Group/Topical: Sustainable Engineering Forum

(529b) A Thermodynamic Analysis of the Photosynthesis Reaction by Assigning Photons a Carnot Temperatue

Craig A. Griffiths[1], Michelle Low[2], Heinrich Wilfrid Kok[2], Diane Hildebrandt[3] and David Glasser[4], (1) COMPS, University of the Witwatersrand, Johannesburg, South Africa, (2)Centre of Materials and Process Synthesis, University of the Witwatersrand, Johannesburg, South Africa, Johannesburg, South Africa, (3)Center of Material and Process Synthesis, University of the Witwatersrand, Johannesburg, South Africa, Johannesburg, South Africa, (4)Centre of Material and Process Synthesis, University of the Witwatersrand, Johannesburg, South Africa, Johannesburg, South Africa

Increasing concentrations in atmospheric carbon dioxide have led to concerns over the use of fossil fuels. These concerns have led to a surge in the development of sustainable alternative fuels. Photosynthesis offers many possibilities since the basic reaction captures carbon dioxide and produces biomass. This biomass can be converted into a various forms of biofuels such as biodiesel. Photosynthetic processes offer the benefit that they are based on sunlight which is a long lasting and freely available energy source. A look at the thermodynamics of a generalised photosynthetic reaction which produces a range of organic compounds shows there is a relationship between the enthalpy (energy) and Gibbs Free Energy (work) over the process. This relationship shows that, for a reversible process, the work and energy needs can be satisfied by a high quality energy source with a Carnot temperature near the temperature of the Sun. By looking at the energy balance over the photosynthetic reaction, the required number of photons can be calculated at different wavelengths, since the energy per photon is determined by its wavelength. A similar calculation can be performed with the Gibbs Free Energy across the reaction and treating the photons as a high quality energy source with a Carnot temperature equal to the surface temperature of the sun. Comparing these results with values obtained from literature discrepancies are observed. The result is that exothermic heat transfer becomes an important issue for real processes.

See more of this Session: CO2 Capture, Control and Sequestration
See more of this Group/Topical: Sustainable Engineering Forum

(529c) Energy Conservation in CO2 Capture by Mechanical Vapor Recompression

Damaraju Phaneswararao, Chemical Engineering, Indian Institute of Technology, Delhi, New Delhi, India, R.V. Janai, Chemical Engg. Dept, Indian Institute of Technology, Delhi, New Delhi, India and Rajat Agrawal, chemical Engg.Dept., Indian Institute of Technology, Delhi, New Delhi, India

Steam and carbon dioxide gas mixture having typical compostions of the gas stream leaving a stripping column in a carbon dioxide removal system based on absorption in a chemical solvent followed by steam stripping, is compressed in diaphragm and reciprocating compressors in order to recover the thermal energy present in it by condensaing the compressed gas in the condensing coils of the reboiler. A pilot plant has been designed with the compressor connected to the desorption column.

See more of this Session: CO2 Capture, Control and Sequestration
See more of this Group/Topical: Sustainable Engineering Forum

(529e) An Integrated Spatiotemporal Modelling, Design and Optimization Framework for the Large-Scale Deployment of CO_2 Capture, Transport and Storage

N.V.S.N. Murthy Konda[1], N. Shah[1], Gert Jan Kramer[2] and Nigel P. Brandon[3], (1)Chemical Engineering Department, Imperial College London, London, United Kingdom, (2)Shell Global Solutions International, Amsterdam, Netherlands, (3)Department of Earth Sciences, Imperial College London, London, United Kingdom

One of the main stumbling blocks to realize large-scale deployment of carbon capture and storage (CCS) is the huge upfront costs involved. In addition, the usual geologically-dispersed nature of the large number of CO_2 point-sources and sinks calls for a perspective that is beyond 'matching each source with the nearest sink' and requires a holistic-systems perspective. Furthermore, since the CO_2 emission mitigation targets are expected to gradually increase over the next several decades, it is important that the CCS network is developed in harmony with the mitigation targets while ensuring that the investments are made optimally (as and when/where they are necessary) to minimising the entire lifecycle costs. Hence, in this contribution we have proposed a comprehensive optimization framework, that is spatially and temporally explicit, to design the least-cost CCS networks and their optimal evolution with time over the next four decades (i.e., until 2050). Such a long-term perspective also helps to optimally place the future fossil fuel-based plants (e.g., power plants and H2 production plants). While spatially explicit CCS networks design is not entirely new, optimization based studies are rather limited and dynamic-model based optimization studies are even more limited in the literature. In this respect, our framework is novel and helps minimize the overall costs.

We have then demonstrated the applicability and usefulness of our approach with a real case study by applying it to design CCS networks for the Netherlands. A recent study (Konda et. al., 2010) has shown that CCS must be an integral part of the Dutch CO_2 mitigation portfolio to comply with the local and regional (i.e., EU level) CO_2 mitigation targets. Further, the availability of a number of large-scale CO_2 point-sources and large storage capacity makes CCS an attractive CO_2 mitigation option for the Netherlands. Potential CO_2 sources considered include 110 power and industrial sources (including refineries and chemical plants). Altogether, these sources emit 80 MT of CO_2 annually and account for more than 50% of Dutch annual emissions. Potential sinks include depleted gas reservoirs (including the Groningen fields) onshore and offshore, enhanced coal-bed methane recovery (ECBM) sites and saline aquifers. Within the Dutch context, depleted oil reservoirs' estimated storage capacity is rather limited and hence they are not considered. Further, due to the uncertainty associated with the availability of the sinks within the Netherlands, we have also considered the possibility of exporting CO_2 to the sinks that are beyond the Dutch borders. These include storage sites in the southern North Sea basin of the UK and the Norwegian part of the North Sea. These are relatively large structures and hence are likely preferred candidates over the smaller fields located in the Netherlands. Further uncertainty analysis is carried out using a scenario-based approach – these include scenarios that are policy-driven with varying commitments. CO_2 is transported by pipelines (both onshore and offshore) and pipeline economies of scale is explicitly accounted for (by piecewise linearization) in our models.

In addition to the Dutch case study, our presentation will encompass a UK CCS case study as well, in which we have proposed a clustered-deployment of CCS networks within the South-East of UK.

Disclaimer: Results/comments are of the authors' and do not necessarily represent the views of the associated organizations/institutions.

Acknowledgements: Authors would like to thank Shell for their financial support. This work benefited from the views of Fred Hage (Linde, The Netherlands).

References: Konda, N. V. S. N. M., Shah, N., Kramer, G.J., and Brandon, Nigel P., "Dutch Hydrogen Economy: A Multi-billion Dollar Question? Reflections from an Optimization Framework," to be presented at PSE Asia 2010, Singapore, July 2010.

See more of this Session: CO2 Capture, Control and Sequestration
See more of this Group/Topical: Sustainable Engineering Forum

SIMULTANEOUS CAPTURE AND MINERALIZATION OF FLUE GAS CARBON DIOXIDE FROM A COAL-FIRED POWER PLANT

Sanil John[a], Katta J. Reddy[a], M.D. Argyle[b], Hollis W. Weber[a], P. Bhattacharyya[a], David T. Taylor[c], and Paul M. Fahlsing[d]

[a]*Department of Renewable Resources, University of Wyoming, 1000 E University Avenue, Laramie WY USA*
[b]*Department of Chemical Engineering, Brigham Young University, Provo, UT USA*
[c]*Department of Agricultural and Applied Economics, University of Wyoming, Laramie WY USA*
[d]*Jim Bridger Power Plant, Point of Rocks, WY USA*

Introduction

Coal reserves are crucial for providing national energy needs because of dwindling petroleum and natural gas reserves and increasing demand for energy. However, flue gas emission from coal-fired power plants is a major source for the release of anthropogenic CO_2 into the atmosphere. Concurrently coal-fired power plants also generate significant quantities of solid residues (e.g. fly ash particles) as by-products. Increasing anthropogenic CO_2 levels, as a consequence of burning fossil fuels, are raising concerns over global warming and climate change. In addition to CO_2 emissions, release of trace amounts of flue gas SO_2, NO_x (NO/NO_2), and Hg into the atmosphere is also critical air quality concerns.

To address the anthropogenic CO_2 problem, multiple CO_2 capture and storage (CCS) processes are proposed [1,2,3]. As a result different CO_2 capture technologies and storage processes are under evaluation. The CO_2 capture technologies include membrane separation technologies, sorbent technologies involving pressure or temperature swing processes, and the use of solvents such as monoethanolamine [4,5,6,7]. The CO_2 storage processes include subsurface pressure injection into geologic strata and reservoirs of saline, oil, and gas. In addition, mineral carbonation, a process of converting CO_2 to stable minerals (mineralization), is also studied extensively. Among different CO_2 storage processes, mineral carbonation is an ideal approach to store CO_2 on a geological time scale [8]. However, CO_2 capture technologies and storage processes have limitations for widespread practical use due to the requirement of separation of CO_2 from flue gas, compression of CO_2, and transportation of concentrated CO_2 to a site where it can be safely stored or used for mineral carbonation. Furthermore, CO_2 separation and capture technologies are severely limited by the flue gas SO_2, because SO_2 is known to affect the performance of amine solvents [9]. In addition to CO_2 and SO_2 air quality concerns, the Hg emissions from coal-fired power plants and its potential for deposition in soils, vegetation, water, and animals also received considerable attention in recent years [10,11].

In a natural chemical weathering process, carbonic acid, which results from the interaction of atmospheric CO_2 and rain water, will convert alkaline earth minerals to carbonate minerals. Similarly, alkaline minerals present in fly ash will also convert to carbonates through a chemical weathering process. In earlier aqueous mineral carbonation (AQMC) studies, CO_2 was bubbled through slurry of industrial residues (e.g., oil shale, coal fly ash) or exposed to a higher CO_2 atmosphere to understand mineral changes associated with the carbonation process [12,13,14]. Some of these studies reported that AQMC process dissolves silicate/oxide minerals and precipitates carbonate minerals in industrial residues. Earlier research also established that the mineral carbonation process will help improve chemical properties of industrial residues and prevent leaching of trace elements (e.g., arsenic, cadmium, lead, and selenium) from land disposal sites into surface water, soils, and groundwater. In addition, the mineral carbonation process also creates a favorable environment in industrial residues for biological processes, because carbonate minerals act as nutrients for microorganisms [15].

Aqueous mineral carbonation of industrial residues is a slow process; hence, an accelerated mineral carbonation (AMC) process was proposed [16]. In this method, oil shale residues with 15-20% of moisture (by weight basis) were exposed to moistened CO_2 pressure (135.8 kPa) for 1 hr. The carbonated solids were subjected to mineral solubility studies. Results of AMC studies were similar to the results of earlier AQMC studies and reported that since AMC uses CO_2, which can be obtained from the combustion process itself, another potential benefit is that it may help to minimize CO_2 emissions into the atmosphere. In a recent study, Dellantonio et al. [17] articulated the role of coal combustion residues in storage of CO_2. Chronological development of AQMC and AMC of industrial residues [e.g., coal fly ash/bottom ash, clean coal technology (CCT) ash, municipal solid waste (MSW) incinerated fly ash/bottom ash, oil shale ash, steel slag, medical solid waste incinerated ash, paper mill ash, and cement kiln dust] and natural minerals [calcium, iron, and magnesium silicate minerals] were published elsewhere [18]. Since AQMC or AMC processes were proposed, CO_2 mineralization studies have been largely investigated at laboratory scale.

The objectives of this research were to: 1) conduct preliminary studies to determine the possibility of simultaneous capture and mineralization of flue gas CO_2 without separation under field conditions; and 2) develop a pilot scale study for further testing to determine the feasibility of proposed AMC process. The preliminary and subsequent pilot tests of the AMC process were conducted at Jim Bridger power plant (JBPP), Point of Rocks, Wyoming. The JBPP burns an average of 21,772 tonnes of coal per day using four units to produce 2,120 MW. All preliminary experiments were conducted on Unit two flue gas stack at approximately 30 m above ground. All pilot scale experiments were conducted in the Unit two ash hopper building.

Preliminary Studies (Patent approval number: PCT/US/20090280046)

Approximately 23 kg of fly ash was collected from the electrostatic precipitators and placed in the fluidized-bed reactor (**Figure 1**). The fluidized-bed reactor was constructed of acrylic (Plexiglas) to allow the bed operation to be observed. The operating portion of the reactor was 0.3 m in diameter and 1.1 m long. A perforated distributor plate with 2×10^{-3} m diameter openings (approximately 307 openings) provided uniform distribution of the flue gas through the ~ 0.6 m deep bed of fly ash above it. A small blower (0.028 m^3/s (STP), HRB600, Republic regenerative blowers) was used to force the flue gas through the fly ash because flue gas from the power plant is near atmospheric pressure and does not provide sufficient pressure to fluidize the bed.

A pleated fabric filter attached to the top flange of the reactor allowed the flue gas to pass through the reactor and return the fly ash to the bed for additional contact with new flue gas. Fly ash samples were collected from the sampling ports at the side of the reactor during experiments at approximately 2, 5, 10, 15, 30, 60, 90, and 120 minutes. The temperature and pressure during the reaction were monitored by mechanical gauges at the top of the reactor. The temperature during the reactions varied between 316 and 327 K, and pressure varied between 88 and 114 kPa.

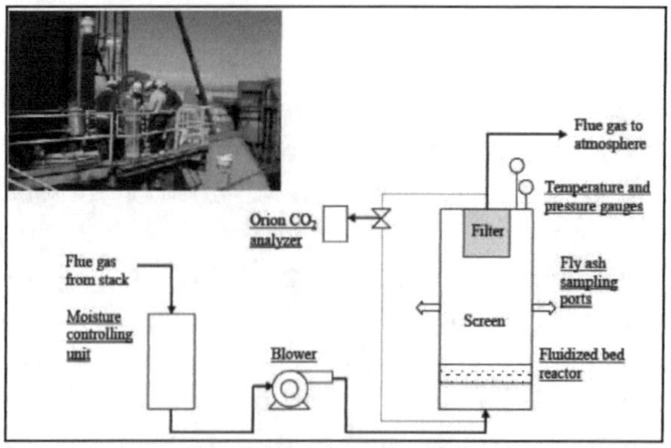

Figure 1 Preliminary experiment set-up for CO_2 capture and mineralization. Inset shows the testing at the plant (Proprietary Information).

An Orion® *plus* IR detector was connected between the blower and outlet, as shown in the Figure 1 to monitor flue gas CO_2 and SO_2 concentrations before and after reaction. The flue gas Hg mineralization was determined by analyzing Hg content of control and flue gas treated fly ash samples. Control fly ash samples were analyzed for particle size, moisture content, pH, total concentration of C, S, and Hg. Fly ash samples were also analyzed for major mineral phases and surface morphology and subjected to trace element fractionation and water solubility studies. Fly ash particle size was estimated using a scanning electron microscope (SEM; JOEL JSM-5800LV). Moisture content of fly ash was measured by heating samples to 373 K in a drying oven (Sargent Analytical Oven) and calculating the difference in mass before and after heating. Flue gas treated fly ash samples were analyzed for pH, C, S, Hg, surface morphology, major mineral phases with elemental composition, metal fractionation, and water solubility studies.

The flue gas CO_2 concentration decreased from 13.0 to 9.6% within few minutes of reacting with fly ash particles (Figure 2a). Conversely, the carbonate content of fly ash increased from non-detectable to 2.34% during the first 2 minutes of reaction (Figure 2b). The pH of flue gas treated fly ash decreased from 12.2 to 10.6 due to the carbonation process. The XRD analysis show calcite ($CaCO_3$) and thaumasite [$Ca_3Si(SO_4)CO_3(OH)_6.12H_2O$] in flue gas treated sample. To verify the formation of these minerals, reacted fly ash was treated with conc. HCl and re-X-rayed. The peaks for calcite and thaumasite were disappeared which indicated the presence of these minerals in flue gas reacted fly ash. These results suggest that flue gas CO_2 is quickly mineralized into carbonates in fly ash. It is well documented that AMC process converts CO_2 into calcite in fly ash and other coal combustion residues [15, 19]. The thaumasite is also reported to form in alkaline industrial residues where carbonate and sulfate are available [20]. The flue gas CO_2 mineralization by the fly ash particles can be explained by the following reactions (here we used CaO as an example):

CO_2 (g) (flue gas) + H_2O (moisture in flue gas) $\leftrightarrow$ $H_2CO_3^0$ (carbonic acid) [1]
$H_2CO_3^0$ (carbonic acid) $\leftrightarrow$ H^+ + HCO_3^- (bicarbonate) [2]
CaO (fly ash) + H^+ + HCO_3^- (bicarbonate) $\leftrightarrow$ $CaCO_3$ (calcite) + H_2O [3]

If we combine equations 1-3, the overall reaction is:
CO_2 (g) (flue gas) + CaO (fly ash) $\leftrightarrow$ $CaCO_3$ (calcite) [4]

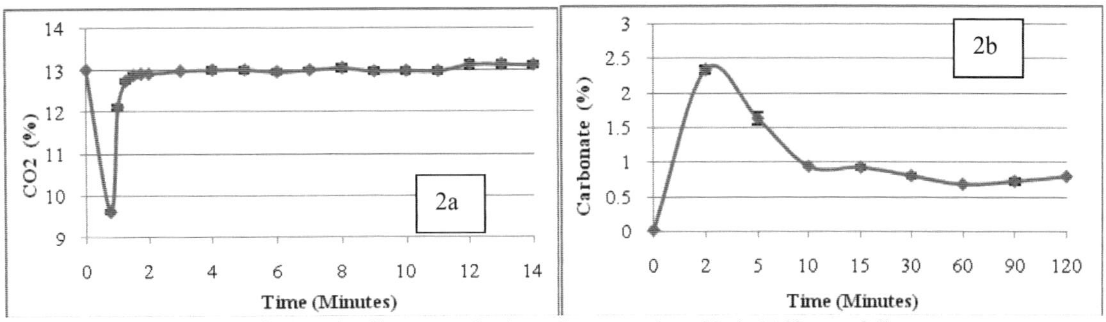

Figure 2 Concentration of CO_2 in flue gas during the reaction (2a). Effect of flue gas treatment on % carbonate content of fly ash (2b). The pH of the flue gas treated fly ash was 10.6. Average of three experiments. Error bar represents standard deviation (SD).

The results of this study also suggest that, in addition to calcite and thaumasite, other carbonate minerals such as dawsonite ($NaAl(CO_3)(OH)_2$) and alumohydrocalcite ($CaAl_2(CO_3)_2(OH)_4 \cdot 3H_2O$) could form in the flue gas reacted fly ash samples. The conditions required for formation of these carbonates include: alkaline pH, source of carbonate, Na and Al, and slightly higher temperatures [21,22]. These conditions do exist in the fluidized bed reactor. The chemical analysis (Figure 3b) of the flue gas treated fly ash and the reference dawsonite material spectrum reported in literature [22] by the EDS are very similar (Na:Al ~ 1:5). Overall, SEM and EDS analysis revealed the formation of distinct new crystal phases with the chemical composition of Al, Fe, Na, Ca, Mg, K, Si, C and S within few minutes of flue gas reaction (Figures 3c and 3d).

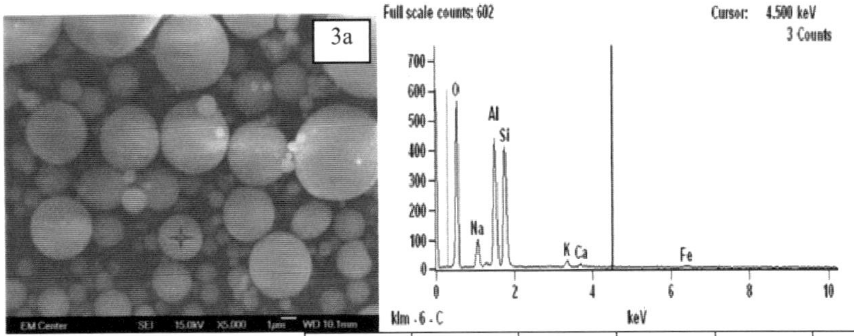

O-K	*Na-K*	*Al-K*	*Si-K*	*K-K*	*Ca-K*	*Fe-K*
59.08	4.43	14.50	16.77	1.56	0.86	2.80

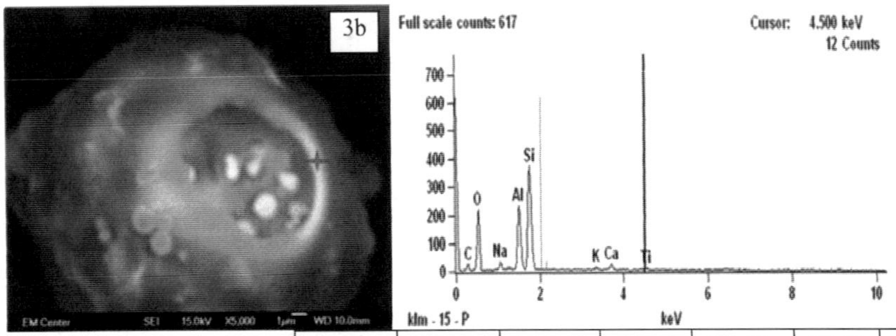

C-K	*O-K*	*Na-K*	*Al-K*	*Si-K*	*K-K*	*Ca-K*	*Ti-K*
15.34	49.73	1.79	9.55	18.88	0.97	2.31	1.44

Figure 3 The SEM images and EDS spectra measured for control fly ash sample (3a) and 2 minutes flue gas treated fly ash samples (3b). The red star in the figures shows the point analyzed.

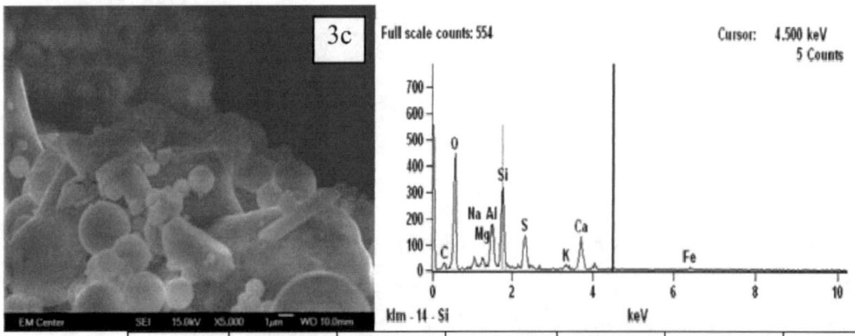

C-K	O-K	Na-K	Mg-K	Al-K	Si-K	S-K	K-K	Ca-K	Fe-K
8.27	57.79	1.78	0.98	4.96	10.17	5.62	0.75	7.25	2.43

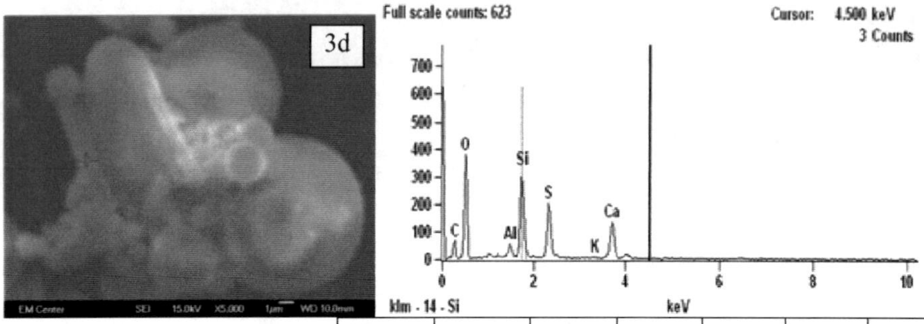

C-K	O-K	Al-K	Si-K	S-K	K-K	Ca-K
18.65	56.51	1.05	8.45	7.20	0.30	7.83

Figure 3 The SEM images and EDS spectra measured for 2 minutes flue gas treated fly ash samples (3c, 3d). The red star in figures shows the point analyzed.

The findings of this study suggest that flue gas CO_2 mineralizes when it reacts with fly ash particles. Two probable mechanisms may explain these observations. First, flue gas CO_2 converting calcium and other oxides in fly ash into carbonate minerals (e.g., calcite, thaumasite, dawsonite, alumohydrocalcite). Second, EDS chemical analysis of the new crystal structures in flue gas CO_2 treated samples shows C along with Al, Fe, Ca, Mg, Na, K, and Si indicating that acidic nature of flue gas (due to nitric and sulfuric acids that were not completely removed in the wet scrubber) probably dissolved amorphous silicate minerals in fly ash, which converted into carbonates. For example, studies have shown that amorphous silicate minerals are much more reactive and soluble in acidic pH [23,24,25].

In addition to the mineralization of flue gas CO_2, the AMC process also mineralized SO_2 and Hg from flue gas (Table 1). The flue gas SO_2 concentration decreases from 107.8 to 15.1 ppmv within two minutes of reaction and stabilized around 22 ppmv. Subsequently S content increased in fly ash from 0.29% to 0.36%. The XRD analysis identified gypsum in flue gas treated sample. The reduction in flue gas SO_2 and increase in S content of fly ash were attributed to the formation of gypsum ($CaSO_4.2H_2O$) in fly ash. The total concentration of Hg in fly ash increased within few minutes of reaction suggesting that AMC process also mineralized Hg from flue. These results suggest that Hg in flue gas probably oxidizes to Hg^{2+} and mineralizes to $HgCO_3$ in fly ash.

Time (min)	SO_2 (ppmv)	Time (min)	S (wt %)	Hg (mg/kg)
0	107.8 ± 0.1	0	0.29 ± 0.01	0.10 ± 0.001
1	5.1 ± 2.0	2	0.23 ± 0	0.22 ± 0.005
2	15.1 ± 5.0	5	0.26 ± 0.03	0.14 ± 0.003
5	27.3 ± 9.6	10	0.30 ± 0.01	0.16 ± 0.001
10	29.4 ± 5.6	15	0.36 ± 0	0.16 ± 0
11	26.6 ± 5.3	30	0.34 ± 0	0.11 ± 0.001
12	23.7 ± 6.8	60	0.30 ± 0.01	0.12 ± 0.007
13	22.8 ± 8.1	90	0.29 ± 0.01	0.13 ± 0.004
14	22.5 ± 8.8	120	0.23 ± 0.01	0.16 ± 0.002

Table 1 Effect of the mineralization process on flue gas and fly ash SO_2 and Hg content

A preliminary economic analysis of the process for 90% capture from a 532 MW power plant yields a mineralization cost of about \$11/tonne CO_2 at a mineralization capacity of 207 kg CO_2/tonne fly ash [26].

Pilot Scale Studies (Patent approval number: PCT/US/20090280046)

Figure 4 A photograph of the pilot process skid installed at the power plant (From right to left, gas analyzer moisture reducing drum, humidifier, and fluidized bed reactor) (Proprietary Information).

Based on the preliminary studies, we designed and developed pilot scale process skid to capture and mineralize flue gas CO_2. The CO_2 capture and mineralization pilot process (Figure 4) consists of three process vessels - a moisture reducing drum (MRD) (0.9 m $\Phi \times$ 1.8 m), a heater/humidifier (0.9 m $\Phi \times$ 1.8 m), and a fluidized bed reactor (FBR) (0.9-1.2 m $\Phi \times$ 3.7 m). Flue gas was withdrawn from the stack and was fed to the MRD at about 0.094 m^3/s. The MRD and the heater/humidifier pretreat flue gas before it enters the FBR. The MRD captures droplets of water entrained in the flue gas to protect the blower placed between the MRD and the heater/humidifier. The heater/humidifier enables control of flue gas moisture and temperature. Approximately 100-300 kg of fresh fly ash was collected from the electrostatic precipitators through ash hopper and placed in the fluidized-bed reactor. The fly ash particles were fluidized by flow of flue gas through a distributor plate in the FBR, ensuring proper mixing and good contact between the fly ash particles and the flue gas. A perforated plate (2.39×10^{-3} m Φ holes) was placed above the distributor plate to minimize collection of ash below the distributor plate. A control valve and a pressure transmitter were used to set the pressure inside the FBR. A particulate removal cyclone in the reactor separated fly ash particles from the exiting flue gas. The pressure drop across the distributor plate, the fluidized bed, and the cyclone were measured by differential pressure transmitters. The temperature at various points inside the humidifier and the reactor were measured by thermocouples. The vessels and piping connecting them are insulated to minimize heat loss through the walls and prevent moisture condensation. The flue gas was continuously monitored by an industrial grade multi-gas analyzer (HORIBA VA-3000). It was connected to the inlet and outlet lines to monitor the real-time concentration of CO_2, NO_x, and SO_2 in the flue gas. Fly ash particles were sampled from the reactor using installed sampling ports. The samples were analyzed for carbonate content by dissolution of the carbonates in an excess of standard acid followed by back titration of the remaining acid with standard alkali.

The experimental conditions of the three runs are summarized in Table 2. The flue gas temperature in the reactor was a function of the temperature of the ash received from the plant hopper and the temperature of the flue gas entering the reactor. The humidifier was set at 323 K for experiment no. 1 but it attained higher temperatures due to heating of the flue gas as it passed through the blower. For experiment no. 2, the humidifier temperature was not controlled and so it depended on the temperature of the flue gas, while for no. 3 and 4 the humidifier temperature was set at 313 and 318 K respectively, by regulating the water flow rate through the humidifier.

Expt. no.	Reactor temperature (K)	Humidifier temperature (K)	Ash quantity (kg)	Reactor pressure (Pa)	Gas flow rate (m^3/s) (STP)
1	321-330	312-328	300	115,142.45	0.094
2	305-331	300-331	200	115,142.45	0.094
3	308-314	313	100	115,142.45	0.094
4	315-320	318	200	115,142.45	0.094

Table 2 Summary of the experiment conditions

The calculated CO_2 conversion cumulative averages and the carbonate content of the fly ash samples are shown in Figures 5a and 5b. The highest conversion is seen around 10 minutes and then there is moderate (Expt. no. 1) or slight (Expt. no. 2 and 3) drop. The CO_2 capture capacity of the amount of ash in the reactor is not fully diminished after 1.5-2 hours of reaction. The maximum $CaCO_3$ content of fly ash was about 1% after about 2 hours of reaction. A sample (~300 mins.) of fly ash obtained after multiple runs shows $CaCO_3$ as high as 2.64%.

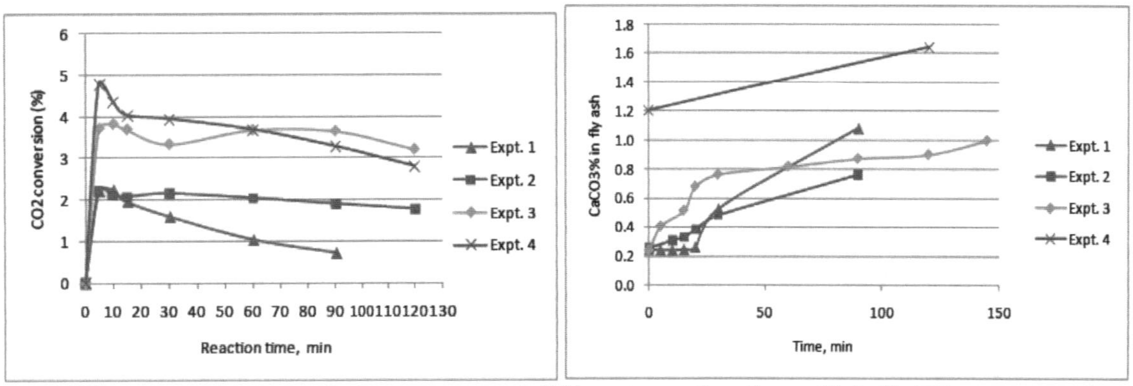

Figure 5 CO_2 conversion cumulative averages at different reaction times (5a). $CaCO_3$ content of various fly ash samples (5b).

Conclusions

We conducted AMC experiments, under field conditions, to capture and mineralize coal-fired power plant flue gas pollutants (e.g. CO_2, SO_2, and Hg) using fly ash particles. Results show that flue gas components can be directly mineralized (without separation) by the fly ash particles using AMC process. The proposed AMC process mineralized flue gas CO_2 into carbonates (e.g., calcite, thaumasite) within few minutes of reaction. However, further research to optimize AMC process parameters (e.g., moisture, temperature, and contact time) to improve the flue gas CO_2 mineralization capacity by fly ash particles will be invaluable. Such information will help to develop scale-up process under industrial conditions and help to minimize emissions of flue gas pollutants into the atmosphere.

References

1. Intergovernmental Panel on Climate Change. Climate Change. The physical science basis: Contribution of working Group I to the Fourth Assessment Report of the IPCC. Cambridge University Press, Cambridge 2007.
2. Pacala S, Socolow RH. Stabilization Wedges: Solving the climate problem for the next 50 years with current technologies. Science 2004;305:968-72.
3. Charles D. Stimulus gives DOE billions for carbon – capture projects. Science 2009;323:1158.
4. Herzog H. Accelerating the deployment of CCS. 8th Annual carbon capture and sequestration conference proceedings, Pittsburg, PA 2009.
5. Reynolds SP, Ebner AD, Ritter JA. New pressure swing adsorption cycles for carbon dioxide sequestration. Adsorption 2005;11:531-6.
6. Atimtay AT. Cleaner energy production with integrated gasification combined cycle systems and use of metal oxide sorbents for H2S cleanup from coal gas. Clean Prod. Processes 2001;2:197-208.
7. Kintisch E. Power Generation: Making dirty coal plants cleaner. Science 2007;317:184-6.
8. Lackner KS. A guide to CO_2 sequestration. Science 2003;300:1677-8.
9. Supap T, Idem R, Tontiwachwuthikul P, Saiwan C. The roles of O_2 and SO_2 in the degradation of monoethanolamine during CO_2 absorption from industrial flue gas streams. EIC Climate Change Technology, IEEE 2006;1-6.
10. López Alonso M, Benedito JL, Miranda M, Fernández JA, Castillo C, Hernández J, Shore, RF. Large-scale spatial variation in mercury concentrations in cattle in NW Spain. Environ. Pollut. 2003;125:173-81.
11. Sullivan TM, Adams J, Milian L, Subramaniam S, Feagin L, Williams J, Boyd A. Local impacts of mercury emissions from the Monticellow coal fired power plant. Environmental Sciences Department, Environmental Research & Technology Division, Brookhaven National Laboratory 2006;BNL-77475-2007-IR.
12. Reddy KJ, Lindsay WL, Boyle FW, Redente EF. Solubility relationships and mineral transformations associated with recarbonation of retorted oil shales. J. Environ. Qual. 1986;15:129-33.
13. Essington ME. Trace element mineral transformations associated with hydration and recarbonation of retorted oil shale. Environ. Geol. Water Sci. 1989;13:59-66.
14. Schramke, JA. Neutralization of alkaline coal fly ash leachates by CO_2(g). Appl. Geochem. 1992;7:481-92.

15. Reddy KJ. Application of carbon dioxide in remediation of contaminants: A new approach. In: Wise DL, Trantolo DJ, Inyang HI, Cichon EJ, Editors. Remediation of Hazardous Waste Contaminated Soils. 2nd Edition, New York: Marcel Dekker Inc; 2000.
16. Reddy KJ, Drever JI, Hasfurther VR. Effects of CO_2 pressure process on the solubilities of major and trace elements in oil shale wastes. Environ. Sci. Technol. 1991;25:1466-9.
17. Dellantonio A, Walter JF, Repmann F, Wenzel, WW. Disposal of coal combustion residues in terrestrial systems: Contamination and risk management. J. Environ. Qual. 2010;39:761-75.
18. Reddy KJ, Argyle MD, Viswatej A. Capture and mineralization of flue gas carbon dioxide. In Baciocchi R, Costa G, Polettini A, Pomi R, editors. Conference proceedings of 2nd Accelerated Carbonation for Environmental and Materials Engineering, University of Rome, Rome, Italy, October 1-3, 2008, p. 221-228.
19. Tawfic TA, Reddy KJ, Gloss SP, Drever JI. Reaction of CO_2 with clean coal technology solid wastes to reduce trace element mobility. Water, Air, Soil Pollut. 1995;84:385-98.
20. Bensted J. Thaumasite-background and nature in deterioration of cements, mortars and concretes. Cem. Concr. Compos. 1999;21:117-21.
21. Golab AN, Carr PF, Palamara DR. Influence of localized igneous activity on cleat dawsonite formation in Late Permian coal measures, Upper Hunter Valley, Australia. Int. J. Coal Geol. 2006;66:296-304.
22. Sirbescu MC, Nabelek PI. Dawsonite: An inclusion mineral in quartz from the Tin Mountain pegmatite, Black Hills, South Dakota. Am. Miner. 2003;88:1055-60.
23. Lindsay WL. Chemical equilibria in soils. 1st ed. New York: John Wiley and Sons; 1979.
24. Drever JI. The geochemistry of natural waters. 3rd ed. New York: John Wiley and Sons; 1997.
25. Yang YF, Gai GS, Cai ZF, Chen QR. Surface modification of purified fly ash and application in polymer. J. Hazard. Mater. 2006;133:276-82.
26. Christensen MH. An economic analysis of the Jim Bridger power plant CO_2 mineralization process. Master's thesis, Dept. of Agricultural and Applied Economics, University of Wyoming. 2010.

(582a) Designing Reforming Catalysts to Condition Biomass Derived Syngas

Kim Magrini[1], Whitney Jablonski[2], Yves Parent[2] and Matthew M. Yung[2], (1)National Renewable Energy Laboratory, Golden, CO, (2)National Bioenergy Center, National Renewable Energy Laboratory, Golden, CO

One approach to the efficient thermochemical conversion of biomass to produce liquid fuels using indirect gasification utilizes a catalytic steam reforming process to remove and reform tars to produce clean syngas which additionally maximizes the use of the biomass carbon and hydrogen content. The catalysts used in the steam reforming reactor can comprise a significant portion of the overall process cost to produce biofuels. In order to reduce the catalyst cost and improve the overall fuel synthesis economics, highly active and stable reforming catalysts are required. To achieve this goal, we are using rational catalyst design to understand, design, and synthesize robust, tailor-made catalysts capable of handling biomass-derived impurities such as tars, polyaromatic hydrocarbons (PAHs), and H2S. Emerging reforming catalysts designed by industrial collaborators are also assessed for reforming performance. Results of catalyst design and reforming performance at laboratory and pilot scales are discussed.

See more of this Session: Catalytic Conversion of Renewable Resources to Synthesis Gases and Pyrolysis Oils
See more of this Group/Topical: Sustainable Engineering Forum

(582b) A DFT Study of Desulfurization and Tar Cracking of Gasifier Effluents Over Ceria-Based Rare-Earth Oxides

Adam D. Mayernick, Department of Chemical Engineering, Pennsylvania State University, University Park, PA, Michael Janik, Department of Chemical Engineering, Pennsylvania State University-University Park, University Park, PA and Kerry M. Dooley, Dept. of Chemical Engineering, Louisiana State University, Baton Rouge, LA

Cleanup of biomass gasifier effluents involves both the removal of sulfur-containing compounds such as H_2S and the cracking of tars. Mixed rare earth oxides of ceria containing lanthanides or transition metal additives are effective H_2S sorbents and tar cracking catalysts. We report density functional theory (DFT +*U*) results used to evaluate the H_2S adsorption capacity and tar cracking activity of pure ceria and ceria doped with a series of lanthanides and group VIIB-VIIIB transition metals. We use ab initio thermodynamics to construct phase diagrams for the surface layers of these rare earth oxides as well their corresponding sulfides and oxysulfides as a function of H_2S and O_2 partial pressures. Our results illustrate the relationship between oxide reducibility and H_2S sorbent capacity, and quantify changes in the surface activity for sulfur adsorption as a result of doping. We also evaluate the effects of transition metal additives on C-H and C-C bond activation, and discuss our results in the context of designing improved rare earth oxide catalysts for simultaneous desulfurization and tar cracking of gasifier effluents.

See more of this Session: Catalytic Conversion of Renewable Resources to Synthesis Gases and Pyrolysis Oils
See more of this Group/Topical: Sustainable Engineering Forum

(582c) Development of a Catalytic Reactor for Conditioning of Biomass Generated Producer Gas and Evaluation of Selected Commercial Catalysts

Ajay Kumar, Hasan K. Atiyeh, Danielle Bellmer and Raymond L. Huhnke, Biosystems and Agricultural Engineering, Oklahoma State University, Stillwater, OK

Gasification of biomass is one of the promising conversion technologies to produce fuels, chemicals and power from renewable resources. Removal of tar is one of the most challenging barriers for commercial-scale use of producer gas generated from gasification of biomass. Tar can be reduced to an acceptable level by cracking in presence of catalysts at high-temperature. This method can also increase carbon monoxide and hydrogen compositions which are low in biomass generated producer gas. In this project, we have designed and developed a catalytic reactor specifically for evaluating commercially-available catalysts for conditioning producer gas which is supplied from a lab-scale gasifier with a biomass feedrate of 3-7 kg/h. In our prior study, selected commercially-available catalysts were evaluated in decomposing toluene as a model tar compound. In this project, the catalysts and reaction conditions will be further evaluated using producer gas with tar generated by gasification of switchgrass as the feedstock. Results are expected to show the effects on selected catalysts, shape of catalysts, and reactor temperature on reducing tar and improving producer gas composition.

See more of this Session: Catalytic Conversion of Renewable Resources to Synthesis Gases and Pyrolysis Oils
See more of this Group/Topical: Sustainable Engineering Forum

(582d) Comparison and Characterization of Bimetallic, Ni-Based Catalysts for Conditioning of Biomass-Derived Syngas

Matthew M. Yung[1], Kimberly A. Magrini-Bair[2], Whitney S. Jablonski[2] and Jessica Olstad[2], (1) National Bioenergy Center, National Renewable Energy Laboratory, Golden, CO, (2)National Renewable Energy Laboratory, Golden, CO

Due to the decreasing availability of low-cost petroleum, efforts to develop technology for transportation fuel production from alternative and renewable sources are being actively researched. Biomass is one resource that has the potential for significant fuel production based on feedstock availability. Thermochemical conversion of biomass to syngas has the potential to be used as a cost-effective and renewable source of transportation fuels, displacing petroleum consumption for this purpose. In order to efficiently utilize the carbon in the biomass and protect downstream processing equipment, the light hydrocarbons and tars in the gas stream following gasification should be converted into additional synthesis gas through steam reforming. To this end, we have designed and evaluated several Ni-based tar reforming catalysts for the conditioning of biomass derived syngas. Catalysts of varying composition are evaluated through reaction experiments for hydrocarbon steam reforming in the presence of hydrogen sulfide, using a model syngas that is representative of the composition that would be created through biomass gasification. Catalyst characterization through several techniques (including XRD, TPR, SEM/EDS, and laser Raman spectroscopy) are also discussed.

See more of this Session: Catalytic Conversion of Renewable Resources to Synthesis Gases and Pyrolysis Oils
See more of this Group/Topical: Sustainable Engineering Forum

(582e) Autothermal Oxidative Pyrolysis of Alternative Hydrocarbon Feedstocks

Christine M. Balonek, Joshua L. Colby and Lanny D. Schmidt, Department of Chemical Engineering and Materials Science, University of Minnesota, Minneapolis, MN

Traditional pyrolysis units are heated externally, requiring large energy inputs, and produce gases, solid chars, and complex liquid mixtures. We have developed and demonstrated an oxidative pyrolysis reactor able to run autothermally with negligible solid char formation. The heat necessary for pyrolysis is generated internally, eliminating the heat transfer issues associated with external heating. This process can produce both synthesis gas and hydrocarbon products from alternative feedstock models of biomass and plastic wastes. Solid cellulose and polyethylene (PE) particles were fed individually with air to a Rh-Ce/γ-Al_2O_3 catalytic fixed-bed reactor. Catalyst temperature and product species were studied as a function of the inlet C/O ratio, the ratio of carbon moles in the solid feed to oxygen moles in the O_2 feed. At all C/O ratios reported, the solid feed was completely converted with negligible coke formation at millisecond time scales. Cellulose and PE pyrolysis produced large amounts of synthesis gas at low C/O ratios (0.5 to 1.0). Cellulose pyrolysis produced up to 20 % carbon selectivity to C_{3+} hydrocarbons at C/O of 1.3 (autothermal operation was unsustainable at higher C/O ratios). At similar conditions, PE pyrolysis had 50 % selectivity to C_{3+}, and selectivity continued to increase with increasing C/O ratio to 90 % at C/O = 3.4. This high initial yield of C_{3+} hydrocarbons from cellulose and PE pyrolysis demonstrates the potential for autothermal oxidative pyrolysis to utilize alternative hydrocarbon feedstocks in a fast, clean, and efficient way.

See more of this Session: Catalytic Conversion of Renewable Resources to Synthesis Gases and Pyrolysis Oils
See more of this Group/Topical: Sustainable Engineering Forum

Production of Bio-oil and Hydrogen from Sawdust via Pyrolysis and Catalytic Steam Reforming

Ebrahim Salehi, Jalal Abedi, Tom Harding, and Fakhry Seyedeyn Azad
Chemical and Petroleum Engineering, University of Calgary, Calgary, AB T2N 1N4, Canada
E-mail: jabedi@ucalgary.ca; Tel: 403-220-5594

ABSTRACT
Fixed-bed and fluidized-bed lab-scale reactors were developed to produce bio-oil using pyrolysis of biomass including sawdust and agricultural wastes such as wheat straw. The role of pyrolysis temperature, nitrogen gas flow rate, and condensation temperature on product yield was investigated. The chemical characteristics of the pyrolysis bio-oil and char obtained under the optimal operational conditions were identified by elemental and calorific analyses. The bio-oil was analyzed using chromatographic and spectroscopic techniques. Pyrolysis temperature had the most significant effect on the bio-oil yield. In addition, an experimental system was designed and constructed to make hydrogen from produced bio-oil in a catalytic steam reforming micro-reactor. Gas chromatographic technique was used to determine the concentration of hydrogen and the other components in produced flue gas from the catalytic micro-reactor. The effect on the hydrogen yield of Ni-based catalysts with different Ni loading was investigated.

Keywords: sawdust; fast pyrolysis; hydrogen; bio-oil, catalyst

1. Introduction

Concerns related to the depletion of fossil fuel reserves as well as environmental issues like global warming make hydrogen, in combination with fuel cells, an attractive alternative energy source. The conventional method of hydrogen production is steam reforming of fossil fuels which results in emission of green-house gasses. Renewable sources, such as biomass, may be used as alternative feedstocks for hydrogen production. Biomass for producing energy has a special place among all other renewable energy sources, and it is estimated to contribute 10-14% to the world's total energy supply [1-7].

To produce hydrogen from biomass, two main technologies are steam gasification and catalytic steam reforming of pyrolysis oil (bio-oil) [8]. The second method has an advantage over steam gasification. In this technique, the first step is to apply fast pyrolysis to bulky and low density biomass and thus convert it into an easy-to-transport intermediate called bio-oil which has a high volumetric energy density. The next step is to steam reform the produced bio-oil into a combustible gas mixture including hydrogen. In addition, bio-oil can be used as a feedstock in conventional petroleum refineries, as well as for the production of high value chemicals [5].

Pyrolysis of biomass is a process in which biomass is subjected to heat in an oxygen-free environment resulting in its conversion to char, liquid, and noncondensable gases. The steam reforming of the bio-oil can be simplified as the catalytic steam reforming of the oxygenated organic compounds (CnHmOk) [9-11], followed by water–gas shift reaction (WGS):

$$C_nH_mO_k + (n-k)H_2O \leftrightarrow nCO + (n+m/2-k)H_2$$

$$nCO + nH_2O \leftrightarrow nCO_2 + nH_2$$

Therefore, the overall process can be represented as follows:

$$C_nH_mO_k + (2n-k)\ H2O \leftrightarrow nCO_2 + (2n+m/2-k)H_2$$

This study aimed to develop first a fixed-bed, and then a fluidized-bed pyrolysis system to produce bio-oil from sawdust. Bio-oil was then steam reformed in another developed experimental apparatus using various Ni-based catalysts made in-house.

2. Experimental

2.1. Feedstock

The feedstock used for pyrolysis was sawdust provided by a waste wood recycling company. The sawdust was sieved into fractions of particle diameter < 590 μm. Before the experiments, the sample to be used was dried overnight at 100°C in an oven. A commercial bio-oil supplied by BTG Company, The

Netherlands, was used in catalytic steam reforming tests. Table 1 shows the elemental analyses and the calorific value of the biomass and bio-oil samples.

2.2. Fixed-bed pyrolysis

The fixed-bed pyrolysis system developed consisted of a stainless steel tubular reactor (1.9 cm I.D., 2.2 cm O.D., and 50 cm length) which was placed in a tubular electrical furnace. In a typical experiment, a sample of 10 g of sawdust was placed in the reactor and nitrogen was used as the carrier gas. After heating the reactor, the biomass particles decomposed to char, non-condensable gas and volatile materials. The gas and the volatile materials passed through two sequential condensers maintained at -10°C by a bath circulator. The schematic diagram of the fixed-bed pyrolysis system may be found in another publication [1].

Table 1. Characteristics of feedstocks

Properties	Bio-oil	Sawdust
Elemental analysis (wt%)[a]		
Carbon	45.83	47.37
Hydrogen	7.15	6.03
Nitrogen	0.87	2.40
Oxygen[b]	46.15	44.2
Heating value[c] (MJ/kg)	17.50	19.05
Water content wt (%)	23.10	-
Density (kg/m3)	1255.00	-

[a] dry basis

[b] by difference

[c] using bomb calorimeter

As discussed in our previous work [1], it was recommended that future tests be performed under fast pyrolysis conditions in a fluidized-bed system. Changing the pyrolyzer from the fixed-bed to a fluidized-bed system allows all feedstock particles to be heated faster and more homogenously which results in higher bio-oil yield and different quality.

2.3. Fluidized-bed pyrolysis

Figure 1 shows a picture of the developed fluidized-bed fast pyrolysis system. This system was designed to operate at temperature between 400-500°C, atmospheric pressure, and with a 100 g/hr biomass feed rate. The system comprised mass flow controllers to control and measure fluidizing gas (nitrogen), fluidized-bed reactor and plenum, biomass feed system, cyclone char collection system for char separation from gas stream, vapor condensation system for bio-oil recovery, instrumentation for data acquisition and control, and heating system to provide heat for the system.

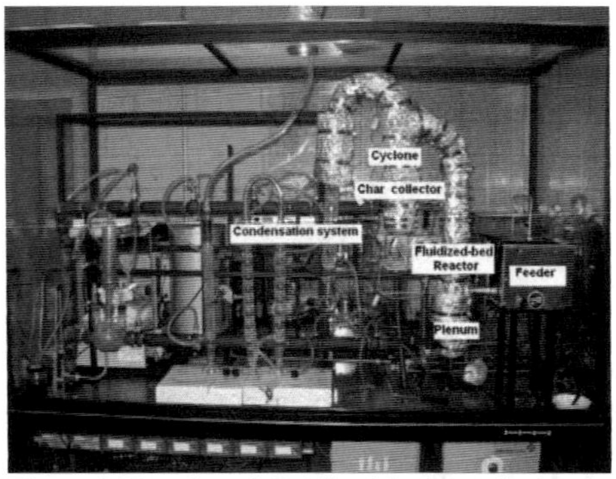

Figure 1. Developed fluidized-bed pyrolysis system

The reactor and the cyclone were sized to provide residence time of less than 1 second for gas phase before reaching the condenser. At the end, the product gas is cooled in a series of water-glycol glass condensers to recover bio-oil from gas steam.

2.4. Catalytic steam reforming

Because at the point of starting steam reforming experiments there was not enough produced bio-oil from the pyrolysis system, a commercial bio-oil was used. Figure 2 shows a picture of the developed catalytic steam reforming system. The heart of the system is the catalytic micro-reactor which is a stainless steel fixed-bed rector. Two syringe pumps, one for water injection and the other one for bio-oil injection, provide feeds to the reactor on top of the catalyst bed. Nitrogen is used as the carrier gas. Before running an experiment, catalyst is reduced in the medium of nitrogen and hydrogen (50:50 vol%) at the temperature of 850 °C.

Figure 2. Catalytic steam reforming system

Ni-based catalysts which are very common for the steam reforming of natural gas were prepared in our lab in four different Ni loading (5.6%, 10.7%, 14.1%, and 18%) on alumina. Down stream of the catalytic steam reforming reactor, the product gases, after passing a glass condenser to recover excess steam, were analyzed by an on-line micro-GC capable of detecting H_2, N_2, CO, CO_2, CH_4, C_2H_6, C_3H_8, and C_4H_{10}.

3. Results and Discussion

In fixed-bed pyrolysis experiments, three operational parameters including reaction temperature, condensation temperature, and nitrogen gas flow rate were selected to investigate their effects on pyrolysis product yields. Maximum bio-oil yield of 45% could be achieved. GC-MS analysis was performed allowing detection of ten components in the produced bio-oil. Results indicated that bio-oil is a complex mixture of organic compounds containing a wide variety of aromatics and oxygenated compounds, such as carboxylic acids, phenols, and aldehydes [1].

Regarding the fluidized-bed pyrolysis system, more than 30 experiments have been done to date. Maximum bio-oil yield of ≈50% has been achieved. The char and gas yields were15% and 35%, respectively. During these initial tests challenges involving the char separation system and bio-oil recovery system have been gradually overcome. Now the main objective is to determine the optimal operational conditions to obtain the highest bio-oil yield and to investigate the effects of operational parameters on the bio-oil quality.

Catalytic steam reforming test results have shown yield of 73% for hydrogen on 14% Ni/Alumina. Compared to the rest of the Ni-based catalysts, the 14% Ni loading provided higher hydrogen yield while the temperature was set at 950 °C and steam to carbon ratio was 5. The main problem regarding catalytic steam reforming of bio-oil is coke deposition on the catalyst. In the next step, the plan is to investigate the effect of temperature and steam to carbon ratio on the hydrogen yield, and also to prepare noble metal catalysts for the bio-oil steam reforming to improve the hydrogen yield.

4. Conclusion

In this study, fast pyrolysis experiments of sawdust were carried out in fixed-bed and fluidized-bed pyrolysis systems. Bio-oil yields of 45% were achieved in the fixed-bed system and a 5% increase in yield to 50% was obtained in the fluidized-bed system. For the pyrolysis phase, the next step is to investigate the effect of operational parameters like biomass particle size, pyrolysis temperature, and vapor residence time on bio-oil yield and quality. Results of catalytic steam reforming show the hydrogen yield of 73%. Improving Ni-based catalysts by adding other metals like Mg, and preparing noble metal catalysts have potential to help prevent or at least decrease coke deposition.

Acknowledgement

Financial support from the Natural Sciences and Engineering Research Council of Canada (NSERC), the Agricultural Funding Consortium of Alberta, Alberta Innovates Bio Solutions, and the Institute for Sustainable Energy, Environment and Economy (ISEEE) at the University of Calgary are gratefully acknowledged.

References

[1] E. Salehi, J. Abedi, T. Harding, Energy & Fuels, 23 (2009) 3767–3772.
[2] N. Ozbay, A. E. Putun, E. Putun, Int. J. Energy Res. 30 (2006) 501.
[3] A. E. Putun, B.B. Uzun, E. Apaydin, E. Putun, Fuel Process. Techno. 87 (2005) 25.
[4] S.Sensoz, D.Angın, Bioresour. Technol. 99 (2008) 5492.
[5] H.J. Park, J.I. Dong, J.K. Jeon, Y.K. Park, K.S. Yoo, S.S. Kim, J. Kim, S. Kim, Chem. Eng. J., 143 (2008) 124–132.
[6] G. Chen, J. Andries, H. Spliethoff, Energ. Convers. Manage. 44 (2003) 2289.
[7] M. Krzesinska, M.B. Pilawa, S.Pusz, Biom. Bioenerg. 30 (2006) 166.
[8] A.C. Basagiannis, X.E. Verykios, Inter. J. Hydrogen Ener. 32 (2007) 3343 – 3355.
[9] A. C. Basagiannis, X. E. Verykios, Catalysis Today 127 (2007) 256–264.
[10] Z. Wang, Y. Pan, T. Dong, X. Zhu, T. Kan, App. Catal. A: General 320 (2007) 24–34.
[11] C. Rioche, S. Kulkarni, F. C. Meunier, J. P. Breen, R. Burch, App. Catal. B: Environmental 61 (2005) 130–139.

(591a) Anisotropic Diffusion in Concentrated Colloidal Dispersions: The Evanescent Diffusivity

James W. Swan, Division of Chemistry and Chemical Engineering, California Institute of Technology, Pasadena, CA and John F. Brady, California Institute of Technology / Division of Chemistry and Chemical Engineering, Pasadena, CA

We employ an analogy to traditional dynamic light scattering to describe the inhomogenous and anisotropic diffusion of colloid particles near a solid boundary measured via evanescent wave dynamic light scattering. Following this approach, new expressions are derived for the short-time self and collective diffusivities of a colloidal suspension of arbitrary volume fraction. Accelerated Stokesian dynamics simulations are used to calculate this evanescent-wave-diffusivity in the limit of large and small scattering wave numbers for a wide range of evanescent penetration depths. We show that at high volume fractions and larger penetration depths, the confinement has little effect on the dynamics of the suspension moving parallel to the wall. However, near and normal to the wall, the diffusivity shows a strong dependence on penetration depth for all volume fractions. This is due to the lubrication interactions between the particles and the boundary. These results are novel as well as comprehensive with respect to the range of penetration depth and volume fraction.

See more of this Session: Fundamental Research in Transport Processes
See more of this Group/Topical: Engineering Sciences and Fundamentals

(591b) Nonequilibrium Statistical Theory for Colloidal Dynamics

Shuangliang Zhao and Jianzhong Wu, Department of Chemical and Environmental Engineering, University of California, Riverside, Riverside, CA

This paper presents a theoretical development of non-equilibrium dynamical density functional theory (DDFT) for colloidal system. We consider the colloidal system as Brownian one, of which the dynamics properties is described by physically important time-dependent quantities including local density, local momentum and local kinetic energy etc.. To address the volatile structure of colloidal system, the dynamical two-body correlation is also included. We set up time-evolution equations for those quantities, and then close the resulting coupled differential equations via Maximum-entropy approach. Comparing to other theories, besides the inclusion of time-dependent correlation, our theory is developed in the pure framework of non-equilibrium dynamics, and no relation is mapped from equilibrium statistical theories of liquids. We give the numerical calculation criteria upon our development, and finally point out this development can be extended to higher level involving more dynamical quantities.

See more of this Session: Fundamental Research in Transport Processes
See more of this Group/Topical: Engineering Sciences and Fundamentals

(591c) Diffusion of Small Molecules in Rubbery Polymers

Liborio Ivano Costa, Giuseppe Storti and Massimo Morbidelli, Dchab-Icb, ETH Zürich, Zürich, Switzerland

The rate of transport of small molecules in polymers, such as monomers, oligomers, plasticizers or additives plays a key role in both polymer reaction, polymer processing and in applications where sorption/desorption kinetics are involved (e.g. drug release from polymer matrixes). Despite that, and the numerous models proposed in past years, to date no fully predictive theory exists to evaluate diffusion coefficients in polymers.[1] Here we show that the Free Volume (FV) concept may be used as a bridge between kinetic and thermodynamic theories: assuming a correspondance between the FV and the unoccupied volume of pure components as given by Lattice Theories (e.g. the Sanchez-Lacombe molecular theory [2]) a new equation is obtained for the calculation of self diffusion coefficients in multicomponent mixtures.[3] A parameter evaluation scheme is proposed, which is substantially simpler compared with the conventional Vrentas-Duda approach,[4] even without losing generality. The theory is compared with experimental data of self diffusion coefficients for several solvent/polymer binary mixtures above the mixture glass transition temperature in wide ranges of physical conditions showing that good correlative ability is achieved in all cases. [1] L. Masaro, X. X. Zhu, Prog. Polym. Sci. 1999, 24, 731. [2] I. C. Sanchez, R. H. Lacombe, Macromolecules 1978, 11, 1145. [3] L. I. Costa, G. Storti, J. Polym. Sci. Part B: Polym. Phys. 2010, 48, 529. [4] J. S. Vrentas, J. L. Duda, J. Polym. Sci. Polym. Phys. Ed. 1977, 15, 403.

See more of this Session: Fundamental Research in Transport Processes
See more of this Group/Topical: Engineering Sciences and Fundamentals

(591d) Diffusion in Fluids with Large Mean-Free-Paths: Non-Classical Behavior Between Knudsen and Fickian Limits

Gregory Aranovich and Marc Donohue, Chemical & Biomolecular Engineering, Johns Hopkins University, Baltimore, MD

Diffusion in fluids is analyzed at non-classical conditions, intermediate between the Knudsen and Fickian limits. The fluid is considered in the framework of the diffusion evolution equation involving expansions of the density distribution in powers of displacement and time. The standard truncation of these expansions results in the classical model of diffusion; however, higher-order terms lead to a departure from classical behavior. This has not been studied or discussed adequately in the literature previously. Here, we present an exact solution of the diffusion evolution equation without truncation of the density expansions. This solution illustrates limitations in the classical truncations and demonstrates non-classical effects due to large mean-free paths, L. In particular, this new solution shows that, at large L, there are significant quantitative deviations from classical diffusion profiles. In addition, this solution demonstrates a dramatic change in the diffusion mechanism from the state where the molecular motions are predominantly ballistic to one of molecular chaos. This has implications for fundamentals of fluids between the Knudsen and Fickian limits, and for a variety of fields where evolution of a system includes random, multi-scale displacement of particles, such as nanotechnology, vacuum techniques, turbulence, and astrophysics.

See more of this Session: Fundamental Research in Transport Processes
See more of this Group/Topical: Engineering Sciences and Fundamentals

Transport Properties for Lennard-Jones Particles

Sun Ung Kim, Department of Chemical Engineering, Department of Mechanical Engineering, University of Michigan, Ann Arbor, MI and Charles W. Monroe, Department of Chemical Engineering, University of Michigan, Ann Arbor, MI

Classical kinetic theory[1-6] provides a reliable method to obtain transport properties of multicomponent dilute gases that saves time and cost compared to experiments. In the theory, transport coefficients, including thermal conductivity, viscosity, and diffusivity, are given as algebraic functions of the collision integral $\Omega_{ij}^{(l,s)}$ arising from Chapman and Enskog's solution of the Boltzmann gas equation. For example, binary diffusion coefficients can be evaluated by $\mathcal{D}_{ij} = 3kT/\left(16n\mu_{ij}\Omega_{ij}^{(1,1)}\right)$. The particles are modeled as a group of colliding pairs, which obey the classical momentum and energy conservation laws with the following parameters: (1) the separation distance r, which is initially infinity; (2) the distance of closest approach r_m; (3) the initial relative speed g, which follows a Maxwell-Boltzmann distribution; (4) the Lennard-Jones (6-12) potential[3] $\phi(r)$, which embeds the characteristics of gas molecules; and (5) the impact parameter b, which is r_m when $\phi(r)$ is zero.

For the evaluation of transport properties, the following three factors should be calculated in consecutive order: (1) the deflection angle $\chi(b,g)$; (2) the reduced collision cross section $Q^{(l)}(g)$; and (3) the collision integral $\Omega_{ij}^{(l,s)}$. First, the deflection angle is given as $\chi(b,g) = \pi - 2\theta(r_m)$, where

$$\theta(r) = b\int_r^{\infty} \frac{dr/r^2}{\sqrt{1 - b^2/r^2 - \phi(r)/g^2}}. \qquad (1)$$

Particle trajectories during a collision event can be drawn with Eq. 1. Integration of $\chi(b,g)$ is difficult, however, because the integrand includes singularities. These can be removed analytically, allowing calculation of the deflection angle χ to high accuracy.

The trajectory map shown in Fig. 1 arises from a contour plot of the deflection angle $\chi(b,g)$ in terms of dimensionless b and g values. Four lines, demarcating deflections ($\chi = 0$), loops ($\chi = \pi$), orbits ($\chi = -\infty$), and inflections, divide the map into six regions. Representative trajectories are also included in each region, showing how Fig. 1 helps to understand the topology of binary Lennard-Jones interactions.

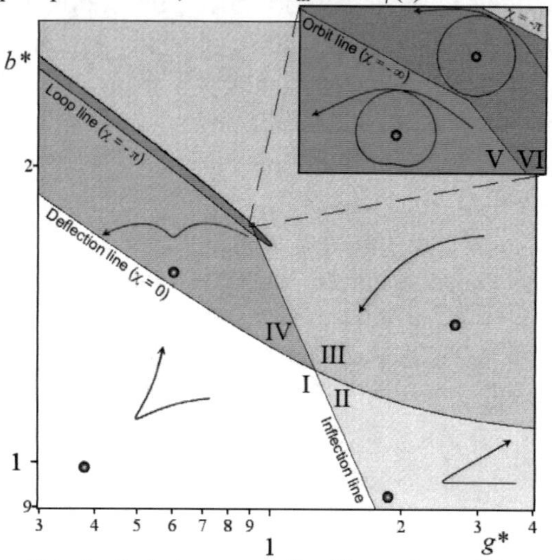

Fig. 1. Trajectory map for binary Lennard-Jones interactions. I: collision with inflection; II: collision without inflection; III: deflection without inflection; IV: deflection with inflection; V: loop with inflection; and VI: loop without inflection.

The trajectory map informs the numerical procedures used when integrating Eq. 1. By comparing the grid independent χ value to the recent previously reported value[6], we could see that our computer program is working properly as well as there are critical regions on the map. In regions IV, V and VI, to estimate χ with the same accuracy as other regions, the number of mesh points should be increased to guarantee the accuracy of the collision integral $\Omega_{ij}^{(l,s)}$.

The ongoing process is to evaluate the reduced cross section $Q^{(l)}(g)$, which is necessary to obtain $\Omega_{ij}^{(l,s)}$, and ultimately the transport coefficients. The reliability will be examined by comparing the numerical results with previously reported experimental values. Once the code is successfully created, a variety of transport coefficients can be evaluated using simple algebraic relations.

References

[1] S. Chapman and T. Cowling, 2nd Ed. Cambridge, 1960.
[2] J. Hirschfelder, R. Bird, E. Spotz, *J. Chem. Phys.* 16 (1948) 968-981.
[3] J. Hirschfelder, C. Curtiss, R. Bird, Wiley, New York, 1954.
[4] E. Akhmatskaya and L. Pozhar, *USSR Comput. Maths. Math. Phys.* 26(2) (1986) 185-190.
[5] U. Storck, *ZAMM* 78 8 (1998) 555-563.
[6] F. Sharipov and G. Bertoldo, *J. Comp. Phys.* 228 (2009) 3345-3357.

(591f) Transport of Brownian Particles Confined to a Weakly Corrugated "Channel"

Xinli Wang, Department of Chemical and Biomolecular Engineering, The Johns Hopkins University, Baltimore, MD and German Drazer, Department of Chemical and Biomomecular Engineering, The Johns Hopkins University, Baltimore, MD

We investigate the average velocity of Brownian particles driven by a constant external force and constrained to move in two-dimensional, weakly-corrugated channels. We consider both the geometric confinement of the particles between solid walls as well as the soft confinement induced by a periodic potential. Using perturbation methods we show that the leading order correction to the marginal probability distribution of particles in the case of soft confinement is equal to that obtained in the case of geometric confinement, provided that the (configuration) integral over the cross-section of the confining potential is equal to the width of the solid channel. We then calculate the probability distribution and average velocity in the case of a sinusoidal variation in the width of the channels. The reduction on the average velocity is larger in the case of soft channels at small Peclet numbers and for relatively narrow channels and the opposite is true at large Peclet numbers and for wide channels. In the limit of large Peclet numbers the convergence to bulk velocity is faster in the case of soft channels. The leading order correction to the average velocity and marginal probability distribution agree well with Brownian Dynamics simulations for the two types of confinement and over a wide range of Peclet numbers.

See more of this Session: Fundamental Research in Transport Processes
See more of this Group/Topical: Engineering Sciences and Fundamentals

(591g) The Effect of Solute-Solute Interactions On Hindered Diffusion of SDS Micelles in Agarose Gels

Wyatt J. Musnicki[1], Nathan Lloyd[1], Stephanie R. Dungan[2] and Ronald J. Phillips[1], (1)Chemical Engineering and Materials Science, U.C. Davis, Davis, CA, (2)Food Science & Technology, Chemical Engineering and Materials Science, University of California, Davis, Davis, CA

The hindered diffusion of ionic micelles in gels is not well understood, and yet plays an important role in numerous applications including foods, drug delivery, and environmental technologies. We have used holographic interferometry to measure the diffusivity of sodium dodecyl sulfate (SDS) in solution and in agarose gels at various SDS, agarose, and sodium chloride concentrations. We have also measured aggregation numbers of the micelles, both in solution and gels, by using time-dependent fluorescent quenching. For SDS micelles, we find the rate at which the micelle diffusivity increases with micelle volume fraction to be dependent both on gel concentration and ionic strength. The fluorescence measurements show a slight increase in the aggregation number of the SDS micelles when in agarose gels, but that increase does not explain the observed changes in rates of diffusion. Those changes can be very significant: at micellar volume fractions of ~3%, solute-solute interactions can more than double the diffusion coefficient. At moderate to high ionic strength, the diffusion data are in quantitative agreement with predictions of a theory that accounts for volume exclusion by the gel fibers, and electrostatic and hydrodynamic interactions between the micelles. At the lowest ionic strengths studied, for which salt concentrations were less than 0.03M, the measured rates of diffusion were significantly higher than the theory predicts. This difference could be a result of the assumption of pairwise-additive solute-solute interactions in the theory.

See more of this Session: Fundamental Research in Transport Processes
See more of this Group/Topical: Engineering Sciences and Fundamentals

(599a) Solubility of E. Coli Endotoxins and Removal From Metallic Biomaterials Using Water-in-CO_2 Microemulsuions

Pedro J. Tarafa, Dept. of Chemical Engineering, University of South Carolina, Columbia, SC and **Michael A. Matthews**, University of South Carolina, Columbia, SC

Bacterial endotoxins have strong affinity for metallic biomaterials because of surface energy effects. Conventional depyrogenation methods may not eradicate endotoxins and may compromise biological properties and functionality of metallic instruments and implants. We evaluated the solubilization and removal of E. coli endotoxin from smooth and porous titanium (Ti) surfaces and stainless steel lumens using a water-in-CO_2 microemulsion phase. The microemulsion system in the liquid CO_2 region (298 K and 27.6 MPa) with strong mixing removed all the endotoxin below detection levels. This suggests that formation of water-in-CO_2 microemulsions penetrates and dissolves endotoxins from all the tested substrates. The successful removal of endotoxins from metallic biomaterials with compressed CO_2 is a promising cleaning technology for biomaterials and reusable medical devices.

See more of this Session: Materials and Biomaterials Synthesis and Processing with Compressed or Supercritical Fluids I.
See more of this Group/Topical: Engineering Sciences and Fundamentals

(599b) Application of Supercritical Fluid Anti-Solvent Processing to Coat and Disperse a Poorly Soluble Compound On Fast Flo® Lactose

Cecile N. Forness[1], Said Saim[2], Kenneth Koenig[1], Jocelyn Gunn[1], Dongmei Qiang[1], John A. Smoliga[1], John Robson[1] and Stephen Horhota[1], (1)Boehringer Ingelheim Pharmaceuticals, Inc., Ridgefield, CT, (2) Collegium Pharmaceutical, Cumberland, RI

Supercritical CO_2 (SC-CO_2) is used as an anti-solvent for a one-step process which coats a carrier (lactose) with a poorly soluble active pharmaceutical compound, poly vinyl pyrrolidone (PVP) and sodium lauryl sulfate(SDS). The product is a rapidly dissolving, free-flowing powder suitable for direct compression into a tablet dosage form.

In this process, a methanol solution containing the poorly soluble drug, PVP and SDS is sprayed onto a stirred bed of lactose (35 °C, 2000 psig). The rapid uptake of methanol by supercritical CO_2 results in deposition of the Drug, SDS and PVP. X ray powder diffraction results of the Drug/ SDS/ PVP product show that the drug is present in a crystalline form and scanning electron microscopy shows fine drug particles dispersed within the SDS and PVP.

Tablets prepared from the Drug/ SDS/ PVP coated on lactose carrier via SCF process show similar dissolution at the 45 minute time point compared to tablets prepared by a traditional high shear granulation/ fluid bed drying process using jet milled drug and SDS with PVP and lactose. However, the tablets prepared by granulation show a decreasing dissolution over the 90 to 120 minute time period, whereas for the SC-CO_2 process the tablet dissolution is maintained through 120 minutes. For the SC-CO_2 process, the tablet dissolution and the crystal morphology of the drug are unchanged after approximately 16 months storage at room temperature. These data suggest the SCF deposited product can increase the solubility of the drug which could result in increased bioavailability.

See more of this Session: Materials and Biomaterials Synthesis and Processing with Compressed or Supercritical Fluids I.
See more of this Group/Topical: Engineering Sciences and Fundamentals

Solubility of Chitosan in Supercritical Carbon Dioxide and Modeling Using SAFT

Requel Carvallo, Temptrol, Tampa, FL, Keyur S Patel, Chemical Engineeering, University of South Florida, Tampa, FL and Aydin K. Sunol, Chemical Engineering, University of South Florida, Tampa, FL

DMSO-CO2 system

Experimental cloud point for the system DMSO-CO_2 was collected in a temperature range of 314.2 °K to 328.2 °K and pressures from 89 to 133.5 bars. The isotherms presented in Figure 1 are consistent with those reported by Rajasingam 2004, Gonzalez 2002 and Kordikowski 1995.

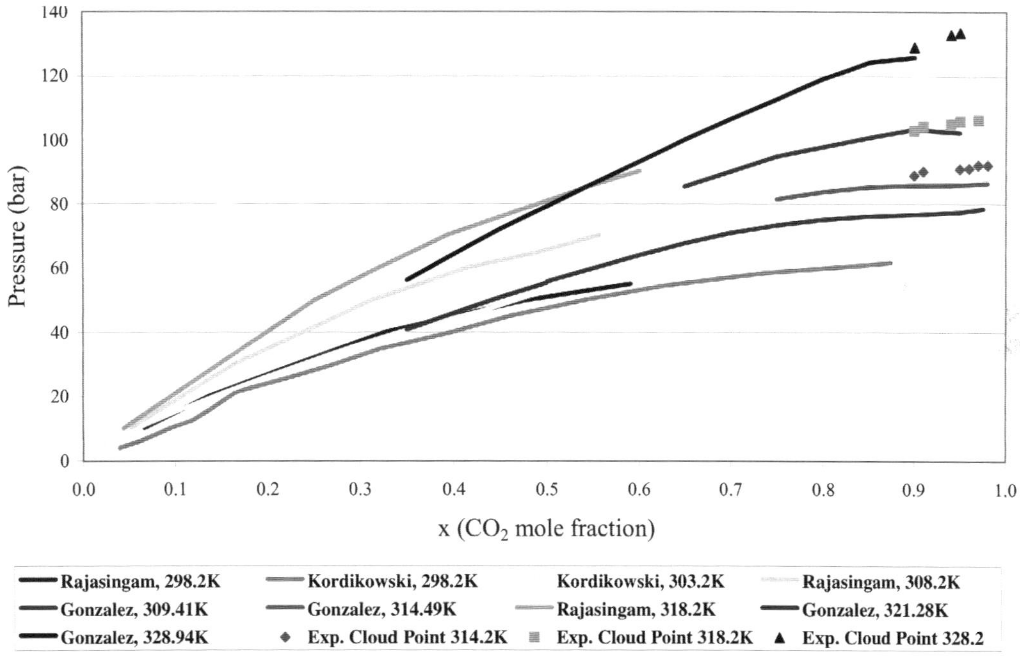

Figure 1. Cloud Point Experimental Data for the System DMSO- CO_2

CO2-DMSO-Citosan system

In Figure 2, at constant temperature the Chitosan solubility increases proportionally with the pressure; and at constant pressure the solubility decrease with temperature increments. This is a general effect that is suitable to the short-range attractive forces between solvent and solute. The solubility of a low-volatile solute, like Chitosan, is enhanced as the density of the volatile solvent (CO_2) increases, as a consequence of temperature drop in an isobaric process or pressure rise in an isothermal process.

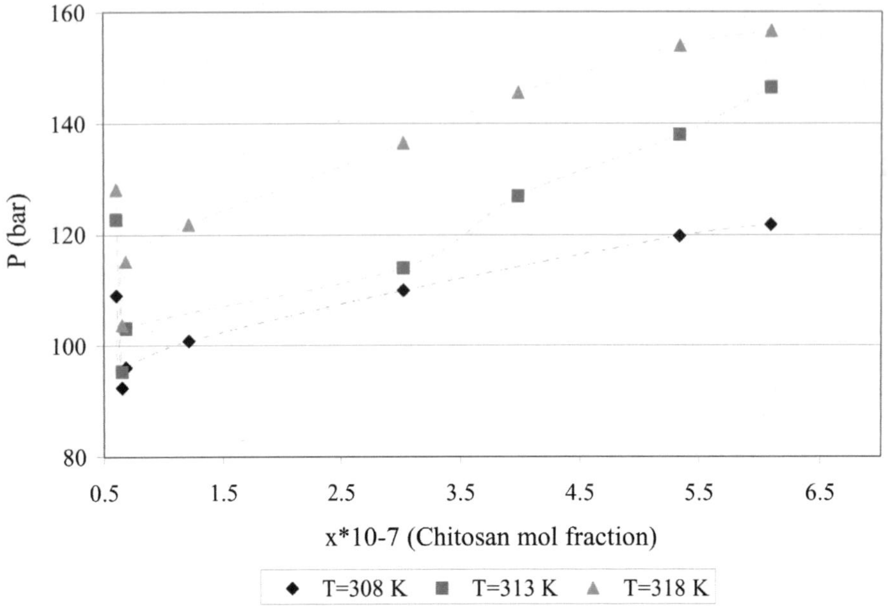

Figure 2. Cloud Point Experimental Data for the System Chitosan-DMSO- CO_2.

References

Rajasingam R., Lioe L., Pham Q. T. and Lucien F. P. "Solubility of carbon dioxide in dimethyl sulfoxide and N-methy-2-pyrrolidone at elevated pressure." *Journal of Supercritical Fluids*, 31, (2004): 227-234

Gonzalez V., R. Tufeu and P. Subra. "High-pressure vapor-liquid equilibrium for the binary systems carbon dioxide+dimethyl sulfoxide and carbon dioxide+dichloromethane." Journal of Chemical Engineering Data, 47 (2002): 492-495.

Kordikowski A., A.P. Schenk, R.M. Van Nielen and C.J. Peters. "Volume expansions and vapor-liquid equilibria of binary mixtures of a variety of polar solvents and certain near critical solvents." Journal of Supercritical Fluids, 8, (1995): 205-216.

Measurement of the Solubility of Poly(methyl methacrylate)-Methyl Methacrylate System in Supercritical CO2 and Their Modeling Using SAFT

Ratka Damnjanovic[1], Keyur S Patel[2] and Aydin K Sunol[1], (1)Chemical Engineering, University of South Florida, Tampa, FL, (2)Chemical Engineeering, University of South Florida, Tampa, FL

1. Experimental setup

There are two different experiments that were conducted using the same experimental setup. First, the methyl methacrylate-CO_2 equilibrium was studied, by determining the bubble point of the mixture at different temperatures. The second system, PMMA-CO_2-MMA, was studied by determining the cloud-point data at different temperatures. The solubility is observed in a vertical high pressure phase monitoring system.

When the cell is full, the CO_2 is shut off, and the initial values of the temperature and pressure in the system are recorded. Then, the mixture in the cell is preheated up to about 105-110 °C, just above the glass transition temperature of the polymer. This is needed in order to avoid the nucleation of the polymer as a result of incompatibility issues related to the density differences between the CO_2 and the PMMA in the system. Once the system is preheated to this temperature, the mixer is turned on to agitate the solution until it becomes a homogeneous optically transparent mixture. The cell assembly is then lowered in its horizontal position, and the heater is switched off. The desired temperature at which the first cloud point would be measured is preset on the controller. The system cools down slowly until the temperature reaches the set point. At this point of time the mixture in the cell may have turned milky or cloudy, due to phase separation in the system. This is expected behavior and it will remain that way until the pressure is increased to bring the system back into the one phase region. Thus, once thermal equilibrium is maintained in the cell at the pre-set temperature, the contents of the cell are compressed into the one phase region where the mixture in the cell becomes transparent. Once in the one-phase region, the pressure is lowered rapidly by moving the piston backwards i.e. increasing the volume until the first appearance of a cloud in the cell. The pressure at which the first cloudy look appears is recorded, and that is the cloud point of the system at the given temperature. After completing the measurement at a given temperature, the cell temperature is then stabilized at a new value, and the experimental procedure is repeated.

An average cloud point is then calculated from the multiple measurements. In this work, cloud point data was measured in ~15°C increments i.e. decrements, and the above procedure was repeated, thus creating a pressure-temperature (P-T) cloud point curve at fixed PMMA, MMA, and CO_2 concentrations.

2. Thermodynamic modeling

The experimental bubble point and cloud point data for CO2-MMA and CO2-MMA-PMMA systems were modeled using PC-SAFT equation of state. PC-SAFT is molecular based equation of state where molecule is modeled as a chain of hard spheres. The repulsive and attractive forces between and within chain molecules are modeled in terms of residual Helmholtz free energy. Each non associating pure component can be described by three parameters in the PC-SAFT equation of state: effective number of segments within the molecule m, temperature independent segment diameter σ and temperature-independent dispersion energy of attraction between two segments u^o/k. Pure component parameters for CO2, PMMA and MMA are listed in Table 1.

Component	u^o/k (K)	σ (A)	m	Literature source
CO_2	169.21	2.7852	1.417	[1]
MMA	208.0	8.700	5.670	[2]
PMMA	240.0	9.560	2532	[3]

3. CO2-MMA system

The experimental bubble point data at three different temperatures are used to fit the binary interaction parameter. The iterative search for optimal interaction parameter is performed using Nelder-Mead simplex algorithm [4]. For the first iteration, the initial guess for the bubble point calculations are computed from phase stability calculations. For subsequent iterations of regression, the computed bubble points provide good initial guess for bubble point calculations. The experimental data are best described using the kij=0.089 with absolute average percentage deviation (AAPD) in bubble point pressure equal to 2.7. Figure 1 shows comparison between experimental and calculated bubble point pressures at different temperatures. The predicted bubble point pressure for kij=0.089 is slightly lower than experimental pressures at 40 C and higher at 105.5 C.

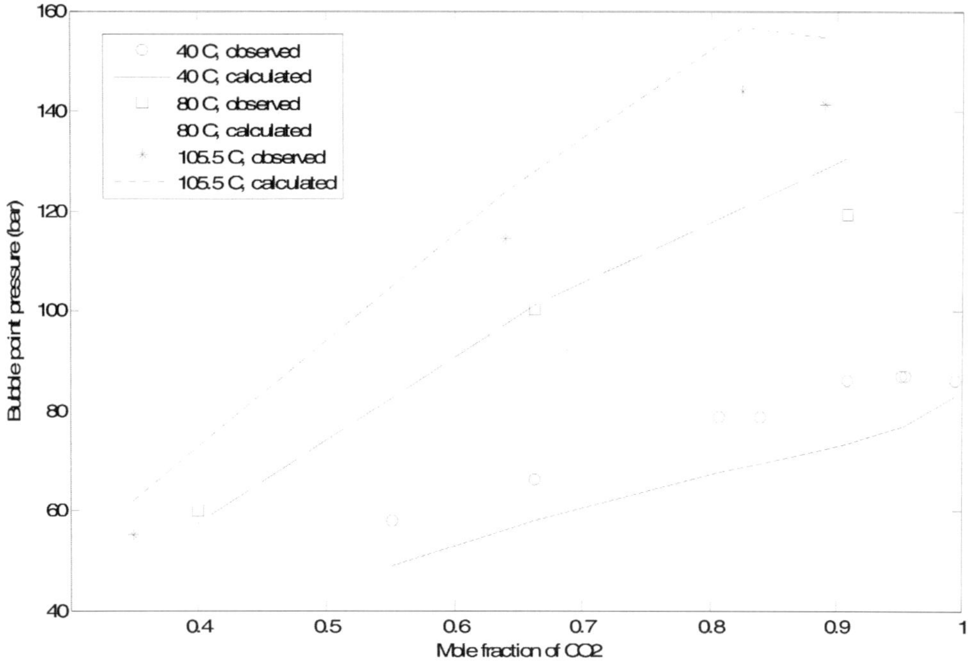

Figure 1. Bubble point pressures of CO2-MMA mixtures at different temperatures and feed compositions with PC-SAFT equation of state for kij = 0.089

References:

[1] J. Gross, G. Sadowski, Perturbed-chain SAFT: an equation of state based on a perturbation theory for chain molecules, IEC Res. 40 (2001) 1244–1260.

[2] F. Becker, M. Buback, H. Latz, G. Sadowski, F. Tumakaka, Cloud-point curves of ethylene-(meth)acrylate copolymers in fluid ethene up to high pressures and temperatures—experimental study and PC-SAFT modeling, Fluid Phase Equilib. 215 (2004) 263–282.

[3] M. Gornert, G. Sadowski, Phase equilibrium measurement and modeling of the PMMA/MMA/CO2 ternary system, Journal of Supercritical Fluids. 46 (2008) 218-225.

[4] Lora M., McHugh M. A., Phase behavior and modeling of the poly(methyl methacrylate)-CO_2-methyl methacrylate system, *Fluid Phase Equilibria* 1999, *157*, 285-297.

[5] Kirby, F. C., McHugh, M. A., Phase Behavior of Polymers in Supercritical Fluid Solvents, *Chem. Rev.* 1999, *99*, 565-602.

(617a) Calculation of Sugar Yields From High Solid Hydrolysis of Lignocellulosic Biomass

Yongming Zhu[1], Marco Malten[2], Mads Torry-Smith[1] and Hui Xu[1], (1)Novozymes North America, Franklinton, NC, (2)Novozymes A/S, Bagsvaerd, Denmark

Accurate determination of sugar yields is important for the evaluation of an enzymatic hydrolysis process. Calculation of sugar yields at high solids is complicated primarily by the varying liquid volume and liquid density owing to solid solubilization and sugar liberation. Ignoring these factors can lead to significant errors in yield calculation. In this paper, a mathematical method was developed for yield calculation in high solids hydrolysis. The information needed in the calculation includes the composition of the substrate, initial solid loading, initial liquid density, and the sugar concentrations before and after hydrolysis. All of these parameters are measurable with conventional procedures. The method was validated experimentally for a range of different substrates and solid loading up to 30 wt%. Compared to the other methods found in the literature, this method is easy to use and represents good accuracy.

See more of this Session: Reaction Kinetics and Transport Fundamentals for Biomass Conversion I
See more of this Group/Topical: Sustainable Engineering Forum

(617b) Diffusion-Adsorption Measurements of Solutes in Cellulose Fiber Beds Using Magnetic Resonance Imaging

David M. Lavenson, Chemical Engineering & Materials Science, University of Calfiornia Davis, Davis, CA, Emilio J. Tozzi, Chemical Engineering and Materials Science, University of Calfiornia Davis, Davis, CA, Michael J. McCarthy, Food Science & Technology, University of Calfiornia Davis, Davis, CA and Robert L. Powell, Chemical Engineering and Material Science, University of California-Davis, Davis, CA

Effective mixing and transport of cellulosic suspensions for enzymatic conversion to glucose is an important biofuels process bottleneck. Most recent literature studies have highlighted the lack of efficient mixing strategies available for combining substrate and enzymes at the initial stages of high solids enzymatic hydrolysis. Design of new energy efficient mixers requires quantitative information about rheology and diffusion in concentrated slurries. We investigated diffusion and adsorption of cellulose suspensions using magnetic resonance imaging. MRI allows for spatially resolved measurements of concentration distributions in highly concentrated suspensions. Paramagnetic tracers of various molecular sizes were used as diffusing solutes in suspensions of various fiber types. Effective diffusion coefficients are compared with predictions of a simple diffusion-adsorption model which accounts for the system porosity, tortuosity, and adsorption.

See more of this Session: Reaction Kinetics and Transport Fundamentals for Biomass Conversion I
See more of this Group/Topical: Sustainable Engineering Forum

Breaking the Barrier to Obtain High Yield on Enzymatic Hydrolysis Using Low Enzyme Loadings

Yoshiki Ueno, Makoto Ikeo, Naohisa Inoue, and Daisuke Taneda
Technology Development Department, JGC Corporation
Higashiibaraki-gun, Ibaraki Pref., Japan

Abstract

JGC Corp. has been developing a method of enzymatic hydrolysis of cellulosic biomass through an approach focused on development of the process. During this effort, we found that by hydrolyzing cellulose without agitation, the enzyme loading can be reduced without decreasing the glucose yield. Moreover, the application of this hydrolysis method allowed more than a 10 fold reduction in the enzyme loading. The studies indicated that there were changes in the characteristics of cellulase, such as adsorption, desorption, and inactivation, when this method was applied. These changes may contribute to the more efficient hydrolysis under the low enzyme loading, and provide insights on further development of the process.

Introduction

Ethanol is one of the biofuels that has been attracting attention as a substitute for fossil fuels. Up to the present, most ethanol has been produced from sucrose and starch by fermentation ; using feedstocks such as sugarcane, beet, and corn. On the other hand, those feedstocks are also used as food. In order to meet an expected increase in the demand for both energy and food in the future, it is necessary to find alternative feedstocks for the ethanol production that will not adversely affect world food resources. To solve this issue, lots of fundamental research and demonstration projects, whose objectives are to develop technologies to produce ethanol from cellulosic biomass such as wood, corn stover, and wheat straw, have been carried out all over the world [1] - [3].

The main issue when utilizing cellulosic biomass is the conversion of cellulose, the main component of cellulosic biomass, to glucose. Enzymatic hydrolysis is one of the technologies that enable this conversion. The cost of the enzymes is a critical issue for the commercialization of the enzymatic hydrolysis technology, because a large amount of enzyme (cellulase) is required to obtain a high glucose yield, and commercial cellulase is expensive.

JGC Corp. has developed a method of enzymatic hydrolysis that enables the reduction of the enzyme costs through an approach focused on the process development. We have found a method that achieves a high glucose yield using low enzyme loadings, which will be introduced in this report. The method can be scaled-up easily, and does not depend on the particular enzyme used.

Materials and Methods

Substrates

Filter paper (No.2, Toyo Roshi Kaisha, Tokyo, Japan) and pulp made from broad-leaved trees were used as substrates. Each substrate was shredded by a paper shredder. The size of the shredded substrate was approximately 15mm-length, 2mm-width, and 1mm-thickness.

Enzymes

A commercial *Tricoderma reesei* cellulase (Cellulase SS), purchased from Nagese ChemteX (Osaka, Japan), was used in this study. This Cellulase SS contained 120 mg of protein per ml as measured by the Bio-Rad protein assay (Bio-Rad Laboratories, Hercules, CA), using Bovine Gamma Globulin (BGG) as a standard protein. The hydrolytic activity of this enzyme was 158 filter paper units (FPU) per gram of protein as measured by the procedure described in the literature [4].

Enzymatic Hydrolysis

Enzymatic hydrolysis experiments were performed in 500ml Erenmeyer flasks placed in a 50 deg-C incubator. The agitation of the flask was controlled using a rotary shaker, and two conditions, 115 rpm (mixing condition) and 0 rpm (static condition) were tested. The original enzyme reagent was diluted in 50mM sodium-acetate buffer (pH 5.0) containing 0.02 w/v% sodium azide to inhibit contamination. The substrate concentration was 10 w/v%, which means that 100ml of the diluted enzyme was used against 10 g of substrate, in all of the experiments.

Analytical Method

HPLC (Shimadzu, Kyoto, Japan) was used for the analysis of monosaccharides. Samples taken during the experiment period were centrifuged under conditions of 13,000rpm and 15 minutes. Then the supernatants were filtrated by disposable filters (Millipore), the pore size of which was 0.45 micro-meters. The filtrated supernatants were used for the HPLC analysis.

Results and Discussion

Hydrolysis of cellulose under the mixing and static conditions

The influence of the enzyme loading on the sugar yield was investigated with respect to the mixing condition. Figure 1 shows the change in the glucose concentration over time. With the exception of enzyme loadings of 158 FPU/g cellulose and 79 FPU/g cellulose, the lower enzyme loadings resulted in lower glucose concentrations. This relationship is difficult to explain from the perspective of conventional catalytic reaction. In a chemical reaction, the catalyst loading generally doesn't affect the final yield of product, although it affects the reaction rate. However, in the case of the enzymatic hydrolysis of cellulose, influences of the enzyme loading on both the yield and the reaction rate have been reported repeatedly, although enzymes work as catalysts [5] - [9]. This phenomenon is a major issue when commercializing the enzymatic hydrolysis technology, because reduction of the enzyme loadings is necessary to reduce the enzyme cost. This barrier has already been identified, but no one has been able to find a solution to it.

However, we found that high glucose yield was achieved at the low enzyme loading when the enzymatic hydrolysis was conducted without agitation. Figure 2A shows the experimental results under both the mixing and the static conditions. The final glucose concentrations obtained from both conditions were the same for an enzyme loading of 79 FPU/g cellulose. However, when the enzyme loadings were reduced to 7.9 FPU/g cellulose, the final glucose concentration obtained under the static condition became higher than that obtained under the mixing condition. Moreover, the final glucose concentration reached the same level as that achieved by hydrolysis with the enzyme loading of 79 FPU/g cellulose. This phenomenon will be called "The Static Effect" from here on.

The results of The Static Effect were also observed in the cases where pulp was used as a substrate (Figure 2B), and where different commercial enzymes were used. (Data are not shown.) In Figure 2B, the sugar concentration indicates the total value of glucose and xylose since the hydrolyzate produced from pulp contains both of these monosaccharides.

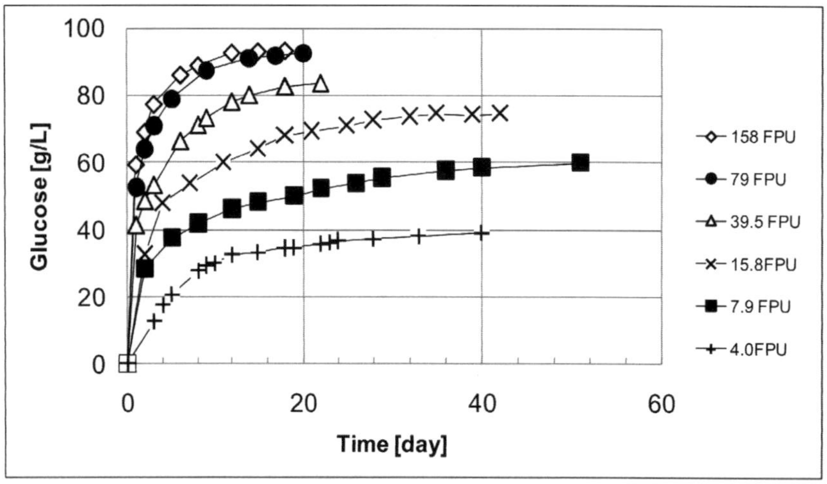

Figure 1. Effect of enzyme loadings (FPU/g cellulose) on enzymatic hydrolysis of cellulose under mixing (115rpm) condition

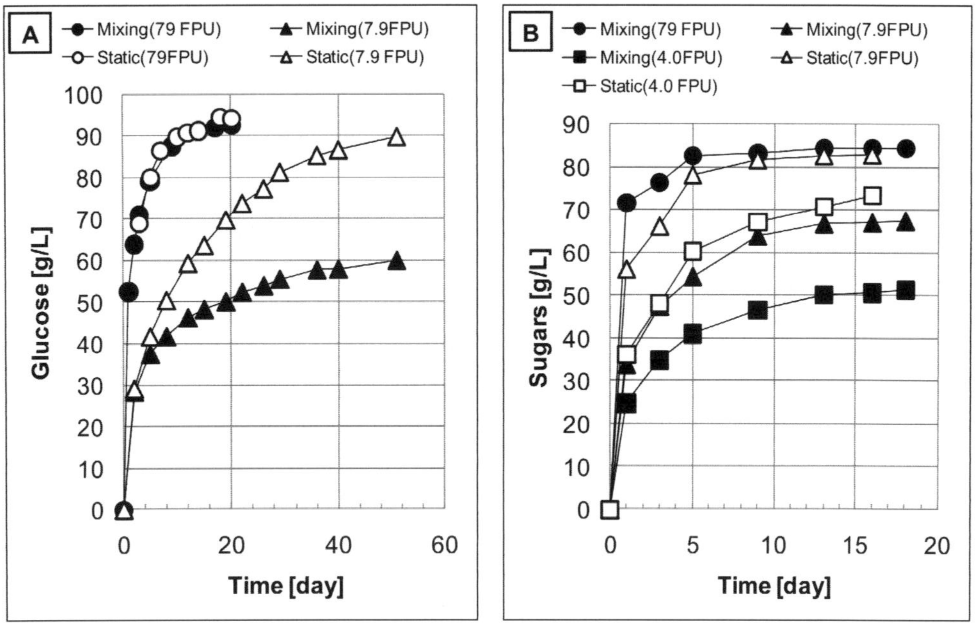

Figure 2. Comparison of enzymatic hydrolysis of cellulose at several enzyme loadings (FPU/g cellulose) under mixing (115rpm) and static condition : (A) filter paper, (B) pulp

Behavior of cellulase under the static condition

Figure 3 shows the changes in the enzyme concentrations when the loading condition was 79 FPU/g cellulose of enzyme. Although the final glucose concentrations of the two experiments were the same (Figure 2), the trends in the enzyme concentrations in the hydrolyzate were different. During the hydrolysis under the mixing condition, the enzyme concentration decreased to 25% of the initial concentration by the adsorption of enzyme on the substrate at the beginning of the reaction, and it didn't change for the succeeding few days. At the end of the reaction, the enzyme concentration in the hydrolyzate was about a half of the initial concentration. However, under the static condition, the enzyme concentration decreased only to 40% of the initial concentration at the beginning, and by the end of the reaction, it had reached a level of up to 75% of the initial concentration. These results indicated that the hydrolysis of cellulose was effectively conducted under the static condition so that a small amount of enzyme absorbed on substrates and much more free enzyme existed in the hydrolyzate compared with the case of the mixing condition. In addition, microscopic observation revealed an increase in the presence of small substrate particles under the mixing condition. The difference in the final enzyme concentrations might have been caused by the adsorption of the enzyme onto these small substrate particles immediately after desorption from substrates under the mixing condition.

One hypothesis is that this adsorption on to the small particles is non-productive, because the final glucose concentrations of the two experiments were the same, and the decrease in this adsorption under the static condition helps to give rise The Static Effect. The influence of the non-productive adsorption will be more significant under low enzyme loading condition, since there is a smaller amount of the enzyme in the solution to start with.

Another possible cause of The Static Effect is the decrease in the inactivation of the enzyme. The effect of agitation on enzymes has been studied, and it was reported that the shear stress can damage the enzyme and decreases the enzyme activity [10] [11].

When the solution containing cellulase was put into the incubator, precipitation was observed. The precipitate was assumed to be the denatured enzyme. The change in the amount of precipitate over time was investigated under the mixing and the static conditions (Figure 4). The amount of the precipitate in the mixing condition was larger than that in the case of the static condition. This result corresponds with the hypothesis that The Static Effect is caused by reduced inactivation of the cellulase. When the influence of the enzyme concentration on the precipitation was investigated, a larger portion of the initial enzyme precipitated at lower enzyme concentrations. The results of these studies indicate that by putting the enzymatic hydrolysis under the static condition, the amount of inactivation of enzyme through non-productive adsorption and denaturing is reduced. The effect of these reductions becomes more significant under the low enzyme loading, and this is assumed to bring about The Static Effect.

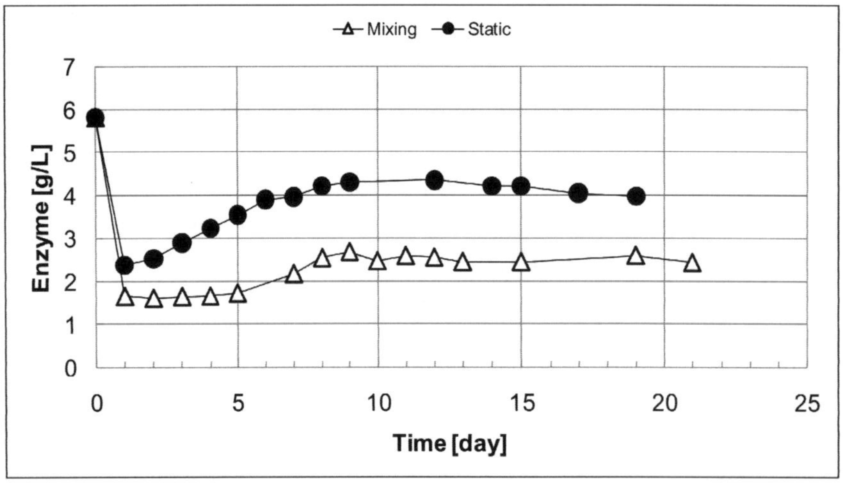

Figure 3. Profiles of enzyme concentration in the hydrolyzate under mixing (115rpm) And static condition : substrate ; filter paper, enzyme loading ; 79 FPU/g cellulose

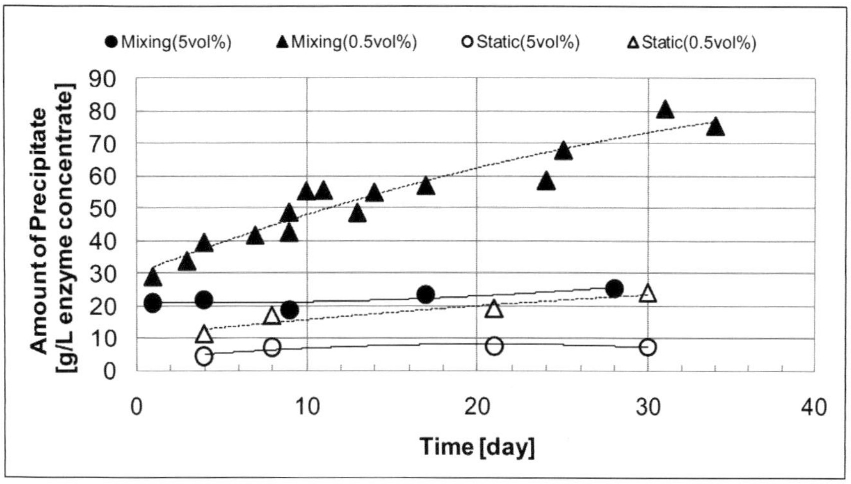

Figure 4. Influence of enzyme concentration and mixing condition (115rpm) on amount of precipitate : substrate was not added

Conclusion

It was found that the static condition is preferable for the enzymatic hydrolysis. Through The Static Effect, hydrolyzate with higher glucose concentration can be obtained with lower enzyme loadings. The study indicates the application of the static method allowed a more than 10 fold reduction in the enzyme loading in the hydrolysis of cellulosic biomass, while maintaining the same level of the glucose yield. This new method is suitable for commercial scale plants, because it obviates the need to mix high solid-content slurry, it reduces the energy consumption required, it uses a simpler saccharification tank, and it lowers the enzyme loading. Moreover, the knowledge acquired from further studies of this method will be applied for the development of microorganisms that will produce a more powerful cellulase.

Experiments using other cellulosic biomass such as rice straw and bagasse are being conducted currently to investigate the preferable pretreatment method.

Acknowledgment

The authors gratefully acknowledge the funding of this work by the New Energy and Industrial Technology Development Organization (NEDO).

References

1. Z. Xiao, X. Zhang, D. J. Gregg, and J. N. Saddler (2004), "Effects of sugar inhibition on cellulases and β-glucosidase during enzymatic hydrolysis of softwood substrates", Appl. Biochem. Biotechnol., Vol. 115, pp.1115-1126.
2. T. I. Georgieva, X. Hou, T. Hilstrom, and B. K. Ahring (2008), "Enzymatic Hydrolysis and Ethanol Fermentation of High Dry Matter Wet-Exploded Wheat Straw at Low Enzyme Loading", Appl. Biochem. and Biotechnol., Vol. 148, pp.35-44.
3. J. S. Tolan (2002), "Iogen's process for producing ethanol from cellulosic biomass", Clean Techn. Environ. Policy, Vol. 3, pp.339-345.
4. T.K. Ghose (1987), "Measurement of cellulase activities", Pure Appl. Chem., Vol. 59, pp. 257-268.
5. W. Sattler, H. Esterbauer, O. Glatter, and W. Steiner (1989), "The effect of enzyme concentration on the rate of the hydrolysis of cellulose", Biotechnol. Bioeng., Vol. 33, pp. 1221-1234.
6. Y. Lu, B. Yang, D. Gregg, J. N. Sadler, and S. D. Mansfield (2002), "Cellulase adsorption and an evaluation of enzyme recycle during hydrolysis of steam-exploded softwood residues", Appl. Biochem. and Biotechnol., Vol. 98-100, pp.641-654.
7. F. Teymouri, L. L. Perez, H. Alizadeh, and B. E. Dale (2005), "Optimization of the ammonia fiber explosion (AFEX) treatment parameters for enzymatic hydrolysis of corn stover", Bioresour. Technol., Vol. 96, pp.2014-2018
8. M. Chena, L. Xia, and P. Xue (2007), "Enzymatic hydrolysis of corncob and ethanol production from cellulosic hydrolysate", Int. Biodeterior. Biodegrad., Vol.59, pp. 85-89
9. C. Tengborg, M. Galbe, and G. Zacchi (2001), "Influence of enzyme loading and physical parameters on the enzymatic hydrolysis of steam-pretreated softwood", Biotechnol. Prog., Vol. 17, pp. 1100-117.
10. S. I. Mussatto, G. Dragone, M. Fernandes, A. M. F. Milagres, and I. C. Roberto (2008)," The effect of agitation speed, enzyme loading and substrate concentration on enzymatic hydrolysis of cellulose from brewer's spent grain", Cellulose, Vol. 15, pp.711-721.
11. S. Mukataka, M. Tada, J. Takahashi (1983), "Effects of Agitation on Enzymatic Hydrolysis of Cellulose in a Stirred-Tank Reactor", J. Ferment. Technol., Vol.61, pp.615-621.

(617d) A Dynamic Model for Cellulosic Biomass Hydrolysis: Validation of Hydrolysis and Product Inhibition Mechanism

Jonas Eeclo[1], Chien-Tai Tsai[2], Ricardo Morales-Rodríguez[1], Krist V. Gernaey[3], Anne S. Meyer[4] and **Gürkan Sin**[1], (1)CAPEC, Department of Chemical and Biochemical Engineering, Technical University of Denmark, 2800 Kgs. Lyngby, Denmark, (2)BioEng, Department of Chemical and Biochemical Engineering, Technical University of Denmark, 2800 Kgs. Lyngby, Denmark, (3)PROCESS, Department of Chemical and Biochemical Engineering, Technical University of Denmark, 2800 Kgs. Lyngby, Denmark, (4)Department of Chemical and Biochemical Engineering, Technical University of Denmark (DTU), 2800 Kgs. Lyngby, Denmark

Production of bioethanol from agricultural waste such as lignocellulosic biomass forms a candidate alternative renewable energy resource. The process of converting biomass to ethanol is complex with many unit operations including the physical pre-treatment and hydrolysis of lignocellulosic material to free simple sugars. One of the challenges in the process is the underlying enzymatic hydrolysis mechanism. The kinetics of multiple enzymes on insoluble substrate is a challenging phenomenon to describe. Although a number of hydrolysis models has been proposed so far in the literature (e.g. Kadam et al., 2004), few or none of the models were independently subject to experimental validation. This sets the aim of this contribution, which is to perform a validation of a dynamic cellulosic biomass model with particular focus on the validation of the hydrolysis and product inhibition mechanisms.

In order to carry out this analysis a number of dedicated experiments were performed. Avicel® and cellobiose were used as substrates. The end product glucose and/or intermediate cellobiose was added to the initial samples in different concentrations, to better identify their inhibition effect. The enzymes used in the experiments were Novozyme 188 and Celluclast 1.5L. Three sets of experiments were performed focusing exclusively on three different conversion reactions:

(i) conversion of cellulose to cellobiose (set 1),

(ii) conversion of cellulose to glucose (set 2) and

(iii) conversion of cellobiose to glucose (set 3).

A total of nine parameters (k1r, k2r, k3r, K1IG2, K2IG2, K3M, K1IG, K2IG and K3IG) involved in the conversion reaction of cellulose to glucose were estimated from the data set 2. The cellobiose to glucose conversion contains three parameters (k3r, K3M and K3IG) which were estimated from the dedicated experimental data (set 3). These three parameters were then used (assumed fixed) when estimating the parameters involved in the reaction for conversion of cellulose to glucose.

Experimental results show that the reaction from cellobiose to glucose goes much faster than initially assumed by the Kadam model (k3r; 3106.7 versus 285.5 and K3M; 2.377 versus 24.3; both result in a reaction that is ±20 times faster, depending on glucose and cellobiose concentrations). This means that inhibition of cellobiose is less critical than originally described by Kadam et al. (2004). Some calculated parameters (k1r, k2r, K1IG and K2IG2) are similar to the values proposed by Kadam et al. (2004) for cellulose to glucose reaction. However due to the fixed parameters of the cellobiose to glucose reaction (which were found faster), some of the model parameters change as well (e.g. faster cellobiose conversion to glucose results in more inhibition for the cellulose to cellobiose conversion). In general, simulation results (using the estimated parameters) fit well for all the experiments.

Overall the model of Kadam et al. (2004) was found to remain largely valid when describing experimental data collected under different initial concentrations of product inhibition (glucose and cellobiose respectively) as inhibitors. One caveat has to be mentioned, that is that some of the model parameters had to be fine-tuned to get a better fit to the experimental data. Nonetheless the results add credit to the reliability of the model and thereby provide confidence in utilising the model for process development and optimisation studies.

References

[1]. Sin, G., Meyer, A. S., Gernaey, K. V., 2010. Comput. Chem. Eng. doi:10.1016/j. compchemeng.2010.02.012.

[2]. Kadam, K.L. Rydholm, E.C. McMillan, J.D., 2004. Biotechnol. Prog. 20, 698-705.

See more of this Session: Reaction Kinetics and Transport Fundamentals for Biomass Conversion I
See more of this Group/Topical: Sustainable Engineering Forum

(617e) Syngas Fermentation: Reaction Kinetics and Pressure Dependencies of the Clostridial P11 Hydrogenase Enzyme

Bradley E. Skidmore, Douglas R. Tree, Jason M. Bray, Ryan A. Baker, Dila R. Banjade and **Randy S. Lewis**, Chemical Engineering, Brigham Young University, Provo, UT

Ethanol usage as a transportation fuel is rapidly increasing in the United States. Production of ethanol from cellulose feedstocks via gasification followed by syngas fermentation offers an approach that mitigates many of the adverse effects associated with production of ethanol from corn. In the syngas fermentation process, the hydrogenase enzyme of the fermentation bacterium supplies electrons to the metabolic pathway, facilitating ethanol production.

Though it is possible for the metabolic pathway to obtain electrons from CO oxidation during syngas fermentation, it is preferred to obtain electrons from H_2 as this leaves the CO available for incorporation into the desired organic chemicals. Processes that depend on CO oxidation to supply electrons greatly reduce their carbon-to-product yield. Hence, understanding the kinetics of H_2 oxidation via hydrogenase may facilitate process improvements resulting in greater H_2 utilization and, consequently, greater ethanol production.

Experiments were performed to measure the dependence of hydrogenase activity from Clostridium P11 on H_2 partial pressure. The data were fit with a variety of published hydrogenase rate laws and the Okura rapid-equilibrium rate law was found to best describe hydrogenase activity in Clostridium P11. A constant that regulates the effect of H_2 partial pressure on hydrogenase activity, K_{H2}, was determined to be independent of electron acceptor and to have a value of 0.31 atm, implying that H_2 must be supplied to syngas fermentation processes employing this bacterium at ~3 atm to maximize hydrogenase activity. K_{BV} and K_{MV}, constants that regulate the effect of benzyl viologen and methyl viologen (the electron acceptors used for the kinetic studies) on hydrogenase activity, were determined to be 1.7-2.4 mM and 10.6 mM, respectively.

See more of this Session: Reaction Kinetics and Transport Fundamentals for Biomass Conversion I
See more of this Group/Topical: Sustainable Engineering Forum

(617f) Secondary Hydrolysis of Pretreatment Liquor Obtained From Continuous High-Solids Dilute-Acid Pretreatment of Corn Stover

Suan Shi, Chemical Engineering, Auburn University, Auburn, AL, Y. Y. Lee, Department of Chemical Engineering, Auburn University, Auburn University, AL, Richard Elander, National Bioenergy Center, National Renewable Energy Laboratory, Golden, CO and Nick Nagle, National Renewable Energy Laboratory, Golden, CO

Dilute-acid pretreatment is one of the most promising process options applicable for various lignocellulosic biomass feedstocks. NREL has developed an efficient continuous pretreatment reactor that can be operated with high-solid charge and with short residence time. Using this reactor, NREL has investigated extensively on pretreatment of corn stover. Because of these unique features, the pretreatment liquors produced from this reactor contain high level of hemicellulose sugars, the concentration surpassing 10 wt. %. On the other hand, the short residence time and low-acid condition limits the hydrolysis of hemicellulose. The pretreatment liquor obtained from this reactor thus contains relatively high amount of xylose oligomers, typically in the range of 10-40% of the total sugar. In order to fully utilize the sugars in the pretreatment liquor, oligomers need to be hydrolyzed to monomers. However, the secondary hydrolysis of this liquor was found to be much more difficult than hydrolysis of xylose oligomers in clean environment. This investigation was undertaken to verify the factors inhibiting the acid-catalyzed oligomer hydrolysis and to establish the comprehensive kinetics recognizing all reactions occurring under the heavily contaminated environment. For this purpose, the secondary hydrolysis was experimentally studied using rapid-heating small-scale batch reactors. The experiments were done for pretreatment liquor and for clean xylose oligomers. The comparison of the results indicated that the a significant fraction oligomers may be bound to lignin fragments, and some of it may also exist in branched form. The kinetic model was set up recognizing different forms of oligomers (linear, branched, and lignin-oligomer complex), and additional reactions of xylose including decomposition, and reversion to xylose dimmers. Model verification was performed using batch hydrolysis data, xylose decomposition data, the data obtained from reversion reaction. The results of the model verification are presented and how the model can be used to determine the optimum operating conditions are discussed.

See more of this Session: Reaction Kinetics and Transport Fundamentals for Biomass Conversion I
See more of this Group/Topical: Sustainable Engineering Forum

(642a) The Constrained Vapor Bubble Experiment – Results From the International Space Station

Arya Chatterjee[1], **Joel L. Plawsky**[1], Peter C. Wayner Jr.[1], David Chao[2], Ronald J Sicker[2], Tibor Lorik[3], Louis Chestney[3], John Eustace[3] and John Zoldak[3], (1)The Isermann Department of Chemical and Biological Engineering, Rensselaer Polytechnic Institute, Troy, NY, (2)NASA Glenn Research Center, Cleveland, OH, (3)ZIN Technologies, Cleveland, OH

The constrained vapor bubble (CVB) experiment is an experiment in fluid science currently operating on the International Space Station. Flown as the first experiment on the Fluids Integrated Rack on the Destiny module of the US part of the space station, the experiment promises to provide new and exciting insights into the working of a grooved micro heat pipe in the zero gravity environment. The CVB consists of a relatively simple setup - a quartz cuvette with sharp corners partially filled with pentane as the working fluid. Along with temperature and pressure measurements, the curvature of the pentane menisci formed at the corners of the cuvette can be determined using optical measurements. This is the first time the data collected in space environment is being presented to the public. The data shows that, while the performance of the CVB heat pipe is enhanced due to increased fluid flow, the loss of convection as a heat loss mechanism in the space environment, leads to some interesting consequences. We present some significant differences in the operating characteristics of the heat pipe between the space and Earth's gravity environments and show that this has important ramifications in designing effective radiators for the space environment.

See more of this Session: Fundamental Research in Transport Processes II
See more of this Group/Topical: Engineering Sciences and Fundamentals

(642b) Relative Dispersion for Turbulent FLOW IN A Channel and Differences BETWEEN Backwards and Forwards Dispersion IN Time

Chiranth Srinivasan, Chemical Engineering, University of Oklahoma, Norman, OK and Dimitrios V. Papavassiliou, School of Chemical, Biological and Materials Engineering, The University of Oklahoma, Norman, OK

Understanding the movement of particle clouds, say in case of pollutant dispersion in the atmosphere or effluent disposal into lakes, is of utmost importance. Two different scenarios arise when studying the motion of a cloud of particles around its center of mass (i.e., turbulent relative dispersion). The first is the case of clouds of particles that arise from a point and move out from the center of mass, which can be termed as forwards relative dispersion, because the particles move forwards in time. The second is the case where particles over a period of time move about and arrive at a particular location, which can be referred to as backwards relative dispersion. This study utilizes a unique approach, which couples a direct numerical simulation of turbulent flow with the tracking of scalar markers in the flow field, termed as Lagrangian scalar tracking, to analyze the effects of relative dispersion. Turbulent flow in an infinite channel with dimensions 4ƒÎh x 2h x 2ƒÎh in the x, y, z directions (where h is the half channel height, and h =150 in viscous wall units) is considered for the study. Four main regions of turbulent channel flow, namely the viscous sub-layer, the transition region, the logarithmic layer, and the outer region of the flow at the center of the channel are examined. Turbulent dispersion statistics are determined at these locations for forwards relative dispersion, by tracking the cloud of markers diffusing from a point, and for backwards relative dispersion, where the statistics of the cloud of markers that arrive at a particular region are calculated. Prior results for the case of isotropic turbulence indicated that backwards relative dispersion is faster than forwards relative dispersion [1], but single particle dispersion data for anisotropic turbulence indicated that there is a dependence on the distance from the wall and on the Prandtl or Schmidt number of the dispersing particles [2]. In this presentation, relative dispersion statistics will be presented as a function of the turbulent flow region and as a function of the particle Prandtl or Schmidt number, ranging from Pr=0.1 to Pr=100.

References cited [1] B.L. Sawford, P.K. Yeung, M.S. Borgas, Comparison of backwards and forwards relative dispersion in turbulence, Physics of Fluids 17 (2005) 095109. [2] Srinivasan, C., and D.V. Papavassiliou, "Backwards and forwards dispersion of a scalar in turbulent wall flows," Int. J. Heat Mass Transf. 53, 1023-1035, 2010.

See more of this Session: Fundamental Research in Transport Processes II
See more of this Group/Topical: Engineering Sciences and Fundamentals

(642c) Scaleout of Pipeline Drag Reduction by Wall-Injected Polyox to a Flat Plate Boundary Layer

Preetinder S. Virk, Andrea M. Pagan, Joubert X. Glover and Benjamin Ruiz-Yi, Department of Chemical Engineering, Massachusetts Institute of Technology, Cambridge, MA

Turbulent drag reduction by wall-injected Polyox solutions was studied by simultaneous axial pressure profile and flow visualization measurements at Reynolds numbers from 50000 to 100000 in a transparent test pipe of ID 16 mm and L/D 430. Friction factors for deionized water solvent adhered to the Prandtl-Karman law for fully-developed turbulent flow in smooth pipes. The P309 Polyox, of MW ~ 8x106 with Nbb ~ 0.5x106 backbone chain links, was injected as a concentrated solution Cinj =1000 wppm through 6 equi-spaced circumferential ports at friction-normalized injection velocities Vinj+ ~ 1 << pipeline mean velocity V+ ~ 25. Downstream polymer concentrations Cline varied from 0.5 to 20 wppm, exhibiting eventual fractional drag reductions DRline from 0.15 to the asymptotic maximum drag reduction MDR ~ 0.80 possible at the present Re. Polymer dissolution into the flow was assessed from the initial increase in fractional drag reduction towards the eventual downstream DRline, which provided a characteristic development distance (L/D)dr and also from flash photography which provided a characteristic distance (L/D)pv where red-dyed strands of the injected solution disappeared. Finally, an attempt was made to "scaleout" the pipeline drag reduction observations to a flat plate boundary layer by treating the latter as an equivalent pipe flow with UCL = Uo, the free-stream velocity, R+ = del.uô/í , with del the thickness and uô friction velocity at the boundary layer trailing edge L, and Type B slip S' taken from pipe flow results at concentration Cbl corresponding to the polymer expenditure rate per unit of wetted area.

See more of this Session: Fundamental Research in Transport Processes II
See more of this Group/Topical: Engineering Sciences and Fundamentals

(642d) Characterization of Transport Phenomena During Flash Pyrolysis of Cellulose

Marco Dressler and Paul J. Dauenhauer, Department of Chemical Engineering, University of Massachusetts, Amherst, Amherst, MA

Flash pyrolysis is a chemical process which allows conversion of renewable energy sources, e.g. corn stover or switchgrass, into fuels, e.g. cellulosic-ethanol. We have set up a pyrolysis experiment to study basic transport phenomena during the ablative flash pyrolysis of cellulose. In our experimental setup a cellulose rod of 10mm diameter and 300mm length is impinged onto a hot spinning metal surface. The whole setup is enclosed in an atmospheric chamber that contains a mixture of gases at prescribed partial pressures and temperature to favor the pyrolytic reactions of cellulose. In our experimental setup, cellulose ablation is controlled via the angular velocity of the rotating plate and the force of impingement of the cellulose rod. Pyrolysis is controlled via the temperature of the spinning plate and the thermodynamic properties of the combustion gas. The aim of this experiment is to explore the parameter space of cellulose ablation (spinning velocity and impingement force) and pyrolysis kinetics (temperature and combustion composition) in order to find optimum values of pyrolysis vapors. The physical and chemical processes in the impingement zone are observed with an optical pyrometer to analyze the chemical kinetics of cellulose decomposition and with a high-speed camera to capture the ablation dynamics. Furthermore, we analyze flash pyrolysis for various modes of the hot surface motion, e.g. start-up of rotation, steady rotation, cessation of rotation, and oscillatory rotation with small and large amplitudes of oscillation. Our experimental studies are paralleled by theoretical investigations based on the physical modeling of transport phenomena and chemical kinetics. For our theoretical investigation we adopt a standard kinetic model of cellulose flash pyrolysis which considers the formation of activated cellulose with a low degree of polymerization as an intermediate product prior to the formation of char, tar, and gas. Moreover, we discuss whether non-Newtonian properties of cellulose and activated cellulose need to be considered for a more reliable transport modeling of flash pyrolysis and whether our experimental observations can be described in the framework of transport theories for chemically stable polymeric liquids. We propose a set of transport equations for flash pyrolysis and we show how it can be adopted to find a set of process parameters that leads to an optimum production rate of pyrolysis vapors at a minimum energy cost.

See more of this Session: Fundamental Research in Transport Processes II
See more of this Group/Topical: Engineering Sciences and Fundamentals

(642e) Numerical Visualization of Heat Flow and Thermal Mixing in Various Differentially Heated Square Cavities Using Bejan's Heatlines

Tanmay Basak and R. S. Kaluri, Chemical Engineering, Indian Institute of Technology Madras, Chennai, India

A comprehensive analysis of heat distribution and thermal mixing in steady laminar natural convective flow in discretely heated square cavities has been carried out via Bejan's heatlines. Heatlines are analogous to streamlines and heat energy flow may be visualized by heatlines silmilar to streamlines which display fluid flow. The trajectories of heatlines indicate direction and magnitude of heat flow and zones of high heat transfer. The heatline approach is implemented to study heat flow in the following three different square cavities which are filled with water (Pr=7): (1) uniformly heated bottom wall (2) distributed heating with heat sources present on central portions of the walls and (3) multiple heat sources on the walls of the cavity. Top wall is maintained adiabatic in all the cases. Galerkin finite element method with penalty parameter has been used to solve non-linear coupled partial differential equations for flow and temperature fields over a range of Rayleigh numbers (Ra = 10^3-10^5). The Galerkin method is further employed to solve the Poisson equation for streamfunctions and heatfunctions. Finite discontinuity exists at the junction of hot and cold walls leading to mathematical singularity. Solution of heatfunction for such type of situation demands implementation of non-homogeneous Dirichlet conditions. Heatlines illustrate that in uniformly heated bottom wall case, the heat from the bottom wall is not adequately distributed to the lower portion of side walls which leads to low temperature in those regions (case 1). In order to improve the heat distribution, the uniform heat sources is divided into three parts and are applied along the central regimes of the walls (case 2). It is observed that, heat distribution and thermal mixing in the cavity is significantly enhanced. However, the lower corner portions are still retained cold. In case 3, multiple heat sources are placed along the walls of the cavity along with heat sources at lower corner regions of the cavity. Heatlines indicate that, the temperature at the core is reduced compared to case 2, but uniform heat distribution results in uniformity of temperature across large area of cavity.

See more of this Session: Fundamental Research in Transport Processes II
See more of this Group/Topical: Engineering Sciences and Fundamentals

(642f) LPG and Ethane Combined Refrigeration Cycle and Offloading Process

Alexandre de França Cordeiro, Thiago Affonso Ferreira Trotta and Carlos Jose Rocha de Oliveira, Special Projects Division, Chemtech Engineering Solutions, Rio de Janeiro, Brazil

As the operation on a process gas facility usually leads to the necessity of storage at low temperatures (e. g: storage of LPG and NGL at refrigerated tanks), the need of cooling utilities is a constant on process gas industry. Currently, the duty required for the cooling process on a plant is achieved by the utilization of a refrigeration cycle (common used utilities are: ethane and propane), those cycles are composed by four steps: compression, condensation of the compressed gas, isenthalpic expansion of the liquid (or isentropic when turbo-expander is applicable), and finally the evaporation of the fluid removing the required energy from the process system. However, as a process plant usually has more than one product that must be refrigerated before the storage, the usage of a refrigeration cycle with an utility may lead to a very large and expensive cycle, or even to the utilization of a cycle for each product, thus resulting in an economically unviable process. Therefore, it is needed to search for an alternative solution that is more economically attractive. A possible solution for this problem is to use one of the products in the refrigeration cycle, to provide the cooling requirement for the other products, and also for the product used on the cycle. As the cooling fluid needs to be stored, this refrigeration process is labelled as an open cycle, which means that the cooling fluid is meant to leave the process at some point on the cycle. This paper studies the applicability and the benefits of this process on a Gas processing plant, which produces Petrochemical ethane, Natural Gas (NG) and Liquefied Petroleum Gas (LPG). The LPG should be stored liquid at atmospheric pressure (which leads to a temperature of -40 °C), and the ethane should be stored liquid at 10 barg. It was used the ethane on an open refrigerated cycle, in order to achieve the storage requirement for both the LPG and the Petrochemical ethane.

See more of this Session: Fundamental Research in Transport Processes II
See more of this Group/Topical: Engineering Sciences and Fundamentals

Heat Exchanger issues within Cryogenic CO_2 Capture process

Christopher S. Bence, Sustainable Energy Solutions

David W. James, Department of Chemical Engineering Brigham Young University

Larry Lin Baxter, Department of Chemical Engineering, Brigham Young University

Key Words, heat exchanger, CO_2 Separation, cryogenic, desublimation, modeling, process

Various methods are being developed to reduce CO_2 emissions from energy and chemical industry processes. A novel concept being explored desublimates CO_2 from a light gas stream such as flue gas by bringing the temperature in the system down to meet the desired CO_2 capture efficiency. This process is called Cryogenic Carbon Capture (CCC).

The CCC process works by compressing ambient temperature flue gas to 6 bar and then cooling the gas prior to entering a turbine. Turbine inlet temperatures can be below the frost point of CO_2, causing desublimation in the heat exchanger. The cold flue gas cools further upon expansion in the turbine, where the remainder of up to 99% of the CO_2 drops out as a solid. After separating the solid and gas streams, solid CO_2 is pressurized to 70 bar. This high pressure CO_2 solid stream and low pressure N_2 gas stream are thermally tied as they are warmed back up to ambient temperature. These cold streams are run counter current to the high pressure flue gas in the heat exchanger between compressor and turbine.

A concern for the CCC process design involves the inability of commercial heat exchanger models to account for proper phase changes that occur in the CCC heat exchangers. An energy balance of the system must account for the actual phase changes that will occur, which is primarily the desublimation of CO_2 and other pollutants rather than condensation as a liquid predicted by most software.

To find the correct temperature profiles inside the heat exchangers a computer model was written to fill the inadequacy of commercially available software. With the thermodynamic properties of the flue gas programmed in, flue gas cooling is modeled by a series of fully defined states equally spaced inside the heat exchanger with a constant enthalpy reduction between them. Each point in the heat exchanger must be a fully defined state to account for phase changes. Figure 1 shows the temperature profile of the incoming flue gas as it is cooled, then shows the thermally tied N_2 and CO_2 streams as they warm up.

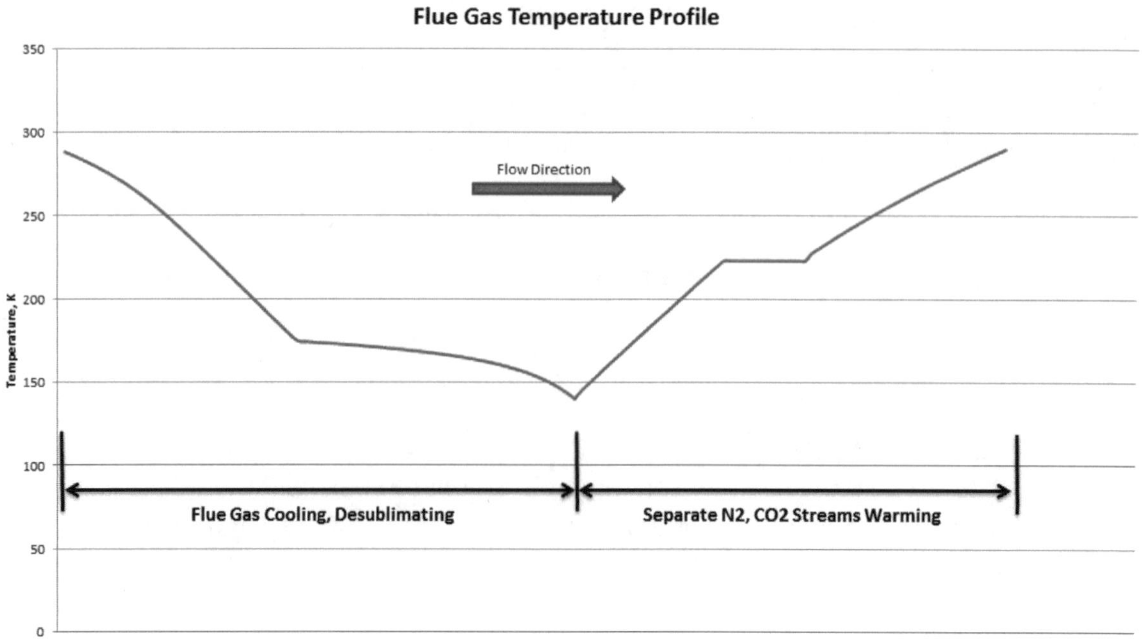

Figure 1. Temperature Profiles as function of enthalpy change.

The shallow slopes on the left side of Figure 1 are due to the latent heats of sublimation of water and carbon dioxide changing phase. The majority of the water drops condenses to liquid above 273 K, while the remainder desublimates as the temperature drops. CO2 starts to desublimate around 175 K with the given flue gas mixture, causing the second kink in the curve. On the right side of Figure 1 is the warming curve of CO2 and Nitrogen. Due to the N2 and CO2 streams being thermally tied, their temperature profiles are forced to take the same shape. The flat portion of the curve is the result of CO2 melting, causing all the energy in this section of the heat exchanger to be consumed as latent heat.

Modeling the desublimating heat exchangers and properly accounting for the locations and amounts of desublimating components is essential to building a heat exchanger that can properly handle solids formation. Figure 2 shows the flow rates of Ice and solid CO2 vs. position in the heat exchanger.

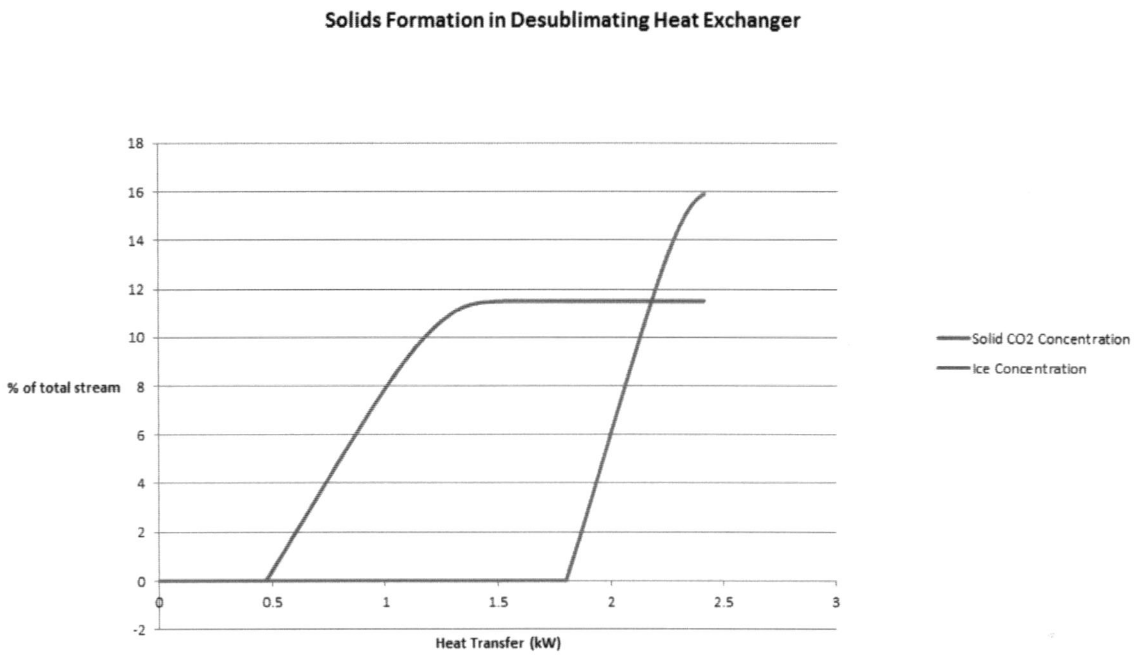

Figure 2. Solids flow rate vs. position in desublimating heat exchanger.

The model is capable of showing the cooling temperature profile for a wide variety of flue gases and can take into account pollutants such as SO2, NO2, and Hg. These trace pollutants are too low in concentration to have a significant effect on the temperature profiles, but knowledge of where they condense inside the heat exchanger is critical to collecting them for removal.

One embodiment of the CCC process uses an external refrigeration cycle to cool the flue gas rather than compressing and expanding it. The inlet to the first heat exchanger is a flue gas composition of 78% N2, 16% CO2, 2% H2O, and 4% O2 at 25 degrees C. The gas flows through the heat exchanger counter current to the evaporating refrigerant, which will be designed to maintain a 10 degree C temperature difference across the length of the heat exchanger. At the exit of the heat exchanger the flue gas leaves at -135 degrees C, by which point over 99% of the CO2 and H2O have desublimated. The latent heat of sublimation has a strong effect on the temperature profile in the heat exchanger, so accurately modeling this behavior is critical to making the proper refrigerant blends.

Due to the non-linear shape of the temperature profiles, a custom refrigeration mixture must be found which has the same shape but is vertically offset from the flue gas cooling in the evaporator, as well as the warming N2/CO2 streams in the condenser.

Of mechanical concern to the CCC process is fouling in heat exchangers from desublimation. Flue gas contacting cold refrigeration pipes will cause CO2 and ice to desublimate on the surface of the pipes, greatly increasing the thermal resistance of the pipes to the point of ineffectiveness. Solids generated in the heat exchanger must be removed at the same rate created. This is a more difficult challenge than condensing heat exchangers where the condensate liquid can easily be moved.

Cryogenic Pollutant Balances Extended Abstract

Rachel Larsen, Cyrus Fox, Bradley Meldrum, and Larry L. Baxter

Abstract

A side benefit of cryogenic carbon capture is the removal of other pollutants such as Hg, HCl, SO_2, SO_3, NO_2, and NO. While cryogenic carbon capture alone is an expensive process, the ability to remove these other pollutants within the same cycle provides an added economic benefit. This process, however, would not be economical by itself if only used to remove these pollutants. Although Hg, HCl, SO_2, SO_3, NO_2, and NO are common pollutants, not much is known about their properties at -135°C. The partial pressures of these pollutants had to be extrapolated from the data found on the DIPPR but give a good estimate of when these pollutants should be removed. The data supports a 99 percent removal of both CO_2 and SO_2 by the time the system has reached temperatures of -135 °C. The process is ready for a larger scale version for more testing to get the technology ready for plant scale operation.

Introduction

Cryogenic Carbon Capture, or CCC, is a Post Combustion Carbon Capture technology that uses thermodynamics to remove carbon dioxide from coal burning power plants. When CO_2 is less than 20% of a gas mixture, it will begin to desublimate once temperatures are below -80°C. This unique property can be used to separate CO_2 from the rest of the flue gas by creating a mostly pure solid that can be separated from the rest of the gases easily. A removal greater than 99 percent can be reached once temperatures reach -135°C. This is a much greater removal than many other technologies that are being studied, and the CO_2 that is removed is much purer than other removal systems.

Alternative options that are being researched include oxy firing, amine cycles, and cryogenic carbon capture. Oxy firing burns the coal with pure oxygen and recycles the flue gas so that there is a relatively pure CO_2 stream that can be captured and stored.. This technology is the furthest developed and has had a successful small scale plant, but there hasn't been a full-scale trial of the technology. While the technology is the most advanced, it is very expensive and would require a major overhaul of the way that the power plant is designed. It would be difficult and expensive to add to existing plants. They are expensive both in terms of money and energy because they require a large Air Separation Unit that has to cool and separate large quantities of air to get enough oxygen to burn in a good ratio. Also, they must reach colder temperatures in order to separate oxygen as its boiling point is 54 K.

Amine cycles use large quantities of toxic amines that the flue gas is bubbled through. The Amine absorbs the CO_2, and releases the CO_2 when the amine solution is heated. This is also expensive technology, and potentially hazardous because of the quantities and types of materials being used. Because Amine solvents are also post Combustion Carbon Capture technologies they require less changes of the plant itself, and can be added to the back of the plant.

Cryogenic Carbon Capture does not require toxic chemicals, or a large remake of the rest of the power plant, and is cheaper to build than the other technologies. However, this technology is the least developed of the other choices. We are currently in the lab stage of development.

Cryogenic carbon capture is an effective method to remove CO_2 and has the potential to replace scrubbers and other flue gas treatment methods. This technology becomes much more economically beneficial with its ability to compliment and even replace some flue gas treatments currently employed. This research has verified that the CO_2 and SO_2 would be removed at the temperatures and percentages that are expected for this process to work. Continued research to verify that other pollutants, especially mercury related species is in the process of being developed.

Theoretical Calculations

The basis of the theoretical calculations assumes that a solid or liquid will only be formed when the vapor pressure of the species is less than or equal to the partial pressure in the gas. Calculations for the vapor pressure from DIPPR provided the temperature dependent correlations to predict the vapor pressure of each pollutant. The partial pressure is then compared to the vapor pressure of each pollutant to calculate the amount of each pollutant that would be theoretically removed. The equations from DIPPR are based on real data that has been collected and verified. However, the data was collected for much warmer temperatures, so the vapor pressures at -135°C were extrapolated and thus may not be completely accurate. They serve as a base estimate for how the components actually behave. The thermodynamics show that the vapor pressure will be extremely low at these temperatures, and indicate that the pollutants will have a greater than 90 percent removal. NO is the exception, as it has a very high vapor pressure even at these cold temperatures, and will not be removed by cooling alone.

Equation 1 states the correlation between vapor pressures and temperature, where *Y* is the vapor pressure; *A, B, C, D,* and *E* are constants specific to the species; and *T* is the temperature.

$$Y = \exp\left(A + \frac{B}{T} + Cln(T) + DT^E\right) \qquad (1)$$

Figure 1, Figure 2, Figure 3, and Figure 4 show how the vapor pressures are dependent on temperature for NO_2, NO, SO_2, and SO_3. As the figures show, the temperatures that we are interested in are not within the range of known data, except in the case of NO. As NO is a liquid at these temperatures, the vapor pressure has been accurately calculated. Because NO is a liquid, its vapor pressure is relatively high, and so will not be directly removed by this process. It could indirectly be removed by this process if it was first converted to NO_2.

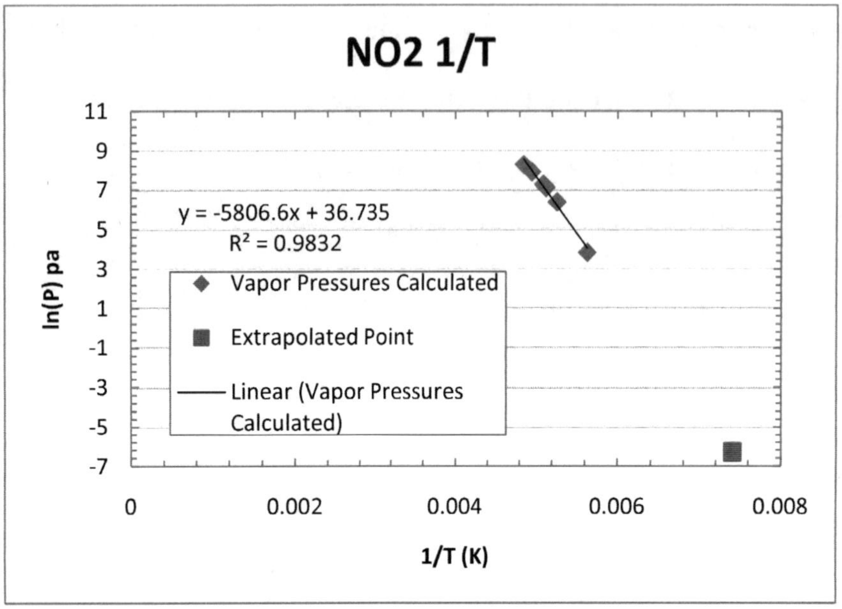

Figure 1 Vapor pressure verses temperature for NO2. For simplicity, the pressure vs. temperature graph is displayed as the natural log of pressure in Pascal verses 1/Temperature in Kelvin. The blue is the data set that the vapor pressure is calculated from and the red is the predicted vapor pressure at -135°C. The vapor pressure for NO2 is predicted to be .00189 Pa.

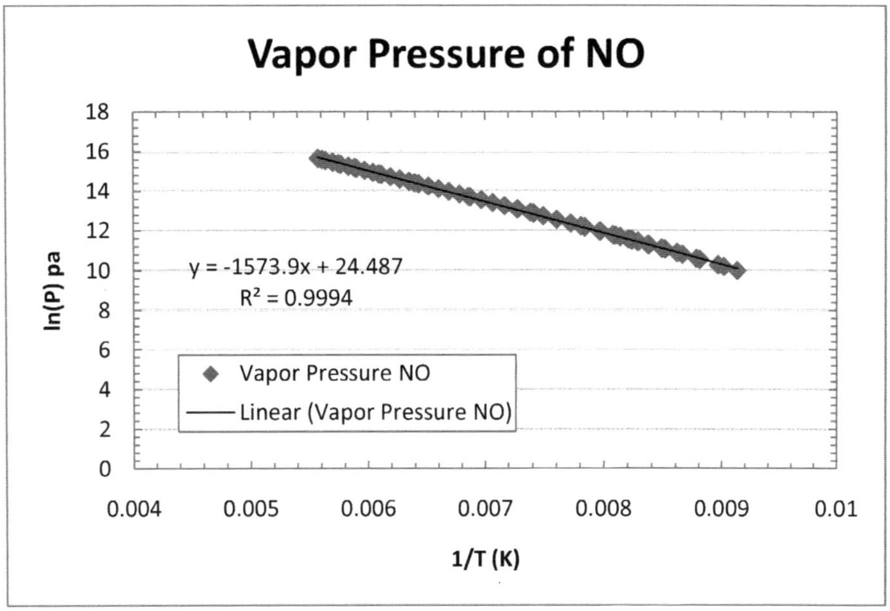

Figure 2 As NO is a liquid at these temperatures; the vapor pressure equations are accurate at these temperatures. The predicted vapor pressure of NO is 409964 Pa. This is rather high, so NO would have to be removed by first converting to NO_2 or using another process.

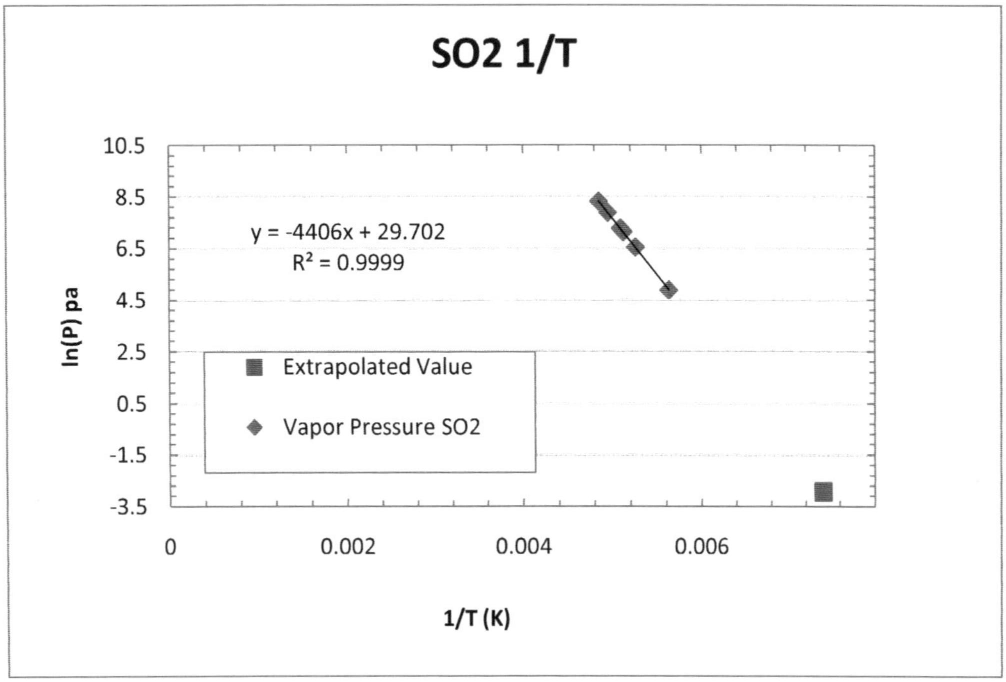

Figure 3 The vapor pressure of solid SO2 at -135°C is predicted to be 0.053 Pa.

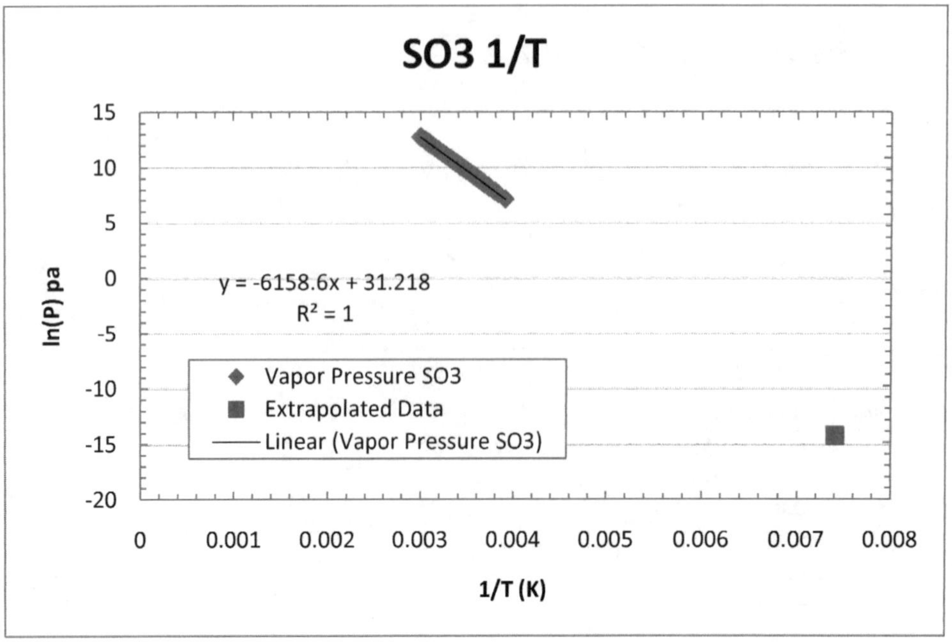

Figure 4 The vapor Pressure of solid SO3 at -135^{0}C is 5.567*10^-7 Pa.

The vapor pressures are used to predict the percent removal of each pollutant by knowing the initial concentrations in the flue gas at room temperature. The gas analyzer calculates concentrations in terms of volume percents. The initial partial pressure can then be calculated using Equation 2 where *pp* is the partial pressure of the species, *y* is the volume percent as a fraction, and *P* is the total pressure of the system. Unless otherwise noted, the system will be assumed to be at atmospheric pressure. The partial pressure is assumed constant until it is equal to the vapor pressure, and the vapor pressure equations account for the temperature dependence. The percent removal can then be calculated using Equation 3, where *I* is the initial volume percent of CO_2, *x* is the volume percent of CO_2 at the point, and *P* is the percent removal of CO_2.

$$pp = y * P \tag{2}$$

$$P = \frac{I-x}{I} * 100 \tag{3}$$

Figure 5 graphically shows the predictions for the removal of CO_2, SO_2, and NO based on temperature using the above equations and atmospheric pressure. NO, as expected from above, will not be removed at any of the temperatures that the experiment reaches. CO_2 will start being removed at -100°C and will reach a 99 percent removal at -134°C. SO_2 will start being removed at -112°C and reach a 99 percent removal at -133°C based off of an original concentration of 100ppm.

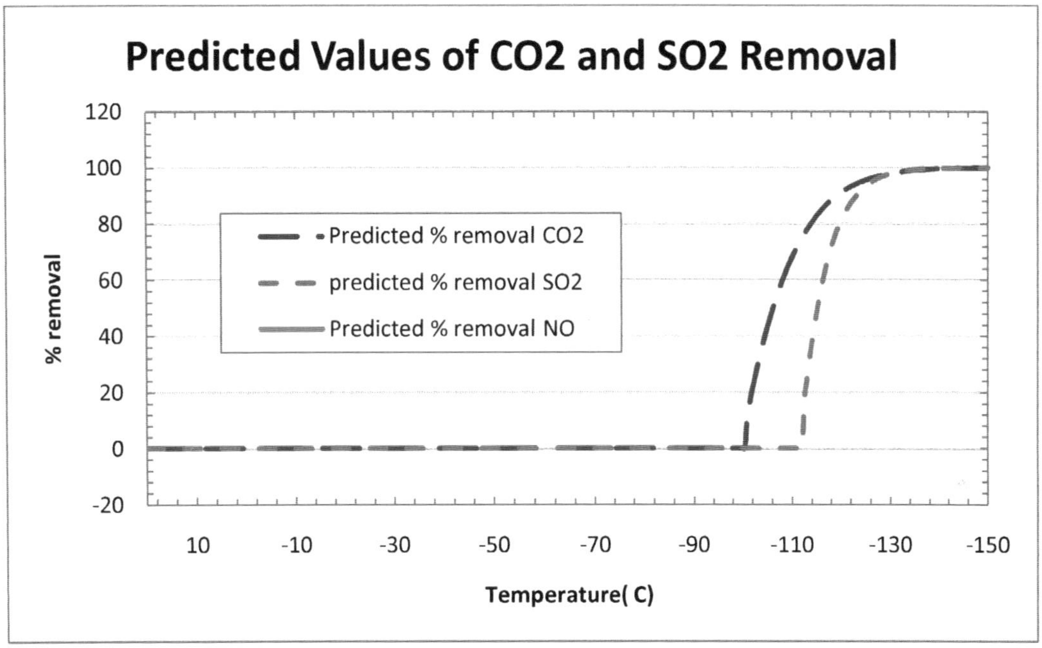

Figure 5 The predicted removal of CO2, and SO2based off of vapor pressures and temperature. SO2 is predicted to begin being removed at -110°C assuming initial concentration is 100ppm. CO2 is predicted to begin being removed at -100°C assuming initial volume 14%.

Experimental Set-Up

The experimental set-up is a source, usually a simulated flue gas, or a coal and methane reactor attached to a heat exchanger then through a Horiba® gas analyzer. Figure 6 displays the basic set-up. The heat exchanger was designed and made by research students. It is made of ¼ inch stainless steel tubing in a snake-like design with valves every 8 inches to allow for thermocouple positioning and gas sampling along the line. The tubing is placed vertically in a Styrofoam box to insulate it and has a small metal grate to hold a pan of liquid nitrogen above the tubing to cool the coils. The liquid nitrogen is filled through a hole in the top of the Styrofoam box during experimentation so that the temperature can remain as constant as possible.

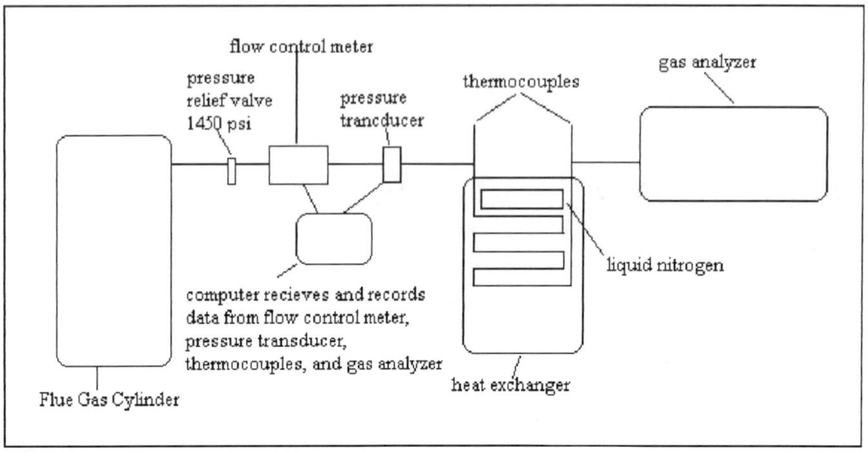

Figure 6 This figure is the basic experimental set-up. Changes and modifications will be duly noted.

Results for Cylinder

A certified cylinder from AirGas® containing by volume 14 percent CO_2, 3 percent O_2, and the balance N_2 was used for this experiment. It is extremely difficult to get accurate measures of the temperature as there are many hot and cold spots because liquid nitrogen is -170°C and the room is 20°C, causing a temperature gradient and uneven cooling. Figure 7 is a comparison of the predicted percent removal and the experimental percent removal. The experimental data varies from the predicted data at the beginning of the removal, but agrees at the point of 99 percent removal at -134°C. As the temperatures get colder, cooling takes longer. This may lead to a more even temperature distribution along the heat exchanger and therefore a more accurate temperature read-out that is closer to the expected data.

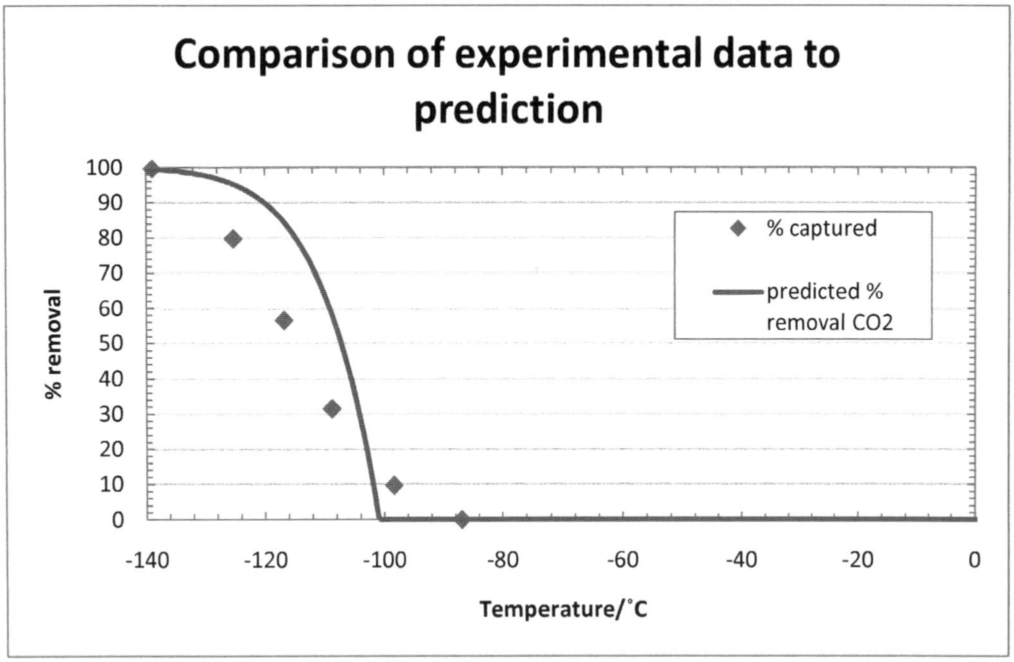

Figure 7 A comparison of the predicted versus experimental data for the removal of CO2. The data may vary from the predicted pattern due to an uneven cooling distribution.

Results for Multi Pollutant Cylinder

Another series of experiments were conducted with a cylinder that contained by volume 14 percent CO_2, 100ppm SO_2, 100ppm NO, and the balance N_2. No oxygen was in the cylinder as NO is unstable in the presence of oxygen at the pressures that the cylinder was under. The same experiment as above was run. This experiment required extra calculations to account for the changing overall volume due to CO_2 being removed. As the gas analyzer measures percent volumes, the percentages of the other species change as CO_2 is removed. We had to do some calculations to account for that change in percentage as it made the SO_2 and the NO artificially increase rather than show that they were removed later or not at all. Figure 8 shows the raw data collected before any corrections were made.

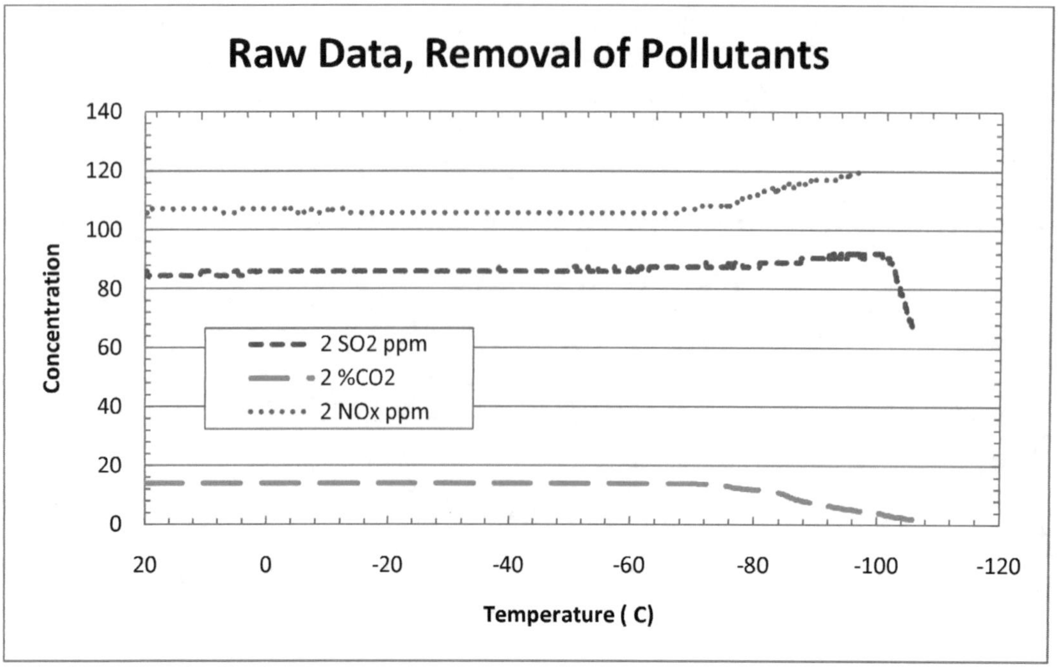

Figure 8 The raw data from the cooling of the artificial flue gas shows an artificial increase of SO2 and NO because the gas analyzer measures percent volume.

The raw data was corrected by using the change in volume of CO_2 to predict the increase of SO_2 and NO assuming only CO_2 was removed. The experimental percent removal of CO2 was used as the basis, so the data matches really well, as seen in Figure 9. The percent removal was calculated using Equations 4, 5, 6, and 7, where I is the initial volume percent of CO_2, x is the volume percent of CO_2 at the point, P is the percent removal of CO_2, R is the remainder of the gas that should not change, F is the fraction by which the values of NO and SO_2 are expected to change based on the percent removal of the CO_2, RI is the initial percent of the gas that is not CO_2, A is the expected amount with the correction of the removal of CO_2, n is the initial amount of species other than CO_2. In this case it would be SO_2 and NO.

$$P = \frac{I-x}{I} * 100 \quad (4)$$

$$R = 100 - P \quad (5)$$

$$F = \frac{RI - R}{RI} \tag{6}$$

$$A = F * n + n \tag{7}$$

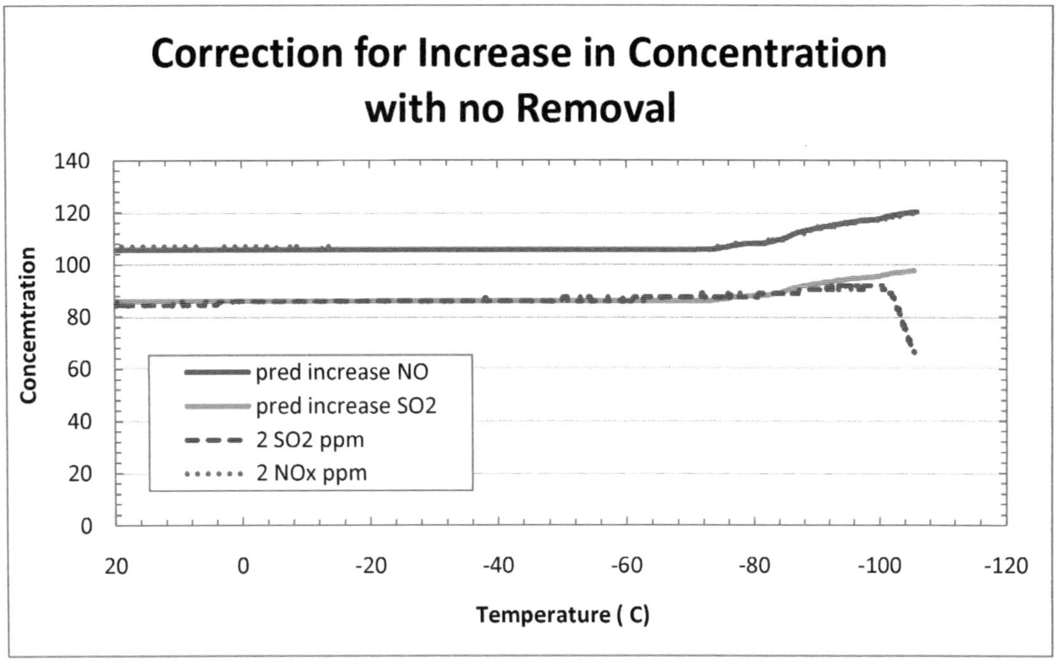

Figure 9 Raw data compared to the prediction of the artificial increase due to the gas analyzer. We expected no removal of NO because of task 1, and the fit agrees. Even though the SO_2 concentration seemed to increase, it was removed.

Using the variation from the expected increase of SO_2 and the measured increase of CO_2, The percent removal for SO_2 and NO was calculated. Comparisons of all the predicted and experimental percent removals are shown in Figure 10. The experimental removal occurred 20°C higher than expected.

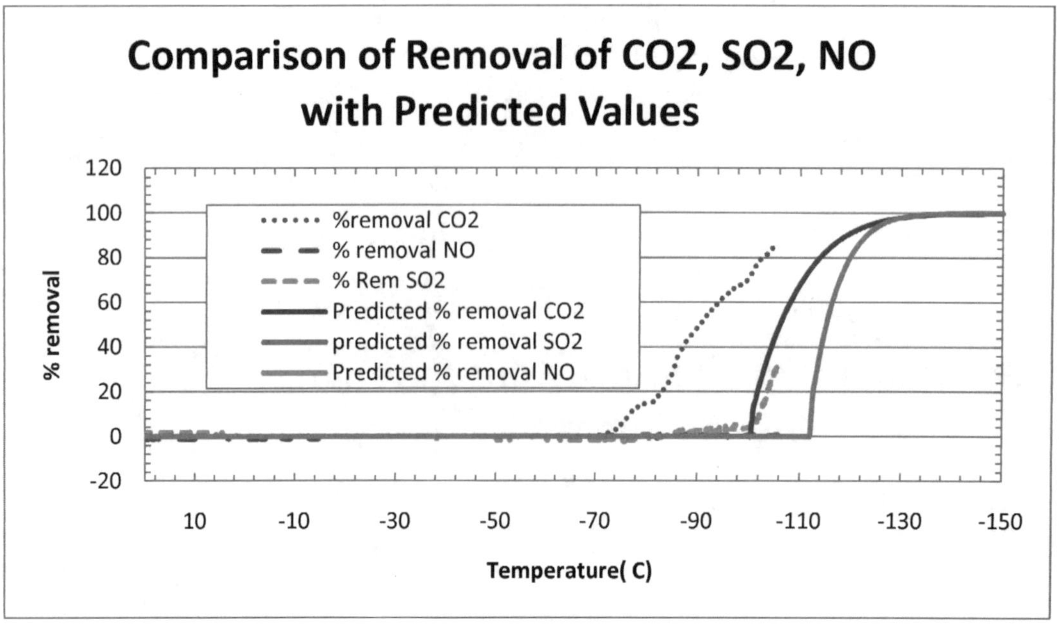

Figure 10 Comparisons of the predicted and experimental percent removals have a similar shape, but happened 20°C earlier than expected.

The difference between the calculated data and the observed data is still a mystery. Most problems that would account for a difference in the data would decrease the temperature rather than increase the temperature at which the CO_2 and the SO_2 would begin to be removed. The experiment seems to lose pressure, and so the gas cannot be cooled more than -115°C. At this time oxygen begins to leak into the gas analyzer and makes further results unreliable. A pressure drop, however, would decrease the temperature at which SO_2 and CO_2 should be removed as it would also decrease their partial pressures. The gas analyzer also has a time lag in reading SO_2, and so the temperatures may not line up with the readings that the gas analyzer is giving. However, this would also decrease the temperature when SO_2 would be removed due to lag time. It should be noted that freezing at higher temperatures is favorable because it will require less energy to extract the pollutants and that means that this process may be even cheaper than was predicted.

The only reasonable explanation as to why the freezing temperatures are warmer would be that the thermocouple is not reading the coldest temperatures that are being reached in the heat exchanger. The thermocouple is placed to read the temperature three inches before it leaves the heat exchanger so the gas does not have time to warm up outside of the heat exchanger before being measured. As the tubing is only ¼ inch in diameter, it is really difficult to control what the thermocouple is reading once it is inside. We are thinking that if we bent the tip of the thermocouple to ensure that it was reading the temperatures closer to the walls of the tubing, the temperatures would be more accurate, especially as the CO_2 and SO_2 are being removed by sticking to the walls.

Experimental Set Up for Real Flue Gas

This is very similar to the previous set up used on simulated canisters. This flue gas is produced by an MFR (multi-fuel reactor) which burns coal by passing it through a methane burner. This coal is pulverized and is burned in forced atmospheric air, refer to Figure 11. A sample line is inserted near the end of the reactor which is pumped out of the reactor, through a desiccant, and an ash filter. Here this flue gas comes in contact with the same heat exchanger and gas analyzer set up as was used in the simulated flue gas (canisters) experiment.

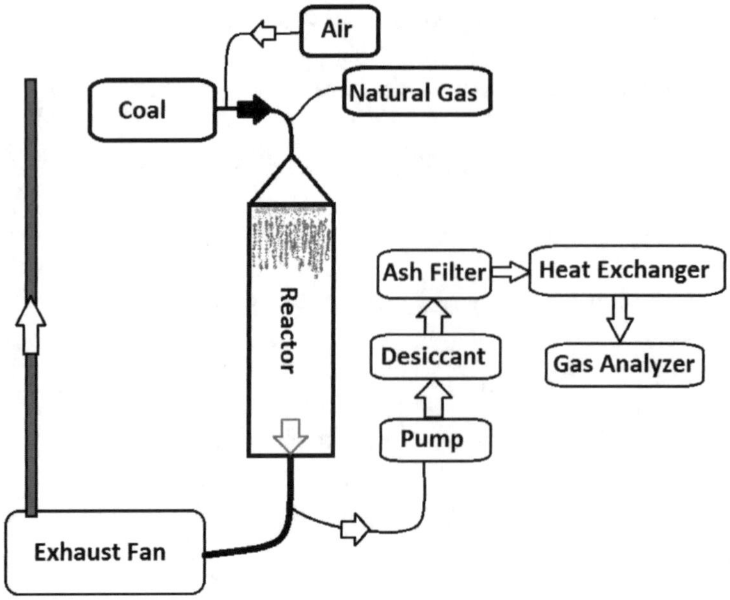

Figure 11 This is a rough outline of the flue gas experiment(I can make another)

The flue gas came from a reactor whose internal temperature at the insulated wall is 700 °C. The flue gas composition is controlled by altering the amount of air and coal entering the reactor. This can be monitored directly on the gas analyzer. The gas analyzer is only capable of reading one stream at a time, and requires a few minutes of steady state flow through it for its numbers to be accurate. At the beginning of each experiment, flue gas would be run through the whole apparatus at room temperature in order to form a solid baseline for the concentrations of pollutants within the gas. Data is taken in intervals at every second.

Initially these test were conducted by continuous cooling, the heat exchanger would be cooled by about 3°C each minute by adding liquid nitrogen. This provided a good picture of CO_2 removal but obfuscated the SO_2 profile due to the time lag of the gas analyzer. To remedy this flaw, test was set up to keep the heat exchanger at constant temperatures of a period of 10-20 minutes. The only flaw with this later set up occurs from flow blockage which occurs from the subliming pollutants which accumulate.

Results for Real Flue Gas

We have begun using real flue gas from a methane and coal reactor to account for all components of the flue gas. We have achieved the best fit of our results for CO_2 removal as seen in Figure 12. However, as the reactor fluctuates more than a cylinder does, good data for the removal of SO_2 has not been achieved. The best data so far for SO_2 is shown in Figure 13, also in Figure 16 and Figure 17 show our first results of SO_2 removal. Note that the data in Figure 13-17 is a continuous cooling curve. We hope to obtain better results for the SO_2 with the same methods used in Figure 12 where the temperature is held at a constant level and the results are averaged for that temperature. We did not achieve a good base for the SO_2 on June 25, se were not able to get an accurate percent removal on that date, but the method appears to get more accurate results than the continuous cooling curve. The reason why the continuous cooling curve method is inaccurate is because there is a time lag in the SO_2 read-outs, and by the time the SO_2 results are read, the temperature is no longer the temperature of the gas that was sampled if the temperature is continually dropping. This would also explain why the continuous cooling curve method fits better at the colder temperatures, because the temperature is not dropping as quickly. To compensate for this time lag, two minutes have been added to the SO2 data as an estimate for the lag.

Removal of both CO_2 and SO_2 occurs on warmer temperatures than were calculated. At -80°C to -90°C reductions of both pollutants becomes evident. Each trial with the live flue gas produces differing results. The trend is certain and shows quite well that nearly all CO_2 and SO_2 are removed at -135°C.

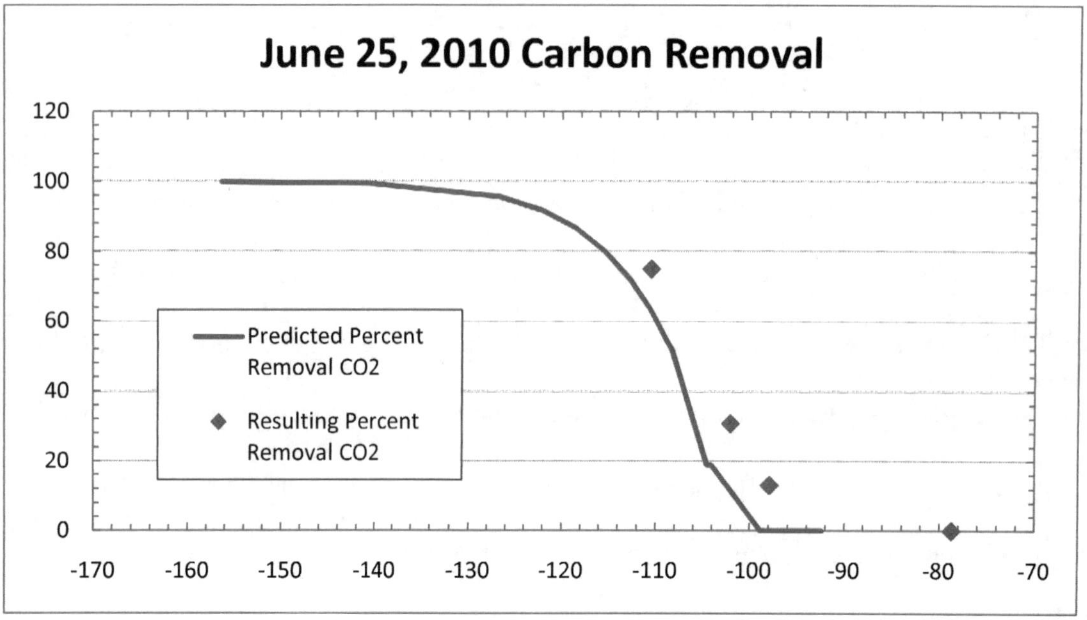

June 25, 2010 Carbon Removal
120
100
80
60
40
20
0
-170
-160
-150
-140
-130
-120
-110
-100
-90
-80
-70
Predicted Percent Removal CO2
Resulting Percent Removal CO2

Figure 12 A comparison of removal of CO2 from real flue gas with predicted removal shows a very close fit. The results were taken on June 25, 2010 by holding the temperature constant at several temperatures and averaging the results rather than a continuous cooling curve.

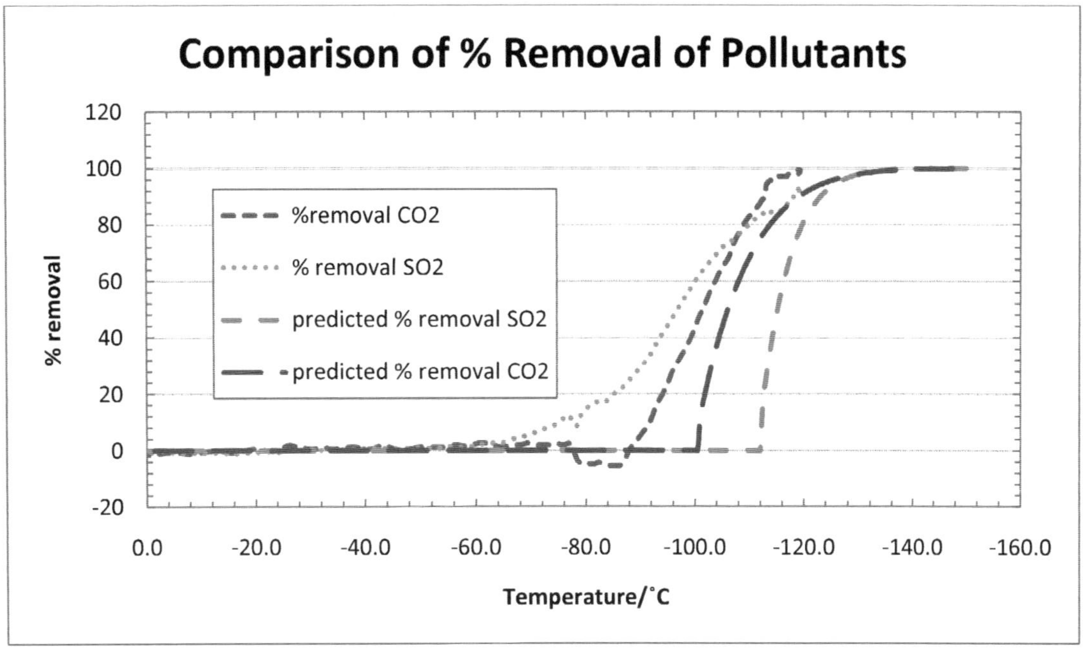

Figure 13 The comparison of the percent removal of pollutants in real flue gas. This figure shows that SO2 is removed before CO2 this may be due to a fluctuation of SO2 in the flue gas as the reactor is not as consistent as a cylinder.

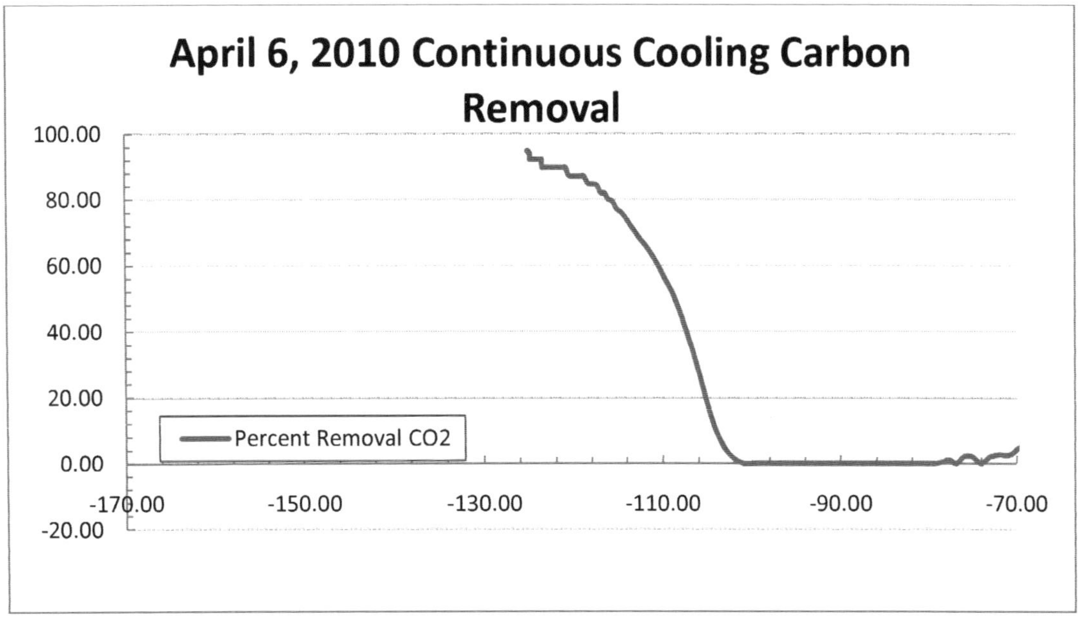

Figure 14First run of the reactor produced this data which confirms highly to predicated values of CO2 removal. This occurred in oxygen rich burning conditions, 15% by volume of this flue gas was O_2.

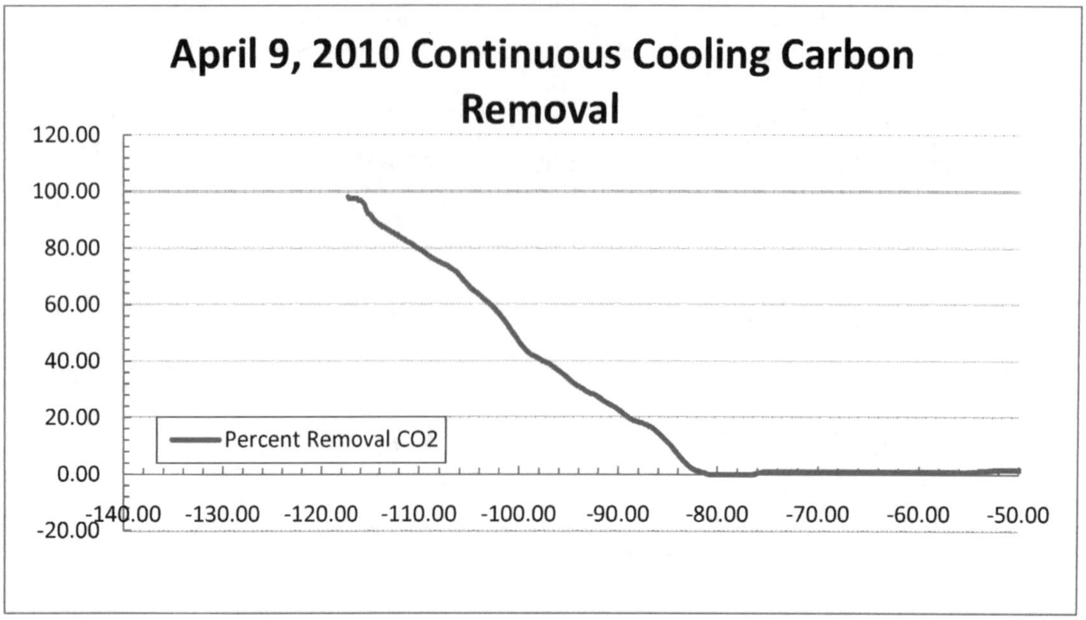

Figure 15Second run of the reactor, flue gas was only 4% oxygen in this run.

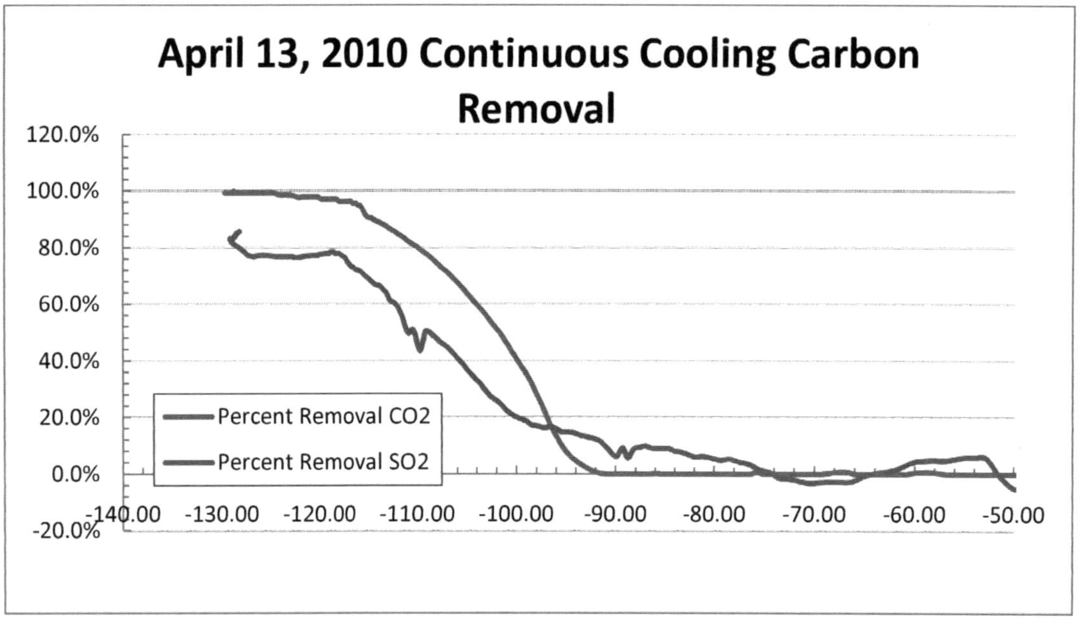

Figure 16By the third run, we were able to obtain SO_2 measurements. A two minute lag time has been added for the SO_2 data to accommodate the gas analyzer's shortcomings. This flu gas was near stoichiometric with flue gas containing .5% by volume O_2.

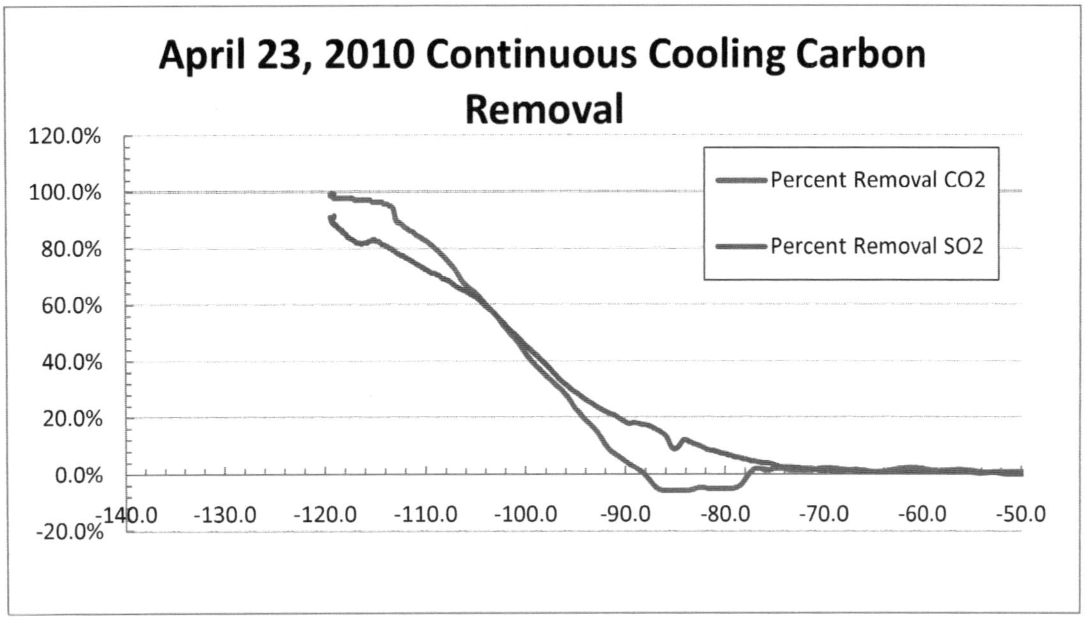

Figure 17 This burn flu gas had the 2% O_2 and also incorporates a two minute lag time on the SO_2 readings.

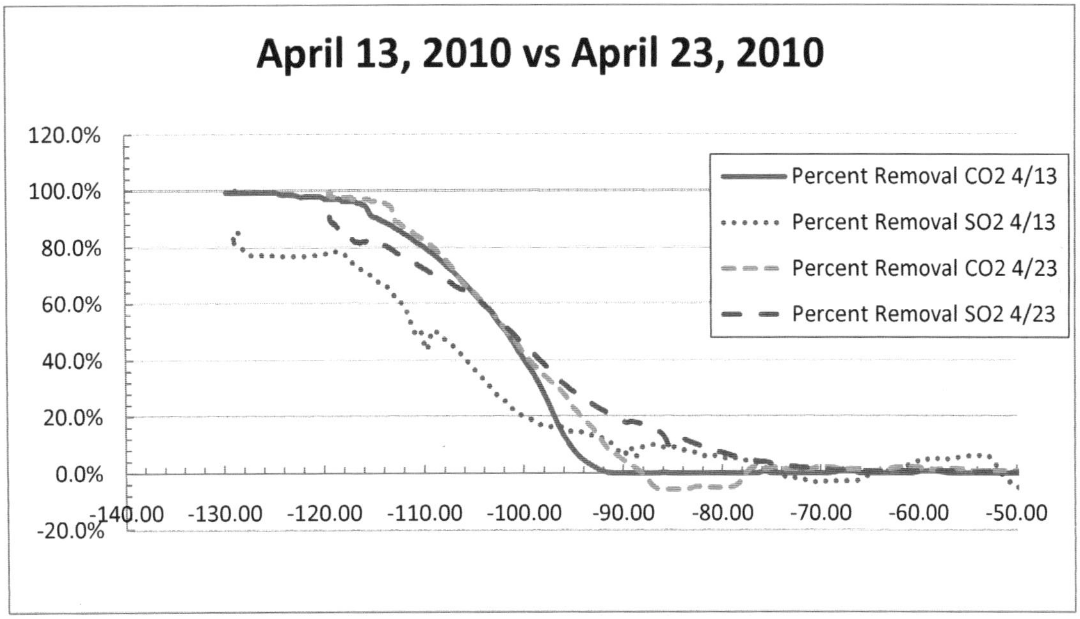

Figure 18This is just a combination of figure16 and 17, the data is strikingly similar.

Conclusion

The experimental results are promising in showing that CO_2 and SO_2 can be removed through excessive cooling due to their physical properties. The exact temperatures at which this happens are still not completely clear. The methods for data acquisition are improving so the temperatures can be more accurate and possibly closer to what we would expect. Continued experiments will help determine exact temperatures and help us to understand how these species all react together in a mixture. It can be concluded from the results that a system that is built to reach temperatures of -135°C would be able to remove 99 percent of CO_2 and SO_2. These results support the use of this process in a larger scale than just the lab setting.

Cryogenic carbon capture is an effective method to remove CO_2 with the potential to replace scrubbers and other flue gas treatment methods. This technology becomes much more economically beneficially with its ability to compliment and even replace some flue gas treatments currently employed.

Acknowledgements

Shawn Kunzler: Helped us learn to use the reactor, made the heat exchanger last summer, and wrote the lab view program.

Dale Tree: Helped us learn how to use the reactor and understand mechanical problems.

Sustainable Energy Solutions (SES): funding

Danish Oil and Natural Gas (DONG): funding and expertise

References

Baxter, Larry. "Cryogenic Carbon Capture Technology." Carbon Capture Journal 10 (2009): 18-21. Print.

Burt, Stephanie, Andrew Baxter, and Larry Baxter. "Cryogenic CO2 Capture to Control Climate Change Emissions." Proc. of Clearwater Clean Coal Conference, Clearwater, Florida. Print.

Dilling, Lisa. "Toward Carbon Governance:Challenges Across Scales in the United States." Global Environmental Politics 7.2 (2007): 28-44. Print.

Gale, Thomas K. "Oxy-Firing for CO2 Mitigation." Southern Research Institute - Home. Southern Research Institute. Web. 27 July 2010. <http://www.southernresearch.org/environmental/oxy-firing.html>.

Hanson, David J. "Legislative Review." Chemical and Engineering News 87.51: 32. Print.

R. L. Rowley, W. V. Wilding, J. L. Oscarson, Y. Yang, R. J. Rowley, T. E. Daubert, R. P. Danner, DIPPR® Data Compilation of Pure Compound Properties, Design Institute for Physical Property Data, AIChE®, New York, NY (2001).

Schreiber, Andrea, Petra Zapp, and Wilhelm Kuckshinrichs. "Eon Starts First Big Carbon Capture Pilot." The International Journal of Life Cycle Assessnent 14.6 (2009): 547-59. Print.

Tomizaki, Kin-ya, Mitsuhiro Kanakubo, Hiroshi Nanjo, Shinkichi Shimizu, Masami Onoda, and Yuichi Fujioka. "C NMR Studies on the Dissolution Mechanisms of Carbon Dioxide in Amine-Containing Aqueous Solvents at High Pressures toward an Integrated Coal Gasification Combined Cycle-Carbon Capture and Storage Process." *Industrial and Engineering Chemistry Research* 49.3 (2009): 1222-228. Print.

Turbine Blade Abrasion Testing for Cryogenic Carbon Capture

Jacob Larsen, Jeff Bean, Grant Evans, Saurav Dhungel, Larry Baxter

Brigham Young University

Abstract:

Consider a turbine running at several thousand rpm being pelted by thousands of tiny solid CO_2 particulates traveling at near sonic speeds. In this paper, we investigate solid CO_2's ability to abrade the surface of a stainless steel 304 sample. The abrasion testing done is in accordance with ASTM standard G 76 – 07[1] with slight variances. We impinge the s.s. sample with a particle stream that is made of a mixture of nitrogen and carbon dioxide gas as well as carbon dioxide solids. After running a series of tests and analyzing the erosion data, we discuss the prospect of utilizing stainless steel 304 as the material for turbine blades in a near-cryogenic turbine expansion chamber based on its ability to resist erosion.

(649a) Effects of Coal Interaction with Supercritical CO2: Chemical Structure

Wei-Yin Chen[1], Guang Shi[1] and Benson B. Gathitu[2], (1)Chemical Engineering, University of Mississippi, University, MS, (2)Department of Mechanical Engineering, Jomo Kenyatta University of Agriculture and Technology, Nairobi, Kenya

Interactions between CO2 and coal have been gaining interests due to its application in geologic sequestration of CO2 and methane recovery in coalmines. Recently, we reported the changes in combustion behaviors, including NO emissions and carbon burnout, and physical structures of coals after treatment. The objective of the current work is to investigate the changes in chemical structure of coals after supercritical CO2 treatment. Lignite and bituminous coal are treated by supercritical CO2 in a batch reactor at 80 and 130 °C, at 2000 to 8100 psig, and with 48 h contact time. The as-received, dried (under vacuum at 100 °C overnight) and treated coals are characterized by temperature-programmed pyrolysis (TPP) with 5 °C/min heating ramp up to 1650 °C followed by online analysis by gas chromatography / mass spectrometer (GC/MS). These coals are also characterized by Fourier-transform infrared spectrometer (FTIR). All raw, dried and treated bituminous coals emit little physically bound CO2 below 200 °C during TPP. All of them emit CO2 with peaks at about 400 and 650 °C with dried and raw bituminous coal treated at 8100 psig emitting highest amount CO2. All lignite samples emit similar amount of CO2 at about 450 °C, which is about ten times that from the bituminous coal samples. CO emission from treated bituminous coal increases dramatically with increasing treatment temperature and pressure during TPP. But such pressure effect is not observed for treated lignite, that emit double the amount of CO that their bituminous coal counterparts do. Drying induces little effect on CO emission during TPP for either bituminous coal or lignite.

FTIR spectra suggest that both drying and treatment significantly enhance the alcoholic/phenolic groups for both bituminous coal and lignite. Both drying and CO2 treatment reduce the ether linkages in bituminous coal and lignite. Carboxyl groups in bituminous coal decrease after drying, but they increase with treatment pressure. For lignite, carboxyl groups do not change significantly with drying and treatment conditions. Carbonyl groups only appear in bituminous coal and they increase significantly after drying and treatment. Drying creates polycyclic aromatic structure in both coals; interestingly, but the newly created polycyclic aromatic structure does not remain in the treated coals.

See more of this Session: Materials and Biomaterials Synthesis and Processing with Compressed or Supercritical Fluids II
See more of this Group/Topical: Engineering Sciences and Fundamentals

(649b) Brownian Particle Coagulation in a Rapidly Expanding Solvent; Example of the "Allowable" Coupling Between Homogeneous Kinetics and Fluid Dilation Rate

Daniel E. Rosner, Chemical Engineering, Yale University, New Haven, CT and Manuel Arias-Zugasti, Vis. Asst Prof.-ChE Dept., Yale University, New Haven, CT

Using recently studied supercritical "anti-solvent" (SAS-) examples for precipitating high-value particles (eg., pharmaceuticals) we suggest1 that the characteristic dilation time [(1/vol)*D(vol)/Dt]^-1 of the swelling solvent can be deliberately made small enough to noticeably reduce the operative coagulation rate "constant", beta. Moreover, we expect that this previously neglected coupling can occur under conditions in which post-nucleation Brownian coagulation must be accounted for in predicting the efficacy of such micron-sized powder production methods. A rational approximate theory for this rate constant "correction factor", beta/beta(0), is proposed1 -------summarized here, with emphasis on the applicable limit of continuum Brownian diffusion-controlled coagulation. We also present a preliminary assessment of the PSD-implications of these "corrections", and indicate possible generalizations. Note that, while P. Curie's principle applied to an isotropic medium of arbitrary EoS does NOT preclude coupling between homogeneous (chemical,...) rate processes and local fluid dilation rate (trace of Def (velocity vector)), this may be the first practical example of such a coupling------ with consequences that can be exploited in the processing of pharmaceutical powders. ______________________________

a Supported at Yale University by NSF Grant CTS 0522944 1. Rosner,DE and Arias-Zugasti M, AIChE J, vol 56 (in press, 2010)

See more of this Session: Materials and Biomaterials Synthesis and Processing with Compressed or Supercritical Fluids II
See more of this Group/Topical: Engineering Sciences and Fundamentals

Particle Encapsulation in Circulating Fluidized Beds

Aydin K. Sunol, and Brandon Smeltzer, Chemical Engineering,

University of South Florida, Tampa, FL

Extended Abstract

Fluidization in supercritical fluids has the potential to offer a unique and energy efficient processing environment for chemical, biological and energy conversion processes. Resulting products, with superior and tunable properties, will allow energy savings and reduce carbon dioxide emission. For decades, particle micro encapsulation, many key reactions, drying, and granulation have been traditionally carried in fluidized beds. However, there are no industrial processes and relevant scale equipment to facilitate fluidization in supercritical fluids. In pilot scale circulating fluidized beds, microencapsulation and functionalization of energetic components have been realized in for synthesis of fuel particles. The particle size range has been extended to nano-scale. The transformational impact of technology for fluidization in supercritical fluids will go far beyond these initial applications in encapsulation.

The potential of the technology is demonstrated in pilot scale using the experimental system shown in Figure 1 while key differences between for particle transport and fluidization in gas, liquid, and supercritical fluid are depicted in Table 1.

Solid fuels such as Ammonium Nitrate (AN) were encapsulated within polymers like Hydroxyl Terminated Poly Butadiene. The pre-polymer and curing agents were deposited from the supercritical carbon dioxide in the circulated fluidized bed. The deposition is usually facilitated through higher operating temperature in the bed (lower polymer solubility region) then extraction/polymer loading chamber (higher polymer solubility region).This retrograde behavior is unique to critical region. The solubility differential allows almost layer by layer deposition. The temperature deposition temperature falls within curing reaction range, too. Micron and sub micron thickness coating of 30-300 micron particles with polymers were achieved within ten minutes. The uniformity of the thickness (surface roughness), measured using AFM (Atomic Force Microscopy), for AN is around 10 nanometers.

In essence the major added benefit of fluidizing with supercritical fluids is that they from homogenous solutions with the components (e.g. coating material), unlike gas counterparts, to allow controllable and rapid deposition and impregnation. Furthermore, the solubility is very sensitive to operating conditions such as temperature pressure composition to allow tunable performance.

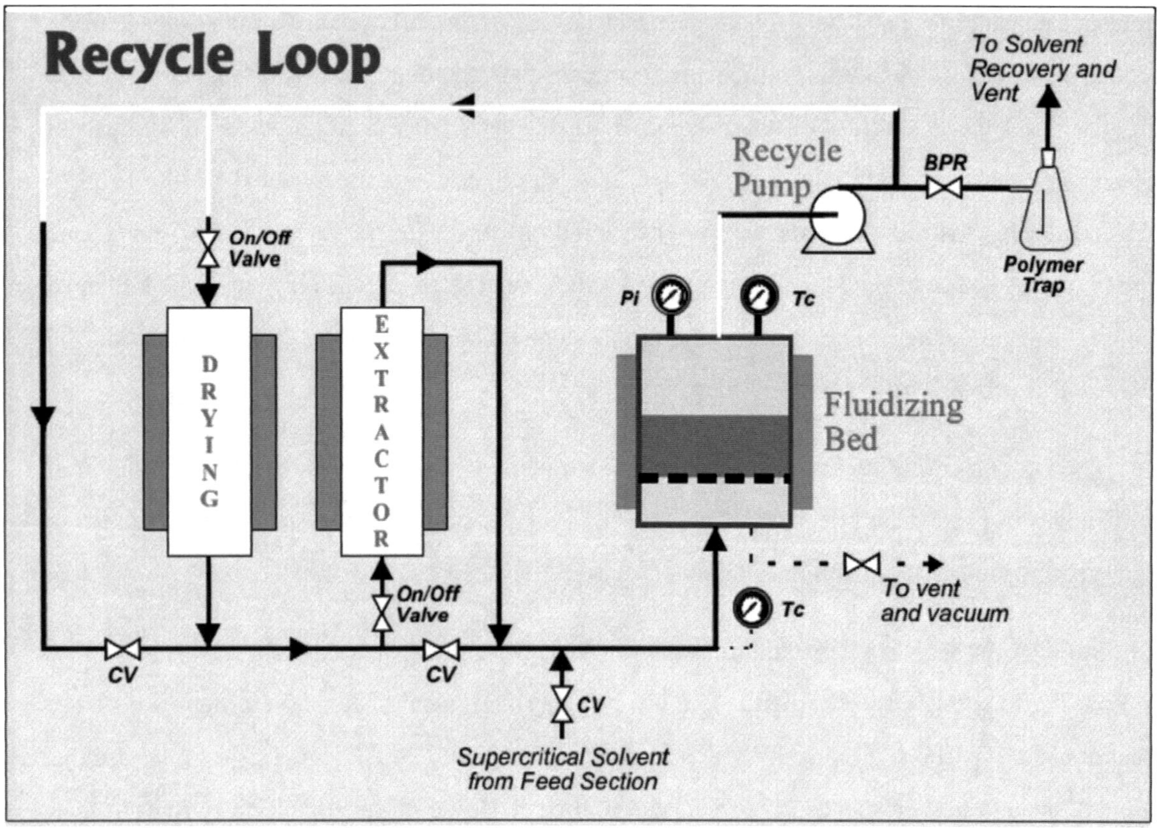

Figure 1. Pilot Circulating Fluidized System for Supercritical Fluid

	Air	Water	SC- CO2
Archimedes (Gravitational Forces/ Viscous F.)	9000	3500	697000
Reynolds (Inertial F/Viscous F) @Min. Fluid.	5.5	2.2	338
Minimum Fluidization Vel. (m/s)	0.12	0.003	0.06
Terminal Velocity (m/s)	2.6	0.08	0.27
Dimensionless Bed Height	20.7	28.2	4.4
Prandtl Number-(momentum dif/thermal dif)	0.7	3.6	3.1
Nusselt Number (convective /conductive)	3.8	3.9	24.8

Table 1. Comparison of Basic Properties of different Fluidizing Media

(649d) Tuning the Precipitation and Fractionation of Nanoparticles in Gas-Expanded Liquids

Steven R. Saunders and Christopher B. Roberts, Department of Chemical Engineering, Auburn University, Auburn, AL

Size-based fractionation of nanoparticles remains a non-trivial task for the preparation of significant quantities of well-defined nanomaterials for certain applications and fundamental studies. Through the use of the pressure-tunable physico-chemical properties of CO_2-expanded liquids, a rapid, precise, and environmentally sustainable size-selective fractionation of significant quantities of ligand-stabilized nanoparticles is possible through simple variations in applied CO_2 pressure (Saunders, S. R. and C. B. Roberts (2009). Nanotechnology (47): 475605). Unfortunately, an applied CO_2 pressure upwards of 35 bar is needed to precipitate dodecanethiol-stabilized gold nanoparticles from a hexane dispersion. However, this work demonstrates that the precipitation and size-selective fractionation of nanoparticles can be tuned by varying the length of the stabilizing ligand, the surface coverage of the stabilizing ligand, or the composition of the solvent. For example, the addition of a certain amount of a liquid antisolvent (e.g. acetone) to a stable dispersion of nanoparticles in hexane can be performed prior to CO_2 addition such that the solvent strength of the liquid mixture (e.g. hexane + acetone) is reduced to near the threshold of precipitation. As such, these liquid solvent mixtures require only an applied CO_2 pressure of 20 bar to induce nanoparticle precipitation, as opposed to the higher pressures necessary for the same nanoparticle dispersion in the neat hexane solvent. Moreover, changing the stabilizing ligand from dodecanethiol to hexanethiol allowed for precipitation and hence fractionation of nanoparticles to occur at applied CO_2 pressures of less than 5 bar. This paper will demonstrate that this technique allows for greater control of the stability of the nanoparticle dispersions allowing for greater precision when processing.

See more of this Session: Materials and Biomaterials Synthesis and Processing with Compressed or Supercritical Fluids II
See more of this Group/Topical: Engineering Sciences and Fundamentals

(649e) Reversible Reactions of CO2 with Amino-Terminated SAM's Employing a Protection/ Deprotection Mechanism for Smart Surface Synthesis

Fiaz S. Mohammed, Chemical and Biomolecular Engineering, Clemson University, Clemson, SC and Christopher L. Kitchens, Chemical & Biomolecular Engineering, Clemson University, Clemson, SC

Self assembled monolayers (SAM's) serve as one of the simplest techniques for control of surface properties and wetabillity. However, the control over the morphology, uniformity and terminal group reactivity of the film is heavily dependent not only on the experimental parameters, but also the availability and presence of different interaction and binding sites. Silicon (Si) substrates were functionalized using solutions of N-(2-aminoethyl)-(3-aminopropyl)-methlydimethoxysilane (AEAPMDS), N-(6-aminohexyl)-(3-aminopropyl)-trimethoxysilane (AHAPTS), Trimethoxysilylpropyl-polyethyleneimine (PEI-Silane) and Aminopropyl-Trimethoxysilane (APS) coupling agents. Due to the presence of the free amines, these agents exhibit an increased number of interactions with the clean Si substrates, leading to difficulties in obtaining large defect free films of uniform morphology and reactivity. This study investigates the reversible reaction of carbon dioxide (CO2) with amine terminated SAMs to provide a non-ionic to ionic change in surface chemistry, thus providing reversible surface properties. We have also investigated the use carbon dioxide (CO2) as a protecting agent to enhance the ability to form uniform SAMs of amine containing moieties.

See more of this Session: Materials and Biomaterials Synthesis and Processing with Compressed or Supercritical Fluids II
See more of this Group/Topical: Engineering Sciences and Fundamentals

(669a) Mechanistic Models for High-Solids Loading Pretreatment and Enzymatic Hydrolysis of Lignocellulosic Biomass

Andrew J. Griggs, Erik M. Kuhn, Xiaowen Chen, James J. Lischeske, Melvin P. Tucker and Jonathan J. Stickel, National Renewable Energy Laboratory, Golden, CO

Conversion of lignocellulosic biomass to fermentable sugars for renewable fuel production remains a formidable challenge for the economically feasible operation of bio-refineries. An improved understanding of the conversion reactions and the development of predictive models will help in overcoming this challenge, especially for high-solids loadings, where mass transfer may limit overall conversion yields. In this work, mechanistically based kinetic models for dilute-acid pretreatment and enzymatic hydrolysis have been developed, which consider the various physical and chemical changes to the feedstock. For dilute-acid pretreatment at high-solids loadings, we have developed a population-balance model, which considers the spatial distribution of reactants and products. The effects of the initial polydisperse particle-size distribution, and its evolution, are captured using a fragmentation-abrasion model. Such a model can aid in interpreting experimental results, determining reaction mechanisms, scale-up of pretreatment processes, and identifying the chemical and structural changes that result in high-xylose yields and enhanced-enzymatic digestibility. For example, when sieved fractions of milled corn stover are subjected to high-solids, dilute-acid pretreatment, the fraction with the largest particle size showed the greatest amount of fragmentation and yielded the most enzymatically digestible substrate, relative to the other sieved fractions. This finding suggests that the amount of fragmentation during dilute-acid pretreatment may be a measure of an effective pretreatment. For enzymatic hydrolysis, the distinct modes of action of the cellulase enzymes on soluble and insoluble substrates are each described by separate rate expressions. Depolymerization of the insoluble substrate by adsorbed enzyme is calculated using distribution kinetics, which tracks the molecular-weight distribution of the cellulose chains, rather than the concentration of individual species, over time. When combined with the method-of-moments, computationally efficient calculations are obtained. Model results throughout enzymatic saccharification are compared to experimental results for cellulose substrates, over a range of solids loadings. These models are capable of examining many of the relevant phenomena occurring during pretreatment and enzymatic saccharification, and aim to guide process development and identify optimal process-operating conditions.

See more of this Session: Reaction Kinetics and Transport Fundamentals for Biomass Conversion II
See more of this Group/Topical: Sustainable Engineering Forum

(669b) Modeling for Identification of Operating Strategies for Continuous Simultaneous Saccharification and Fermentation (cSSF)

Yi Jin[1], Jiacheng Shen[2], Jian Shi[2], Mirvat Ebrik[3] and Charles Wyman[4], (1)Chemical and Environmental Engineering,Bourns College of Engineering, Center for Environmental Research and Technology (CE-CERT) University of California, Riverside, Riverside, CA, (2)Center for Environmental Research and Technology, Bourns College of Engineering, University of California, Riverside, Riverisde, CA, (3) College of Engineering-Center for Environmental Research and Technology, University of California, Riverside, Riverisde, CA, (4)College of Engineering-Center for Environmental Research and Technology, University of California, Riverside, Riverside, CA

Although many economic studies assume continuous Simultaneous Saccharification and Fermentation (cSSF) will be the preferred configuration for commercial applications to achieve higher productivity and eliminate the need for seed fermentors, little information is available to guide process design or optimization. Therefore, in this paper, we focused on developing kinetic models to predict cSSF performance and applying them to better understand potential tradeoffs among enzyme loadings, residence times, yields, organism washout, and organism fermentation features to inform our new experimental research approach. Comparisons are made between batch and continuous processing to understand how end product inhibition and other characteristics differ and their effects on performance. In addition, simplified kinetic models are devised that require fewer measurable parameters to more easily predict the threshold for organism washout and how enzyme loadings impact system performance. Experimental results are compared to model predictions, and opportunities are identified for improving model accuracy. The result will be applied to identify optimal enzyme loadings and residence times for different pretreated feedstocks based on their compositions.

See more of this Session: Reaction Kinetics and Transport Fundamentals for Biomass Conversion II
See more of this Group/Topical: Sustainable Engineering Forum

(669c) Molecular Modeling of Cellulose Pyrolysis Using Molecular Dynamics

Vishal Agarwal, Chemical Engineering, University of Massachusetts, Amherst, MA, George W. Huber, Chemical Engineering, University of Massachusetts-Amherst, Amherst, MA, William C. Conner, Department of Chemical Engineering, University of Massachusetts-Amherst, Amherst, MA and Scott M. Auerbach, Chemistry and Chemical Engineering, Univ. of Massachusetts, Amherst, MA

Biomass pyrolysis is one of the viable routes, which can be used for producing gases and liquids that can replace petroleum-derived products. However, molecular mechanisms that control the physical and chemical transformation of cellulose during pyrolysis is poorly understood. Understanding this can help to control degradation process and minimize unwanted coke formation. Low temperature degradation of cellulose is proposed to proceed via a "active cellulose" stage [1]. However, such transition process and the nature of so called "active cellulose" is poorly understood. We have applied NPT molecular dynamics (MD) using ReaxFF (a reactive force field) [2] to model this transition process. We have used cellulose Iβ as the starting material since it is the most abundant crystalline form in plant material. We have found that ReaxFF can accurately model the crystal structure of cellulose Iβ. The modeled crystal agreed reasonably well with experimental results, including crystal density, unit cell parameters, chain conformation parameters, hydrogen bonds and thermal expansion coefficient at room temperature. The transition from cellulose to "active cellulose" is a physical change involving breaking of only intermolecular hydrogen bonds. This transition occurs at 480 K agreeing well with experimental results [1]. We have computed the changes in cell parameters, hydrogen bonding network, crystal density and chain conformational parameters with temperature. The results have been found to be in good agreement with the experimental results.

References

1. Lin Y., Cho J., Tompsett G. A., Westmoreland P. R. and Huber G. W., "Kinetics and Mechanisms of Cellulose Pyrolysis", J. Phys. Chem. C 113, 2009, 20097-20107.

2. Chenoweth K., van Duin A. C. T. and Goddard A., "Reactive Force Field for Molecular Dynmaics Simulations of Hydrocarbon Oxygen", J. Phys. Chem. A 112, 2008, 1040-1053.

See more of this Session: Reaction Kinetics and Transport Fundamentals for Biomass Conversion II
See more of this Group/Topical: Sustainable Engineering Forum

(669d) Reaction Kinetics of Transesterification of Canola Oil to Biodiesel Catalyzed by Nanopowder Calcium Oxide

Lina Zhao, Zheyan Qiu and Susan M. Stagg-Williams, Chemical & Petroleum Engineering Department, The University of Kansas, Lawrence, KS

The reaction kinetics model of transesterification of canola oil to biodiesel catalyzed by nanopowder calcium oxide was expressed.The reaction mechanism based on Langmuir-hinshelwood model has been proposed. In this proposed mechanism the overall reaction rate is assumed to be controlled by triglyceride adsorption step in the initial period followed by chemical reaction-controlled period. The apparent reaction constants, apparent activation energy and pre-exponential factors for both periods have been calculated based on experimental data. The internal mass transfer does not limit the reaction due to the nanosize of the catalyst particle.

See more of this Session: Reaction Kinetics and Transport Fundamentals for Biomass Conversion II
See more of this Group/Topical: Sustainable Engineering Forum

(669e) Water-Only Flowthrough Pretreatment of Poplar and Birchwood Xylan

Heather L. McKenzie[1], Nancy L. Engle[2], Joshua F. Emory[3], Bruce A. Tomkins[3], Timothy J. Tschaplinski[2], Gary J. Van Berkel[3] and Charles E. Wyman[1], (1)Chemical and Environmental Engineering, Center for Environmental Research and Technology, University of California Riverside, Riverside, CA, (2)Environmental Sciences Division, Oak Ridge National Laboratory, Oak Ridge, TN, (3) Chemical Sciences Division, Oak Ridge National Laboratory, Oak Ridge, TN

The conversion of lignocellulosic biomass to ethanol requires recovery of the sugars contained in the hemicellulose and cellulose. High temperature water-only pretreatment solubilizes primarily sugars and oligomers in hemicellulose prior to the enzymatic hydrolysis of the cellulose in the remaining solids to glucose. The understanding and kinetic models of hemicellulose hydrolysis are generally simplistic and do not fully address the existence of xylooligomers, lignin, and the interaction of biomass components. Furthermore, pretreatment of cellulosic biomass in the batch reactors typically employed makes the collection of time-dependent data challenging. On the other hand, a fixed bed flowthrough reactor facilitates tracking the release of biomass components as a function of time, thus making it a valuable laboratory tool. Analyzing the flow behavior and release patterns of xylooligomers from poplar and xylan would improve the understanding of the relative importance of mass transfer and kinetics during pretreatment. Analysis of void volume and flow behavior of a fixed bed flowthrough reactor will be determined to help characterize mass transfer in the system. Although lignin morphology is known to change during pretreatment, the cause of these changes is not well-known. It has been suggested that lignin depolymerizes (Li et al., 2007) or melts (Donohoe et al., 2008) during pretreatment. In an attempt to clarify the differences between these mechanisms, the release of lignin and lignin derivatives from poplar are followed over a range of reaction times, temperatures, and flow rates. Finally, pretreatment results from poplar and purified xylan are compared to clarify xylan interactions with other poplar components, such as cellulose and lignin.

See more of this Session: Reaction Kinetics and Transport Fundamentals for Biomass Conversion II
See more of this Group/Topical: Sustainable Engineering Forum

(669f) Comparative Pyrolysis Kinetics for One Legume and Two Grass Hays

Joseph J. Biernacki[1], Jessica Murillo[1], Prejeeth Ambuken[1] and C. Pat Bagley[2], (1)Chemical Engineering, Tennessee Technological University, Cookeville, TN, (2)Agriculture and Human Ecology, Tennessee Technological University, Cookeville, TN

While the reactions involved in plant biomass are somewhat understood, there are still unknowns, particularly regarding the details of the kinetic pathways. Furthermore, there are a wide range of biomass feedstocks presently being considered for which there is no kinetic information available. This work compares the pyrolysis of one legume (alfalfa hay) and two grass hays (sudan grass hay and tall fescue). Alfalfa hay is a premium lifestock feed and would likely not be used in the future as a feed for production of biofuel, however, provides a basis upon which to compare the two grasses, which are of poor and very poor nutritional content and are likely candidates for biofuel conversion. Conventional pyrolysis using thermalgravimetric analysis (TGA) was performed on the dried, pulverized feeds. Heating rates of 5, 10, 30 and 50 oC/min were used as a basis. Mass spectroscopy was performed and and pyrolysis rate curves analyzed using a multi-process model wherein pyrolysis is assumed to be a series of discrete independent reactions. The results were used to infer alternative appraoches that account for the dramatic change in pyrolysis efficiency as a function of heating rate.

See more of this Session: Reaction Kinetics and Transport Fundamentals for Biomass Conversion II
See more of this Group/Topical: Sustainable Engineering Forum

(694a) "Bi-Variate Population-Balance Analysis of Ethanol-Fueled Continuous Flow Spray Combustors"

Daniel E. Rosner, Chemical Engineering, Yale University, New Haven, CT and Manuel Arias-Zugasti, Vis. Asst Prof.-ChE Dept., Yale University, New Haven, CT

We formulate/present here results of our recent bi-variate Population Balance (PB-)-based theoretical analysis of the performance of well-(macro-)mixed adiabatic combustors fed by EtOH-containing sprays of prescribed DSD and composition^1-3. Our "pre-bio-fuel" historically relevant/instructive example is the well-documented liquid fuel-cooled V-2 rocket-------which used 75 wt pct EtOH + H2O solution, along with the cryogenic oxidizer O2(L) at a chamber pressure of 15 atm. Of special interest are our predicted combustion intensities (GW/m^3), and efficiencies (EtOH fraction vaporized) at each "vaporization Damkohler number" [ratio of combustor mean residence time to feed-SMD QS-vaporization lifetime]^4. Our droplet evaporation rate-controlled formulation exploits a Quasi-Steady gas-diffusion-controlled and internally well-mixed individual droplet evaporation rate law, and we exploit the formally exact method of characteristics to solve (numerically) the associated first-order PB-PDE governing the joint pdf(m1,m2) exiting such a chamber, where m1 = EtOH mass/droplet, and m2 = H2O mass/droplet. This function also enables predictions of corresponding unconditional DSDs, and the possibly (remotely-) measurable jpdf(diam,droplet temperature). Apart from their intrinsic multi-phase chemical reaction-engineering interest as an instructive, tractable limiting case, our numerically exact formulation/results can also serve as valuable "meter-sticks" for computationally convenient approximate methods (BV-QMOM- and spectral/weighted residuals) to predict such "correlated" bi-variate pdfs in more realistic spray combustor environments. We also demonstrate the inability of a seemingly plausible "pseudo-unary" approximation to capture key features of the performance of combustors fueled by alcohols capable of absorbing their own combustion product : H2O. ______
*Supported at Yale University by Yale NSF Grant : CTS 0522944 1 Rosner, DE , Ind.Eng.Chem Res. Vol 48, 6453-6464(2009) 2 Rosner, DE and Arias Zugasti, M, AIChE J (submitted May, 2010) 3 Rosner, DE and Arias Zugasti, M, AIAA-J Prop.Power, vol 25(3)826-828(2009) 4 Rosner, DE, Arias-Zugasti,M and Labowsky, M., Chem. Eng. Sci. vol 63, 3909-3920(2008)

See more of this Session: Mathematical Modeling in Transport Processes
See more of this Group/Topical: Engineering Sciences and Fundamentals

(694b) Stokes' Second Problem for An Oldroyd-B Fluid in a Porous Half Space

Balram Suman, SP-TD-02, Corning Inc., Corning, NY and Nazish Hoda, Department of Chemical Engineering, University of Michigan, Ann Arbor, MI

A modified form of Darcy's law is used to study the flow of Oldroyd-B fluids in a semi-infinite porous domain bounded by an oscillating plate. A close form analytical expression is obtained using Laplace transform, which suggests that depending upon the choice of natural and forced frequencies markedly distinct velocity fields can be observed. The fluid velocity oscillates in time with a frequency which is same as the plate frequency. However, there exists a phase lag that increases with increasing distance from the plate. At a fixed distance from the plate, the phase-lag increases with increasing Reynolds number, $Re = U(K/\phi)^{0.5}/\nu$ and Wiessenberg number, $Wi = \lambda U^2/\nu$, and decreasing viscosity ratio, $\beta = \lambda_t/\lambda$, where U is the amplitude of plate oscillation, K and ϕ are permeability and porosity of the porous media, ν is the kinematic viscosity, and λ_t and λ are retardation and relaxation times, respectively. This is consistent with our analytical prediction that the phase-lag is pronounced when $Wi\ Re^2 >> \beta^2$. The phase lag also decreases with decreasing Strouhal number, $St = \nu\omega/U^2$ where ω is the plate oscillation frequency, and it vanishes in the limit of $St = 0$. Furthermore, the fluid velocity increases with increasing β and Re, and decreasing distance from the plate. Away from the plate the fluid velocity shows a strong transient effect where the amplitude of oscillation either increases or decreases with time depending upon Re, St, and β. The above findings can be of importance for the fabrication of nano-mechanical devices that are designed to operate at varying induced frequencies.

See more of this Session: Mathematical Modeling in Transport Processes
See more of this Group/Topical: Engineering Sciences and Fundamentals

(694c) Mathematical Simulation of Acid Rock Drainage

surya P. Sunkavalli, Edward M. Trujillo, Marcin Niewiadomski and You Li, Chemical Engineering, university of Utah, Salt Lake City, UT

Acid rock drainage (ARD) occurs in rock piles and tailings from mining operations when sulfide minerals oxidize in the presence of water and air to produce sulfuric acid. The rock piles and tailings are generally unsaturated and water (acid) flow is dictated by annual rainfall and snowfall patterns. The resulting acid leaches through the rock material, reacts with other minerals and can contaminate nearby streams and lakes. There have been numerous mathematical models of weathering and (ARD) over the last few decades, starting with rather simple models to very complicated coupled transport models. The problem is that ARD is a very complex process involving not only mineralogical and geochemical changes but physical and biological changes as well. We have developed a mathematical model of ARD using a modified version of TOUGHREACT that includes all the major mechanisms involved and that can be adapted to any particular mining site and environment. The multi-component, multi-phase, non-isothermal model has been calibrated with laboratory data collected from experiments using a modified standard operating procedure over a 52 week duration with actual mining rock material.

See more of this Session: Mathematical Modeling in Transport Processes
See more of this Group/Topical: Engineering Sciences and Fundamentals

(694d) Analysis, Modeling and Simulation of a Gravity-Independent Porous Membrane Condenser Cooled by a Thermoelectric-Driven Heat Pump

J.M.R. Apollo Arquiza[1], Robert Morrow[2], Ross Remiker[2] and Jean B. Hunter[1], (1)Biological and Environmental Engineering, Cornell University, Ithaca, NY, (2)Orbital Technologies Corporation, Madison, WI

Dehumidification of air in astronaut living quarters and space wastewater treatment systems that involve evaporation or distillation require a microgravity-compatible condenser for recovering liquid water. The porous membrane condensing heat exchanger (PMCHX) provides gravity-independent condensation and phase separation in a single step. No free liquid is produced since the condensate is confined by capillary forces in the porous membrane. Design equations for the PMCHX in microgravity were derived from a theoretical analysis of the principles governing its operation. A PMCHX driven by a thermoelectric (TE) heat pump was fabricated as part of a closed air-loop drying system for astronaut cabin waste. The PMCHX was designed to condense water vapor from the hot, moist air leaving the drying chamber while recovering some of the latent enthalpy to heat the air leaving the condenser before it returns to the drying chamber. Experiments on the PMCHX gave the effect of moist air flow rate, air temperature, and temperature difference between the hot and cold side of the TE on condensate production and thermodynamic efficiency. Results showed that the compact PMCHX system is capable of producing 2 L/day of condensate with a COP of 2.5. A computational model for the condenser system based on the conservation equations for momentum, energy, and moisture applied to the air and porous membrane phases together with the TE heat pump equations has been developed. The resulting differential equations were solved numerically by the finite element method. Good agreement between the simulation and experiment was observed. The model was used to predict the system performance and energy cost per unit of recovered water for a range of operating conditions with different values of cold-side plate temperature, air flowrate, humidity, and temperature.

See more of this Session: Mathematical Modeling in Transport Processes
See more of this Group/Topical: Engineering Sciences and Fundamentals

(694e) Modeling Electrokinetic Remediation of Concrete by G/FEM

Patrick A. Oglesby, School of Chemical Engineering, Purdue University, West Lafayette, IN and M T Harris, Chemical Engineering, Purdue University, West Lafayette, IN

This paper reviews the process of electrokinetic remediation of a concrete disk to remove various contaminants, especially those found in the nuclear industry. A one dimensional model is developed and solved using the Galerkin/Finite element method (G/FEM) with adaptive time stepping.

This method is proven to be quick and reliable in calculating the time and spacial progression of the various species, the pH, and the electric potential variations throughout the system. An adaptive mesh focuses elements on regions of rapid change allowing fewer total elements to be used.

Results are presented for select initial and boundary conditions detailing parameters such as breakthrough times, solute loading removed, and power input required to perform the operations.

See more of this Session: Mathematical Modeling in Transport Processes
See more of this Group/Topical: Engineering Sciences and Fundamentals

(694f) Dead Zones in Reactive Porous Materials

Roger L. York[1], Kaitlin M. Bratlie[2], Larry K. Jang[3] and Lloyd R. Hile[3], (1)Koch Institute for Integrative Cancer Research, Massachusetts Institute of Technology, Cambridge, MA, (2)Chemical Engineering, MIT, Cambridge, MA, (3)Chemical Engineering, California State University, Long Beach, Long Beach, CA

This talk examines the conditions under which a dead zone, or a portion of a catalyst (or any reactive porous material) devoid of reactant, can form in a porous catalyst (material) in which simultaneous reaction and diffusion are occurring. The condition that allows for the existence of a dead zone is defined by a critical Thiele modulus. When the Thiele modulus – the ratio of chemical reaction to diffusion - is greater than the critical Thiele modulus, a dead zone exists. This dead zone can be mathematically defined by a change of boundary conditions. We examine nth order reactions in isothermal infinite slabs, infinite cylinders, and spheres, demonstrating analytical expressions for concentration profiles and efficiency factors for zero order reactions in infinite cylinders and spheres. In addition, we provide analytical concentration profiles and efficiency factors for zero-order reactions in non-isothermal infinite slabs (in the so-called low beta approximation). We also discuss some common errors and misconceptions associated with this phenomenon.

See more of this Session: Mathematical Modeling in Transport Processes
See more of this Group/Topical: Engineering Sciences and Fundamentals

(694g) Modeling Mass and Energy Transport in Thermodynamically Coupled Reactions

Erin Lennon, M. C. Tanzy, V. A. Volpert and A. Bayliss, Engineering Sciences and Applied Mathematics, Northwestern University, Evanston, IL

We develop and analyze models for mass and energy transport in non-adiabatic self-propagating reactions. The reaction-diffusion systems under consideration consist of thermodynamically coupled reactions that may also be chemically coupled. Each reaction can be modeled as having either an exponential or a step-function temperature dependence. Parameters such as thermal conductivity and heat-loss through environmental and chemical sinks are explored in both analytic and numerical approaches. The results are compared with experimental systems and suggest a different set of thermal processes than previously considered. The models are applicable to a wide range of reactions, including the combustion synthesis of nano-scale powders for catalytic applications.

See more of this Session: Mathematical Modeling in Transport Processes
See more of this Group/Topical: Engineering Sciences and Fundamentals

(694h) Nonlinear Charging Dynamics of Porous Electrodes

Martin Z. Bazant[1], Yeqing Fu[2], Todd Ferguson[2] and P. M. Biesheuvel[3], (1)Chemical Engineering and Mathematics, Massachusetts Institute of Technology, Cambridge, MA, (2)Chemical Engineering, Massachusetts Institute of Technology, Cambridge, MA, (3)Department of Environmental Technology, Wageningen University, Wageningen, Netherlands

Porous electrodes have many uses, ranging from energy storage to seawater desalination, but their theoretical description remains challenging. In 1963, Newman and Tobias pioneered the continuum theory of transport and reactions in porous electrodes, based on formal volume averaging, and in the same year De Levie derived equivalent circuit models (RC transmission lines) for linear response. These approaches have been successfully applied for decades, e.g. to supercapacitors and batteries, but various nonlinear effects have yet to be analyzed. Here, we extend the continuum theory of porous electrodes by taking into account double-layer salt adsorption (for capacitive desalination) and pseudo-capacitance (for electrochemical energy storage in nanoparticles). Scaling arguments, analytical approximations, and numerical solutions serve to describe different dynamical regimes, and comparisons are made with experimental data.

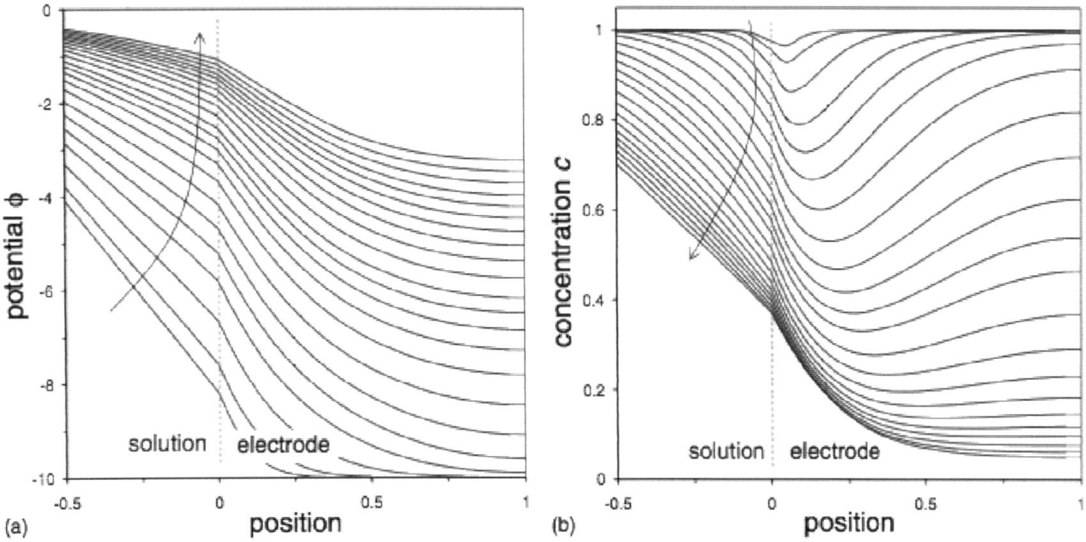

(a) (b)

See more of this Session: Mathematical Modeling in Transport Processes
See more of this Group/Topical: Engineering Sciences and Fundamentals

Upscaling multi-components two-phase flow dispersion with partitioning coefficient

C. Soulaine [a,*] , G. Debenest [a], M. Quintard [a,b]

[a] Université de Toulouse ; INPT, UPS ; IMFT (Institut de Mécanique des Fluides de Toulouse ;

Allée Camille Soula, F-31400 Toulouse, France

[b] CNRS ; IMFT ; F-31400 Toulouse, France

* corresponding author. E-mail adress : cyprien.soulaine@imft.fr

Mass transfer and fluid flow with liquid-vapour phase exchange in porous media appears in a large number of situation such as nuclear safety devices, transport in petroleum reservoirs, aquifer contamination, desalination processes by way of distillation...

This paper comes in the continuity of the works of *Quintard and Whitaker (1994)* [2] and of *Coutelieris and al. (2006)* [1]. In the first paper, a macroscopic model describing a binary mixture is obtained from the pore-scale equations using a volume averaging upscaling method. This theory is extended in the second paper to multicomponent mixtures with a partitioning equilibrium condition at the interface between an immobile non-aqueous liquid phase and a flowing aqueous phase. In this paper, we extend these theoretical results to the case of a two-phase, multicomponent system. The two phases are mobile. In addition, the upscaling methodology is improved with respect to the interface movement.

The pore-scale problem under consideration corresponds to mass transfer of a component A in a two-phase flow system. The liquid (denoted β-phase) and its vapour (γ-phase) flow through a porous medium (σ-phase). It is assumed that the chemical potential equilibrium at the interface is linearised to a partitioning relationship (Henry's law). We have for component A,

$$\frac{\partial \rho_\beta \omega_{\beta A}}{\partial t} + \nabla.(\rho_\beta \omega_{\beta A} \mathbf{v}_\beta) = \nabla.(\rho_\beta D_{\beta A} \nabla \omega_{\beta A}) \ \texttt{in} \ \mathcal{V}_\beta \tag{1}$$

$$\mathbf{n}_{\beta\sigma}.\rho_\beta D_{\beta A}\nabla\omega_{\beta A} = 0 \ \texttt{at} \ \mathcal{A}_{\beta\sigma} \ ; \ \mathbf{n}_{\gamma\sigma}.\rho_\gamma D_{\gamma A}\nabla\omega_{\gamma A} = 0 \ \texttt{at} \ \mathcal{A}_{\gamma\sigma} \tag{2}$$

$$\mathbf{n}_{\beta\gamma}.(\rho_\beta(\mathbf{v}_\beta - \mathbf{w}_{\beta\gamma}) - \rho_\beta D_{\beta A}\nabla\omega_{\beta A}) = \mathbf{n}_{\beta\gamma}.(\rho_\gamma(\mathbf{v}_\gamma - \mathbf{w}_{\beta\gamma}) - \rho_\gamma D_{\gamma A}\nabla\omega_{\gamma A}) \ \texttt{at} \ \mathcal{A}_{\beta\gamma} \tag{3}$$

$$\omega_{\beta A} = H_A \omega_{\gamma A} \ \texttt{at} \ \mathcal{A}_{\beta\gamma} \tag{4}$$

$$\frac{\partial \rho_\gamma \omega_{\gamma A}}{\partial t} + \nabla.(\rho_\gamma \omega_{\gamma A} \mathbf{v}_\gamma) = \nabla.(\rho_\gamma D_{\gamma A} \nabla \omega_{\gamma A}) \ \texttt{in} \ \mathcal{V}_\gamma \tag{5}$$

The volume averaging method is applied to the microscopic equations written above in order to develop a macro-scale model for homogeneous porous media. We consider local mass non-equilibrium, i.e., averaged concentrations are not necessarily linked by the equilibrium interface relationship Eq.4. The resulting macroscopic equations involve dispersion tensors, additional convective transport terms and a mass exchange term that are influenced by the mass transfer process. All these effective coefficients can entirely be determined by three closure problems solved at the unit-cell scale. A closed form of the total mass transfer rate is also proposed. We obtain the following two equations which are linked by a mass transfer coefficient

$$\underbrace{\frac{\partial \epsilon_\beta \rho_\beta \Omega_{\beta A}}{\partial t}}_{accumulation} + \underbrace{\nabla.(\epsilon_\beta \rho_\beta \Omega_{\beta A} \mathbf{V}_\beta)}_{convection} + \underbrace{\nabla.(\rho_\beta \mathbf{E}^*_\beta (H_A \Omega_{\gamma A} - \Omega_{\beta A}))}_{convection\ correction}$$
$$= \underbrace{\nabla.(\rho_\beta \mathbf{D}^*_{\beta\beta}.\nabla\Omega_{\beta A})}_{dispersion} + \underbrace{\nabla.(\rho_\beta \mathbf{D}^*_{\beta\gamma}.\nabla\Omega_{\gamma A})}_{dispersion\ correction} - \underbrace{\dot{m}_A}_{mass\ exchange} \quad (6)$$

$$\frac{\partial \epsilon_\gamma \rho_\gamma \Omega_{\gamma A}}{\partial t} + \nabla.(\epsilon_\gamma \rho_\gamma \Omega_{\gamma A} \mathbf{V}_\gamma) + \nabla.(\rho_\gamma \mathbf{E}^*_\gamma (H_A \Omega_{\gamma A} - \Omega_{\beta A}))$$
$$= \nabla.(\rho_\gamma \mathbf{D}^*_{\gamma\gamma}.\nabla\Omega_{\gamma A}) + \nabla.(\rho_\gamma \mathbf{D}^*_{\gamma\beta}.\nabla\Omega_{\beta A}) + \dot{m}_A \quad (7)$$

In these equations, ϵ_i denote the volume fraction of the i-**phase**, the intrinsic volume average of a ϕ_i variable is defined as

$$<\phi_i>^i = \frac{1}{\mathcal{V}_i} \int_{\mathcal{V}_i} \phi_i d\mathcal{V} \quad (8)$$

where $\mathcal{V}_i$ is the volume occupied by the i-**phase**, and in order to clarify the notation, the average quantities are denoted as $\Omega_{\beta A} =< \omega_{\beta A} >^\beta$, $\Omega_{\gamma A} =< \omega_{\gamma A} >^\gamma$, $\mathbf{V}_\beta =< \mathbf{v}_\beta >^\beta$ and $\mathbf{V}_\gamma =< \mathbf{v}_\gamma >^\gamma$.

The closed form of the mass exchange term of the A-component depends on the effective variables $\mathbf{u}_{\gamma\beta}$, $\mathbf{u}_{\beta\gamma}$ and α which are calculated from the closure problems

$$\dot{m}_A = -\mathbf{u}_{\gamma\beta}.\nabla\Omega_{\beta A} + \mathbf{u}_{\beta\gamma}.\nabla\Omega_{\gamma A} + \alpha(H_A \Omega_{\gamma A} - \Omega_{\beta A}) \quad (9)$$

and we can calculate the total mass transfer rate by

$$\dot{m} = \sum_A \dot{m}_A \quad (10)$$

To validate the approach, we consider the case of the falling film along a vertical fixed wall. In this case, the closure problems can be solved analytically. The obtained effective dispersion coefficients are in good agreement with the falling film theory results and the mass exchange coefficient corresponds to the Lewis and Whitman double film theory. Moreover, we compared with success the macro-scale model with a direct numerical simulation (see Fig.1) of the pore-scale model.

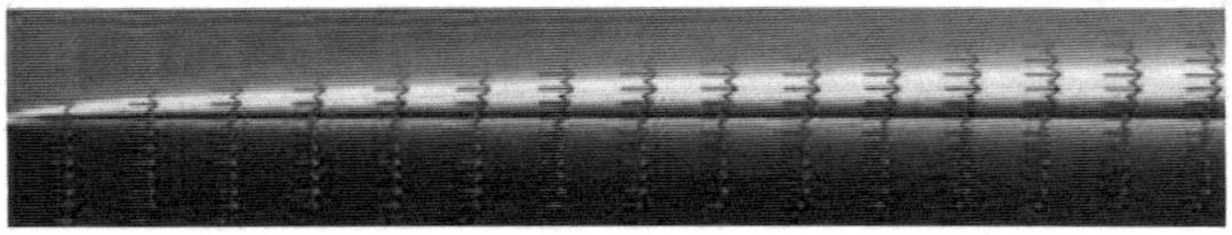

Figure 1: Direct Numerical Simulation

References

[1] F.A. Coutelieris, M.E. Kainourgiakis, A.K. Stubos, E.S. Kikkinides, and Y.C. Yortsos. Multiphase mass transport with partitioning and inter-phase transport in porous media. *Chemical Engineering Science*, 61(14):4650 – 4661, 2006.

[2] Michel Quintard and Stephen Whitaker. Convection, dispersion, and interfacial transport of contaminants: Homogeneous porous media. *Advances in Water Resources*, 17(4):221 – 239, 1994.

(694j) Mathematical Modelling of the Heap Leaching Process

Maik Irrazábal Sr., Chemical Engineering, Pontificia Universidad Católica de Valparaíso, Valparaíso, Chile

This work is focused on the development of a simple mathematical model of copper heap leaching, able to predict the concentrations profiles of sulphate ion (SO4-2), as the leaching agent, and copper ion (Cu +2), both as a function of time. The model is based on the conservation equations applied in the heap, neglecting temperature and deformation effects. The system is considered as an isotropic porous media with porosity homogeneously distributed. An exact solution to this problem will be of utility if the optimal operational parameters are needed. Specifically, the system is considered as a three phase problem, with liquid, solid and gaseous phases, in which the transport problem is modelled considering the components sulphuric acid in liquid phase and copper's ions in liquid and solid phase. The main transport mechanism of Cu+2 through heap is advection through liquid phase. However the small amounts of Cu+2 in the leaching solution do not affect the liquid flow. Therefore, the model assumes that it is possible to uncouple the transport equations to solve the liquid flow independently. Analytical solutions for the liquid phase flow are based on the solution of the Richards' equation, Gardner retention curves, and exponential dependence of the heap permeability with liquid content. This configuration is useful to describe the liquid flow in saturated or unsaturated conditions. In order of achieving an exact solution, techniques, such as linearization, variable transformations, and perturbation analysis are useful in the development of the model. The second part of this work is focused on the study of the effects of several operational parameters (temperature, pH of the leaching agent, heap granulometry, etc.) on the amounts of recovered copper. These experimental data were used for validating the model predictions. Results agree with experiments and other works previously reported.

See more of this Session: Mathematical Modeling in Transport Processes
See more of this Group/Topical: Engineering Sciences and Fundamentals

AUTHOR INDEX

AUTHOR INDEX

AUTHOR INDEX

AUTHOR INDEX